Paolo Gasparini
Gaetano Manfredi
Jochen Zschau

Earthquake Early Warning Systems

Paolo Gasparini
Gaetano Manfredi
Jochen Zschau

(Editors)

Earthquake Early Warning Systems

With 153 Figures

Springer

PROF. PAOLO GASPARINI
University of Napoli
Department of Physics
Via Cintia
80126 Napoli
Italy
E-Mail: paolo.gasparini@na.infn.it

PROF. GAETANO MANFREDI
University of Napoli
Department of Structural Engineering
Via Claudio 21
80125 Napoli
Italy
E-Mail: gaetano.manfredi@unina.it

PROF. DR. JOCHEN ZSCHAU
GeoForschungsZentrum Potsdam
Telegrafenberg E 425
14473 Potsdam
Germany
E-Mail: zschau@gfz-potsdam.de

Library of Congress Control Number: 2007925906

ISBN-13 978-3-540-72240-3 Springer Berlin Heidelberg New York

Springer is a part of Springer Science+Business Media
springer.com

Cover design: deblik, Berlin
Typesetting: camera-ready by doppiavoce, Napoli, Italy
Editorial coordination: doppiavoce, Napoli, Italy
Production: Christine Adolph
Printing: Krips bv, Meppel
Binding: Stürtz AG, Würzburg

Printed on acid-free paper 32/2132/ca 5 4 3 2 1 0

Preface

In the last few decades economic losses due to natural disasters have increased exponentially worldwide and little progress has been seen in reducing their rate of fatalities. This also holds for earthquake disasters and is mainly due to increasing population and industrial density in high hazard and vulnerability areas. Although the prediction of earthquakes is not yet practicable, current technology allows prompt identification of the onset of any dangerous seismic event. Hence *early warning and rapid disaster information systems* are becoming important means for strengthening prevention and social resilience against the adverse effects of major natural events and should therefore become the keystones of disaster mitigation. The term *early warning* is now widely used with various meanings in scientific, economic and sociological communities. Even in the scientific world the term is used in slightly different ways although there is a growing consensus in defining *early warning* as all the action that can be taken during the *lead time* of a catastrophic event. The *lead time* is defined as the time elapsing between the moment when the occurrence of a catastrophic event in a given place is reasonably certain and the moment it actually occurs. Typical lead times are of the orders of seconds to tens of seconds for earthquakes, minutes to hours for tsunamis, and hours to days for landslides, floods and volcanic eruptions.

In more general terms, early warning is the provision of timely and effective information, through identified institutions, allowing individuals exposed to a hazard to take action in order to avoid or reduce their risk and prepare for effective response.

Although the definition of lead time for non-seismic hazards may be ambiguous (the term "reasonably certain" may need a more precise probabilistic definition), for earthquakes the definition is unequivocal as the lead time will start when the first waves are released by the earthquake source. Indeed, the physical basis for earthquake early warning is simple: strong ground shaking is caused by shear-waves and by the subsequent surface waves which travel at about half the speed of the primary waves and much slower than electromagnetic signals transmitted wireless and/or by cable. Thus, depending on the distance of a strong earthquake from the endangered urban area, transmission of information and real-time analysis of the fast primary wave may provide warnings from a few seconds to a few tens of seconds before the arrival of strong ground shaking. This may be used to minimize property damage and loss of life in urban areas and to aid emergency response. When a suitable seismic network is available fast processing methods can be applied to locate an earthquake, determine the magnitude, and estimate the distribution of ground motion (regional approach). At a site or structure equipped with seismic sensors, a site-specific warning is possible using the first low amplitude arrivals (P-waves) to infer the motion due to the following high amplitude shear and superficial waves (on-site approach).

The application of earthquake early warning systems (EEWS) can be very effective in real time risk mitigation, enhancing the safety margin of specific critical engineered systems such as nuclear power plants, lifelines or transportation infrastructures by reducing the exposure of the facility with automated safety actions. The early warning system can be used to trigger the orderly shutdown of pipelines and gas lines to avoid fires, or the shutdown of manufacturing operations to reduce both potential damage to equipment and industrial accidents. Also, personal safety might be enhanced if people were alerted. In addition, the functions of modern society will be less likely to turn chaotic if an early earthquake alert is available and if training of appropriate actions has been performed. Last not least, emergency response teams may be dispatched where they are needed most if maps of strong ground shaking can be provided by the early warning system within a few minutes.

In addition, seismic early warning systems can be of great value in reducing damage and loss due to secondary events triggered by earthquakes. These may include landslides, tsunamis, fires and industrial accidents. The fires that devastated San Francisco after the 1906 earthquake and the tsunami of December 2004 in Indonesia are two classic cases, but in most of the major earthquakes economic losses and human casualties have been enhanced by secondary phenomena.

Despite the above considerations, at present the potential of seismic early warning methods is not fully used. This is not only true for developing countries but also for highly industrialized countries including those of Europe.

Most existing seismological processing methods have not been developed or optimized for real-time or near real-time applications as required for early warning. The development of real-time analysis, modeling and simulation methods, their integration with appropriate facilities for data processing, visualization and rapid information systems and their application to earthquake early warning in conjunction with disaster management is, therefore, one of the major challenges of today's seismology.

All of these issues were raised and discussed during a workshop held in Naples, Italy, on September 23-25 2004, focusing on "Seismic Early Warning for European Cities: toward a coordinated effort to raise the level of basic knowledge". The workshop was organized in the framework of the EC FP 6 SSA Project "Natural Risk Assessment (NaRAs)". Researchers attending the meeting from eight European countries (France, Germany, Greece, Iceland, Italy, Portugal, Switzerland, Turkey), United States, Japan and Taiwan unanimously approved a recommendation submitted to the European Commission, stressing the still unresolved basic questions for full application of earthquake early warning to society's needs and asking for future calls to contain specific reference to seismic early warning methods.

This book is mostly based on the articles that were presented at the workshop. Given the long time needed to collect all of them, they have since been updated. They were written in their final form at the end of 2006.

The short review by Hiroo Kanamori points out the main problems for automatic application of earthquake early warning to real time risk reduction.

One of the basic problems in seismic early warning is the development of real-time algorithms for fast determination of earthquake source parameters and the estimation of their reliability. This includes the problems of real-time event detection and location, real-time fault mapping as well as new approaches for fast magnitude/moment determinations based on strong motion data, modern seismic array technology and the concept of energy magnitude. The latter promises to be extremely useful for estimating the size of mega-events. The scientific and technological challenge is to obtain this kind of information only a few seconds after the first P-wave arrivals. Classic seismic processing tools still need larger portions of a seismogram and are thus not suited to this purpose.

A group of five papers deals with the above problems. In particular, the paper by Stefan Nielsen discusses from a theoretical viewpoint whether reliable information on the size of an earthquake can be obtained from processing the waves released at the onset of a fracture. The paper by Richard Allen discusses the ElarmS system based on the processing of first P-wave arrivals to predict ground motion at different sites. Aldo Zollo and Maria Lancieri use an earthquake database to simulate real-time magnitude determination from the Earthquake Early warning system implemented in the Campania Apennines. They identify the parameters most robustly correlated with moment magnitude. Maren Böse et al. present the PreSEIS (pre-seismic shaking) method they developed and applied to the Istanbul case. The method is based on an artificial neural network and is as fast as the on-site warning approach, because it combines information from several sensors within small seismic subnets with apertures of about one hundred kilometers to estimate source parameters from the first few seconds of seismic recordings. Satriano et al. present an evolutionary method for real-time location based on the equal differential time formulation and a probabilistic approach.

Along with the development of appropriate real-time algorithms, it is crucial to develop a strategy for rapidly communicating the obtained seismic information not only to the disaster managers, but also to other interested parties from civil protection, politics, media, science and the public. The warning time involved in this task may, however, have to be extended to minutes, tens of minutes or more. Of special importance for emergency planners will be the concept of the virtual seismologist, which takes into account pre-existing information to estimate and possibly reduce the uncertainties of source parameter determinations, and which, in particular, can deduce from the source parameter information specific decision support for disaster management, as discussed in the paper by Georgia Cua and Thomas Heaton.

The evolutionary method and the virtual seismologist concept are very useful for providing continuously upgraded real-time alert maps and predicted shake maps within seconds and minutes as well as maps of measured ground shaking within a few minutes after the event. The development of proper attenuation algorithms, as discussed by Vincenzo Convertito et al., is crucial in order to also account for site corrections in such maps. Maps of expected ground motion before a catastrophic event for various scenarios are useful information to design the way effects of ground vibrations on structures can be reduced as well as for fast map calibration once the event occurs. 3D simulations of ground response and the key parameters needed to optimize the probabilistic approach are discussed by Jean Virieux et al.

Earthquake Early Warning Systems are efficient tools in urban areas where a significant portion of the buildings are structurally deficient. In cases where the seismic source zone is clearly known and sufficiently far away, the population can be warned by radio, television, etc. Operation of critical facilities and processes can be stopped. In the case of very short pre-warning times of a few seconds, it is still possible to slow down trains, to switch traffic lights to red, to close valves in gas and oil pipelines, to release a SCRAM in nuclear power plants, etc. Early warning systems can also be used to alarm the population where rapid response is needed. A typical example would be to issue the so-called water alarm, i.e. alarming the population living in the downstream region of a large dam. Early warning systems are useful for facilities and processes, such as nuclear power plants, high speed trains, gas mains and highways, where rapid response can contribute to reduction in the seismic risk.

In addition to such immediate uses, further development of an early warning system may include the implementation of semi-active interfaces with infrastructures that can use the early warning information for real-time risk reduction. For example, construction companies in Japan are developing buildings with semi-active control systems. The buildings can change their mechanical properties within a few seconds to better withstand ground motion. Implementation of this "few seconds engineering" requires careful assessment of the false alarms or "cry wolf" and missed alarm probabilities on the decision chain as discussed in the papers by Grasso et al., and Iervolino et al. A second paper by Iervolino et al. discusses several real-time engineering applications in the light of performance-based earthquake engineering for risk reduction.

Finally in the last part of the volume four different earthquake early warning systems are described, which are already in operation.

The first system to be operative in the world was the UrEDAS (Urgent Earthquake Detection and Alarm System). It was implemented to protect sections of the fast Japanese Railway Systems. The history of seismic early warning since the original idea of J.F. Cooper in 1868, the development of the UrEDAS system and a report of its performance are given by Nakamura and Saita. The same authors also describe a portable device for on-site early warning applications.

The early warning system implemented in Taiwan, described in the article by Wu, is a regional system which can issue an alert after 22 sec from the onset of an event. This gives a lead time of more than 10 sec to locations more than 100 km away from the epicentre, and the application of a novel processing method has the perspective of decreasing the processing time to about 10 sec and the "blind" zone to about 25 km.

The system implemented in Romania was designed to protect mainly Bucharest and some industrial structures from the intermediate depth earthquakes originating in the Vrancea region. Some specific characteristics of the seismic activity (such as the stationary epicentres, the stability of radiation patterns) and a line-of-sight connection between the epicentral area and the capital allowed a simple and robust system to be designed, which is currently being tested to protect a nuclear power plant, as described by Marmureanu et al.

The fourth system, described by Weber et al., is being implemented in the Campanian Apennines, southern Italy, along the fault systems which have been the source of many strong crustal earthquakes in previous centuries (the last occurred in 1980). It is a local network broadcasting the signals to the city of Naples, developed together with the Civil Protection of the Campania Regional Authority.

The systems described in this book do not cover all the existing operational systems. For the sake of completeness, at least two further cases are mentioned in the paper by Iunio Iervolino, Gaetano Manfredi and Edoardo Cosenza in their review of engineering applications, namely the regional system implemented to protect Mexico City and the local system designed for the Ignalina nuclear power plant in Lithuania.

The seismic alert system (SAS) for Mexico City (Mexico) is an EEWS for large earthquakes, which are likely to cause damage in Mexico City and have their source in the subduction zone of the Pacific coast at a distance of about 320 km. The warning time varies between 58 to 74 seconds. Information received from the stations is processed automatically to determine magnitude and is used in the decision to issue a public alert. The Radio Warning System for users disseminates the seismic early audio warnings via commercial radio stations and audio alerting mechanisms to residents of Mexico City, public schools, government agencies with emergency response functions, key utilities, public transit agencies and some industries. During rush hours, approximately 4.4 million people are covered by the system.

The seismic alarm system for the Ignalina nuclear power plant in Lithuania consists of a Seismic Alarm System (SAS) designed to detect potentially damaging earthquakes and to provide an alarm before the arrival of the shear waves at the reactor. Six SAS stations are installed at a distance of 30 km from the power plant forming an array, which is referred to as a seismic "fence". An earthquake with an epicenter outside the fence is detected about 4 seconds before it is "felt" by the reactor. The required time to insert the control rods is 2 seconds. Potentially, the reactor could be shut down before the earthquake arrives. At present, the SAS will only initiate an alarm signal.

A few recent examples of practical use of earthquake early warning information are shortly discussed in the review paper by Kanamori.

We hope that the contents of this book show convincingly that implementation of effective earthquake early warning systems is scientifically and technologically feasible. However, to be really effective any early warning system must include three components:

1. the scientific-technological component that provides information on an impending extreme event,
2. the decision making component that issues a warning, and
3. the response component that ensures an adequate response to the warning.

Today, the main problems in this warning chain occur as a result of inadequate interaction between these different components. This is particularly true for earthquake early warning. Even when the technological means necessary for earthquake early warning, such as seismic instrumentation, computerized systems and telecommunication, are in place, their ability to serve the needs of disaster management and decision makers has only been marginally exploited.

We have the feeling, shared by most of the scientific community, that "end users", such as civil defense organizations, industries and public administrators, react very cautiously to the challenge issued by the scientific community due to the complexity they foresee in activating the second and third components of the earthquake early warning chain. Indeed, sound information and education of the public and officials living in the "protected" area are required in order to produce an effective increase in resilience.

In turn, close interaction between scientists, administrators and the public is the path to follow to take full advantage of the developments offered by science and technology to allow people to continue to live in areas prone to natural hazards with an acceptable level of risk.

Paolo Gasparini
Gaetano Manfredi
Jochen Zschau

This volume was worked out in the framework of the EC FP6 Project No. 511264 "NaRAs" (Natural Risk Assessment) coordinated by AMRA, Napoli.

Contents

List of Contributors

Richard M. Allen
Seismological Laboratory, Department of Earth & Planetary Science, University of California Berkeley, CA, USA

Pierre-Yves Bard
LGIT – Maison des Géosciences, Saint-Martin-d'Hères, France

James L. Beck
Department of Applied Mechanics and Civil Engineering, Caltech

Antonella Bobbio
Istituto Nazionale di Geofisica e Vulcanologia, Osservatorio Vesuviano, Napoli, Italy

Maren Böse
Karlsruhe University, Geophysical Institute, Karlsruhe, Germany

Luciana Cantore
Dipartimento di Scienze Fisiche, Università di Napoli Federico II, Napoli, Italy

Vincenzo Convertito
Istituto Nazionale di Geofisica e Vulcanologia, Osservatorio Vesuviano, Napoli, Italy

Margherita Corciulo
Dipartimento di Scienze Fisiche, Università di Napoli Federico II, Napoli, Italy

Edoardo Cosenza
Dipartimento di Ingegneria Strutturale, Università di Napoli Federico II, Napoli, Italy

Georgia Cua
Swiss Seismological Service, Swiss Federal Institute of Technology (ETH) Zurich, Switzerland

Raffaella De Matteis
Dipartimento di Studi Geologici ed Ambientali, Università degli Studi del Sannio, Benevento, Italy

Martino Di Crosta
Dipartimento di Scienze Fisiche, Università di Napoli Federico II, Napoli, Italy

Luca Elia
Istituto Nazionale di Geofisica e Vulcanologia, Osservatorio Vesuviano, Napoli, Italy

Antonio Emolo
Dipartimento di Scienze Fisiche, Università di Napoli Federico II, Napoli, Italy

Mustafa Erdik
Bogazici University, Kandilli Observatory, Istanbul, Turkey

Massimiliano Giorgio
Dipartimento di Ingegneria Aerospaziale e Meccanica, Seconda Università di Napoli, Aversa, Italy

Veronica F. Grasso
Dipartimento di Ingegneria Strutturale, Università di Napoli Federico II, Napoli, Italy; Visiting Special Student, Caltech

Adrian Grigore
National Institute for Earth Physics, Bucharest, Romania

Thomas Heaton
Department of Civil Engineering, California Institute of Technology, Pasadena, USA

Nai-Chi Hsiao
Central Weather Bureau, Taipei, Taiwan

Giovanni Iannaccone
Istituto Nazionale di Geofisica e Vulcanologia, Osservatorio Vesuviano, Napoli, Italy

Iunio Iervolino
Dipartimento di Ingegneria Strutturale, Università di Napoli Federico II, Napoli, Italy

Constantin Ionescu
National Institute for Earth Physics, Bucharest, Romania

Hiroo Kanamori
Seismological Laboratory, California Institute of Technology Pasadena, CA, USA

Maria Lancieri
RISSC-Lab, Dipartimento di Scienze Fisiche, Università di Napoli Federico II, Napoli, Italy

William H.K. Lee
U. S. Geological Survey (retired), Menlo Park, CA, USA

Anthony Lomax
Anthony Lomax Scientific Software, Mouans-Sartoux, France

Gaetano Manfredi
Dipartimento di Ingegneria Strutturale, Università di Napoli Federico II, Napoli, Italy

Alexandru Marmureanu
National Institute for Earth Physics, Bucharest, Romania

Gheorghe Marmureanu
National Institute for Earth Physics, Bucharest, Romania

Claudio Martino
Dipartimento di Scienze Fisiche, Università di Napoli Federico II, Napoli, Italy

Hormoz Modaressi
BRGM, ARN, Orléans, France

Yutaka Nakamura
System and Data Research Co. Ltd.

Stefan Nielsen
Istituto Nazionale di Geofisica e Vulcanologia, Roma, Italy

Annalisa Romeo
Dipartimento di Scienze Fisiche, Università degli Studi di Napoli Federico II, Napoli, Italy

Jun Saita
System and Data Research Co. Ltd.

Claudio Satriano
RISSC-Lab, Dipartimento di Scienze Fisiche, Università di Napoli Federico II, Napoli, Italy

Tzay-Chyn Shin
Central Weather Bureau, Taipei, Taiwan

Ta-liang Teng
Southern California Earthquake Center, University of Southern CA, Los Angeles, CA, USA

Jean Virieux
Géosciences Azur, Valbonne, France

Emanuel Weber
Istituto Nazionale di Geofisica e Vulcanologia, Osservatorio Vesuviano, Napoli, Italy

Friedemann Wenzel
Karlsruhe University, Geophysical Institute, Karlsruhe, Germany

Yih-Min Wu
Department of Geosciences, National Taiwan University, Taipei, Taiwan

Aldo Zollo
RISSC-Lab, Dipartimento di Scienze Fisiche, Università di Napoli Federico II, Napoli, Italy

1 Real-time Earthquake Damage Mitigation Measures

Hiroo Kanamori

Seismological Laboratory, California Institute of Technology Pasadena, CA, USA

Abstract

Some reflections on real-time earthquake information and early warning methods application to risk mitigation are discussed. A list of application and the recent obtained results are discussed. The main seismological problems related to the implementation of the method are outlined.

1.1 Introduction

Real-time earthquake damage mitigation refers to a practice with which we rapidly determine immediately after a significant earthquake the source parameters and the estimated distribution of shaking intensity and distribute the information to various users. The users include emergency services officials, utility companies (electric, water, gas, telephone etc), transportation services, media, and the public. This information will be useful for reducing the impact of a damaging earthquake on our society.

In most cases, it takes a few minutes to hours to process the data and when the information reaches the users, the damage may have already occurred at the user site. In this case, the information is called the post earthquake information. This information is important for orderly recovery operations in the damaged areas.

In contrast, if the data processing and information transfer can be done very rapidly (i.e., within 10 sec), the information reaches some sites before shaking starts there. In this case, the information is called "Earthquake Early Warning" (EEW). This concept has been around for more than 100 years, but it had not been put in practice until recently for technical and practical difficulties. In Japan, in conjunction with the operation of the high-speed bullet train (Shinkansen) in the 1960s, a warning system for impending ground shaking after a nearby large earthquake was implemented. This system was later extended to UrEDAS (Nakamura 1988, Nakamura and Saita 2007, this issue) which led the subsequent developments for earthquake early warning methodology for more general purposes.

1.2 Post-Earthquake-Information and Earthquake Early Warning

Post earthquake information was routinely issued by various organizations such as the U.S. Geological Survey (USGS) and the Japan Meteorological Agency (JMA), and has been widely used. In California, a project to distribute post earthquake information to various users began in the 1990s. This project was called CUBE (Caltech-USGS Broadcast of Earthquakes) and aims at not only just distribution of earthquake information but also better communication between the providers of the information (e.g., universities and government agencies) and the users. The spirit of this project was inherited by the ShakeMap (Wald et al. 1999, Fig. 1.1). ShakeMap is a map showing the distribution of ground-motion parameters which is produced automatically within a few minutes to an hour after a large earthquake. At present, ShakeMap is used widely by the USGS and other agencies as the basic information for taking emergency measures after a damaging earthquake. For this type of information to be useful, it is important to have close interaction between the providers and the users. One-way communication has only limited utility. In the CUBE project, the interaction was promoted by regularly scheduled meetings to discuss the effective use of real-time information, and in the event of large earthquakes, the feedback from the users concerning how accurately, rapidly and effectively the information was sent to the users and how the information was actually used for emergency operations. This feedback was extremely important for development of CUBE.

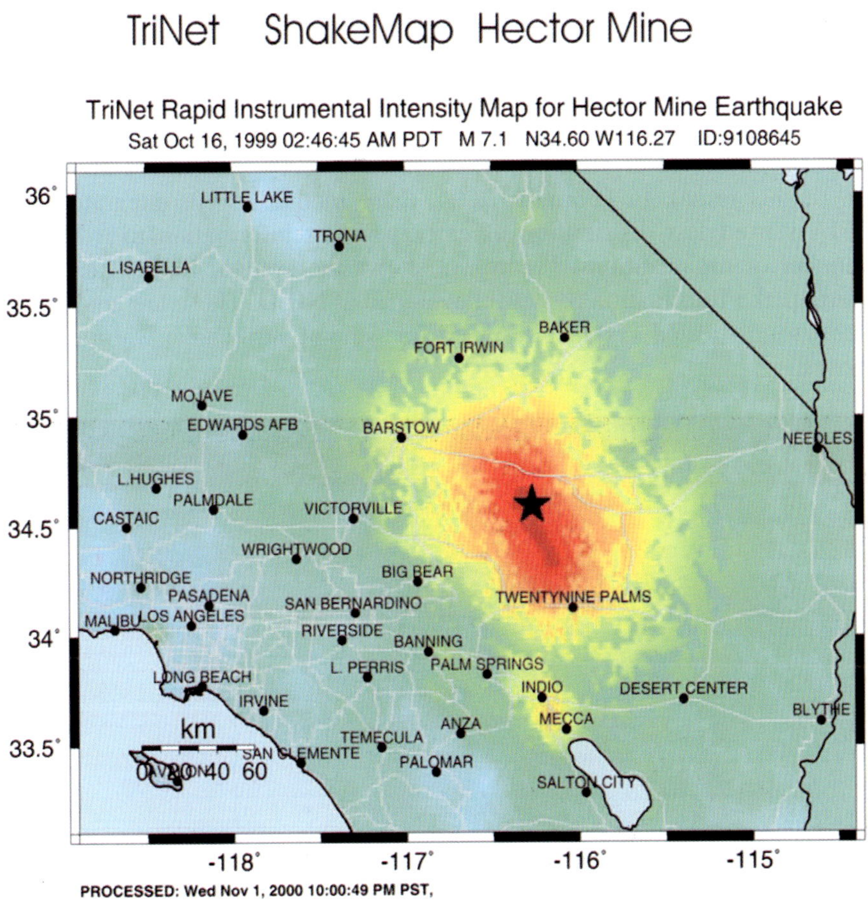

PERCEIVED SHAKING	Not felt	Weak	Light	Moderate	Strong	Very strong	Severe	Violent	Extreme
POTENTIAL DAMAGE	none	none	none	Very light	Light	Moderate	Moderate/Heavy	Heavy	Very Heavy
PEAK ACC.(%g)	<.17	.17-1.4	1.4-3.9	3.9-9.2	9.2-18	18-34	34-65	65-124	>124
PEAK VEL.(cm/s)	<0.1	0.1-1.1	1.1-3.4	3.4-8.1	8.1-16	16-31	31-60	60-116	>116
INSTRUMENTAL INTENSITY	I	II-III	IV	V	VI	VII	VIII	IX	X+

Fig. 1.1 ShakeMap for the 1999 Hector Mine (M_W=7.1, California) earthquake. ShakeMap shows the intensity distribution computed automatically from the observed ground motion, and is usually distributed to the users within a few minutes to 1 hour after an earthquake.

The rapid progress of modern seismological practice, information processing, and data telemetry in recent years has made it possible to produce similar information in a matter of a few seconds, instead of a few minutes, after a large earthquake. This progress made earthquake early warning a realistic goal. To date, several warning systems are practically used in Japan (Shinkansen), Mexico, and Taiwan. In Japan, various methods were developed in the 2000s at JMA, Railway Technical Research Institute, and National Research Institute for Earth Science and Disaster Prevention (NIED) (Horiuchi et al. 2005, Tsukada et al. 2004, Nakamura and Saita 2007, this issue). In February, 2004, these methods were integrated and JMA started test distribution of early warning information to a limited number of organizations. Figure 1.2 shows the system, REIS (Real-time Earthquake Information System) developed at NIED. These are among the most sophisticated systems designed for general earthquake warning purposes.

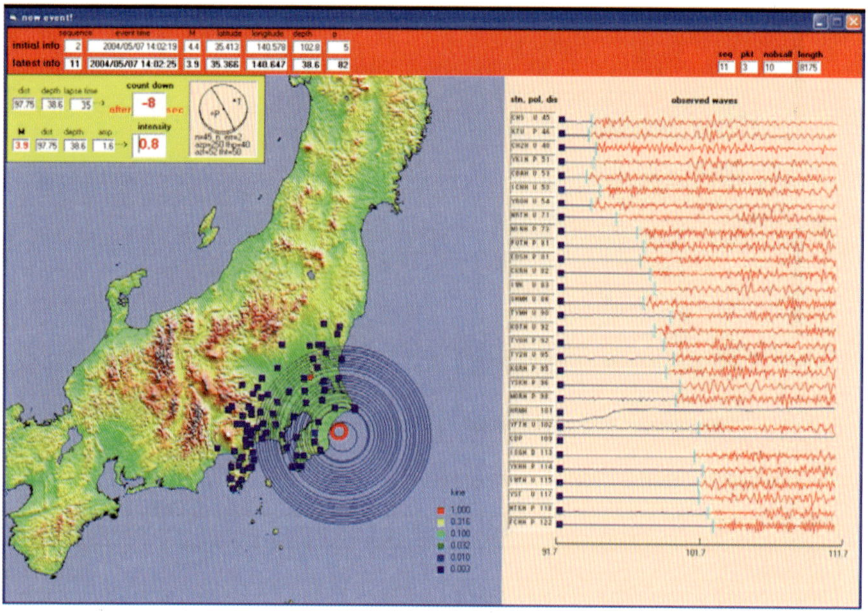

Fig. 1.2 This map was produced with REIS after an earthquake offshore of Boso, Japan. The wave front propagating from the epicenter is shown on the map. The estimates of the magnitude, intensity and the arrival time of strong ground motion at the receiver site are displayed. The seismograms at different locations are shown on the right. The users can tell when they should expect the onset of the strong shaking. (Courtesy of Dr. S. Horiuchi).

At present, the research is focused on how to best utilize the early warning information issued by these modern systems. It is still unknown how these systems will perform for very large earthquakes with large source dimensions, and for earthquakes at short distances (less than 30 km), but this development attracted public attention to the practical use of earthquake early warning. For example, a research group at Nagoya University is conducting an active interdisciplinary research on the practical use of earthquake early warning information. The interdisciplinary approach involving seismologists, engineers, social scientists, and emergency management personnel is critically important for successful implementation of early warning information in the future.

1.3 Implementation and Associated Problems

In the following, we list a few recent examples of practical use of earthquake early warning information.

1. UrEDAS has long been used for controlling the speed of Japanese Shinkansen after a large earthquake. During the recent Chuetsu earthquake in Japan (October 23, 2004, M_W=6.6), a UrEDAS located in the epicentral area issued an warning at 1 sec after the P-wave arrival at the site, which resulted in power shutdown and activation of emergency brakes on the train moving at a speed of 200 km/h near the epicenter (Fig. 1.3, Nakamura 2005). In this case, the train eventually derailed (no casualty) a few seconds later, and some media made somewhat negative reports to the effect that early warning was a failure. However, this view seems to be missing the point. It is remarkable that the early warning system worked as it is supposed to in such a short time. The system is not intended to prevent derailment; it is designed to slow down the train to minimize the impact of strong ground motion.
2. Motosaka et al. (2006) reports an experiment involving an elementary school in Sendai, Japan, to practice emergency exercises in response to an earthquake early warning to be issued by the JMA system. In this area, magnitude 7 earthquakes (Miyagi-Oki earthquakes) are known to occur offshore once every approximately 30 years. Motosaka et al. (2006) demonstrates the merit of such warnings for large offshore events.
3. Kanda et al. (2006) reports the use of the JMA early warning system in the construction site of a high-rise building in Yokohama. For the safety of the workers at the construction site, when an earthquake early warning is received at the site the workers are immediately notified the possibility of impending strong ground motion so that they can take proper safety measures such as stopping the elevators at the nearest floor and setting the tower cranes in a safe position.

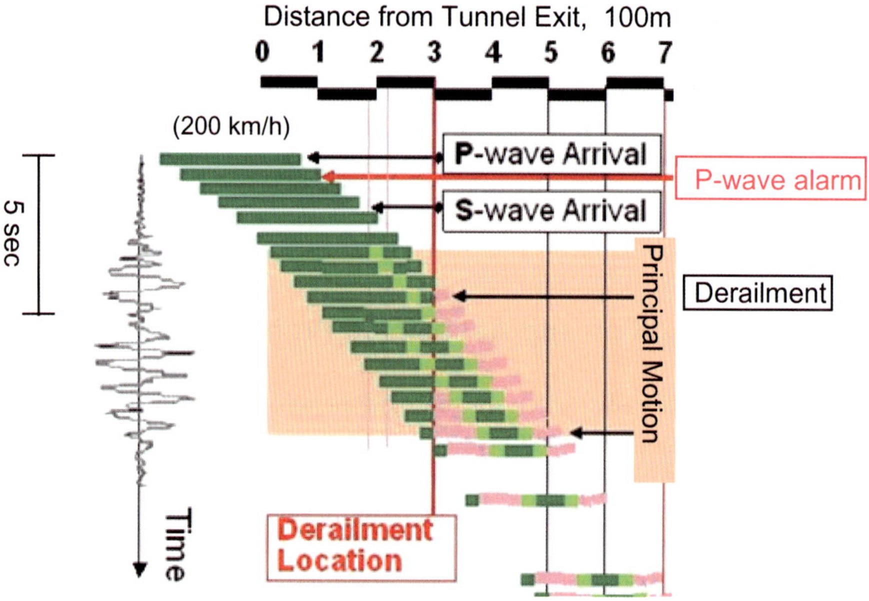

Fig. 1.3 Schematic diagram showing how UrEDAS worked during the 2004 Chuetsu, Japan, earthquake (M_W=6.6). The locations of the train are shown by green bars. The horizontal axis indicates the distance from the exit of a tunnel and the vertical axis is the time. A seismogram at a location near the train is shown along the vertical axis to indicate the ground motion. An alarm was issued at about 1 sec after the arrival of P wave. The ground motion was not very large at this time, and it took a few more seconds before the maximum motion occurred. The derailed cars are indicated by pink. (Courtesy of Dr. Y. Nakamura).

In addition to various technical issues, the overall reliability, the impact of false alarms and missed alarms, and the associated liability are among the issues being vigorously discussed these days. Needless to say that these issues are important, but at present when not many early warning system are in operation, it is somewhat difficult to fully understand its utility. It is probably most important at this point to accumulate more experience by testing various real-time systems for practical applications. Since we need to deal with complex earthquake processes and even more complex societal problems, it would be inevitable to encounter some difficulties associated with false alarms, missed alarms, and the resulting chaotic social responses. Accordingly, it would be better to start with practical use of earthquake early warning information for applications where false alarms and missed alarms will not cause catastrophic consequences. Introduction of any completely new concepts and methodology inevitably involves some risk. Nevertheless, considering the extremely serious impact of a large earthquake on modern metropolitan areas, introduction of effective short-term damage mitigation measures is desirable. Now that the technical feasibility has been demonstrated, it is most important to start exploring the effective use of earthquake early warning.

1.4 Basic Research on Seismology and Earthquake Early Warning

Besides its practical importance, earthquake early warning is an interesting subject of basic seismological research. After an earthquake has occurred, the wave propagation process is essentially governed by the crustal structure and the wave equation, and the uncertainty is expected to be fairly small. This is in contrast to the traditional earthquake prediction, in which the process is governed by many factors such as the distribution of stress, strength, the extent of interaction between different parts of the crust etc, and the prediction is inevitably very uncertain. In earthquake early warning, if the displacement field at the early stage can be measured accurately, its future development can be estimated fairly accurately using the wave equations and the known (at least approximately) crustal structure. To proceed with this method effectively, we need extensive research on the physics of earthquakes and on wave propagations in three-dimensionally heterogeneous media. Thus, the problem of earthquake early warning is not only an important practical problem but also an interesting scientific problem. Earthquake early warning may be one of few problems in which relatively accurate short-term predictions can be achieved. In most seismological problems, accurate short-term predictions are difficult because of the many unknown elements involved.

References

Horiuchi S, Negishi H, Abe K, Kamimura A, Fujinawa Y (2005) An Automatic Processing System for Broadcasting Earthquake Alarms. Bull Seism Soc Amer 95;708-718

Kanda K, Nasu T, Miyamura M, Kobori T, Takahashi M, Nagata T, Yamaya H (2006) Application of earthquake early warning system to construction sites. Proceedings of 4th world conference on structural control and monitoring, UC, San Diego, 11-13 July 2006

Kikuchi M (2003) Real-time Seismology. University of Tokyo Press, pp 222

Motosaka M, Fujinawa Y, Yamaguchi K, Kusano N, Iwasaki T, Satake A (2006) Application of early warning system for disaster prevention in schools using real-time earthquake information. Proceedings of the 8th U.S. National Conference on Earthquake Engineering, April 18-22, 2006, San Francisco, California, USA, paper no. 719

Nakamura Y (1988). On the urgent earthquake detection and alarm system (UrEDAS). Presented at Ninth World Conf. Earthq. Eng., Tokyo

Nakamura Y (2005) Earthquake early warning and derailment of Shinkansen train at the 2004 Niigataken-Chuetsu earthquake (in Japanese). Proceedings of Earthquake Engineering Symposium of the Japanese Society of Civil Engineers, August 23, 2005, paper no. 115

Nakamura Y, Saita J (2007) UrEDAS, the Earthquake Warning System: Today and Tomorrow. In: Gasparini P, Manfredi G, Zschau J (eds) Earthquake Early Warning Systems. Springer

Tsukada S, Odaka T, Ashiya K, Ohtake K, Zozaka D (2004) Analysis of the envelope waveform of the initial part of P-waves and its application to quickly estimating the epicentral distance and magnitude. Zisin 56:351-361

Wald DJ, Quitoriano V, Heaton TH, Kanamori H, Scrivner CW, Worden CB (1999) TriNet "ShakeMaps": Rapid generation of peak ground motion and intensity maps for earthquakes in southern California. Earthquake Spectra 15:537-55

2 Can Earthquake Size be Controlled by the Initial Seconds of Rupture?

Stefan Nielsen

Istituto Nazionale di Geofisica e Vulcanologia, Roma

Abstract

It has been argued that the dominant period T_p derived from the initial seconds of a seismogram, hence only depending on the initial phases of earthquake rupture, seems to scale with the final size of the earthquake. We provide a physical interpretation for the observed scaling and explain how the final earthquake size could be controlled by the initial phase of rupture.

2.1 Introduction

What are the chances that an initially small rupture continues to propagate and turns into a large earthquake? Propagation or arrest of earthquake rupture is ultimately controlled by the energy balance between the work of frictional breakdown and that of elastic stress (Aki 1979). Despite the above simple statement, the problem is not trivial: it can be shown that under given initial conditions, the balance strongly depends on the rupture history and modality. In particular, the energy flow essentially differs if the fracture propagates in the form of a large crack or a fracture pulse of variable size (Nielsen and Madariaga 2003). The strength of an earthquake barrier can be defined in terms of friction parameters and stress, but its capacity to stop rupture will critically depend on dynamic properties and essentially, on the characteristic length Λ of a fracture pulse (or in the case of a crack, its radius). As a consequence, the probability that a starting fracture will continue to propagate depends on the size Λ of the fracture pulse. For an average rupture propagation velocity v_r, the rise time can be defined as $T_r \approx \Lambda / v_r$, hence the probability of continued propagation should depend on the rise-time. It has been argued that a dominant period T_p can be derived from an earthquake seismogram; although T_p is derived from the initial seconds of the seismogram and hence only depends on the initial phases of rupture, it seems to scale with the final size of the earthquake

(Allen and Kanamori 2003; Olson and Allen 2005). These intriguing results immediately raise the question of causality: a physical justification should be found for this apparent predetermination of the earthquake size. We propose that T_p is linked to risetime T_r, and show how in T_r the initial phases of the rupture may affect the final size of rupture.

2.2 Statement of the Problem

The scope of early warning studies is to anticipate as much as possible the response to a potentially destructive event. The size of an earthquake should be determined as soon as possible in order to trigger a proper response, and reducing the delay of a few seconds only may be determining for the success of final size of an earthquake (in the probabilistic sense), even before the rupture propagation has ended. Note that we are not discussing the properties of the slow, quasi-static nucleation phase, but the early phases of dynamic rupture acceleration and advancement.

An earthquake is triggered on a fault at time t_0. At time t_1 the rupture has already propagated to a finite size A (Fig. 2.1), radiating a wavefiled which is captured by one or more seismographs in the vicinity of the fault at time t_3. Do the properties of rupture at time t_1 (and of the wavefield recorded at time t_3) carry some information on the probability that the fracture will continue to propagate until it reaches a final size B? If the answer is positive, which physical model of rupture is in agreement with such a statement?

2.3 Fracture, Barriers and Energy Concepts

The probability Π_p that an earthquake rupture continues to propagate is complementary to the probability Π_s that the rupture stops, i.e., Π_p=1- Π_s. In other words, we have to investigate the *stopping dynamics* of earthquake rupture in order to understand what mechanism could control the final size of rupture. Earthquake propagation stops when the rupture runs into a sufficiently strong barrier, so the first step is to quantify barrier strength.

The relative barrier strength should be defined in terms of energy balance, and several classical studies have treated the question of rupture propagation or arrest in term of fracture energy. The concept was originally developed by Griffith (1921). when describing the conditions under which a static crack becomes unstable and starts to grow. It was subsequently developed to more complex situations, including several cases of dynamic propagation. In all cases, the problem is essentially described as the balance between the loading conditions or fracture-driving force, on the one hand, and the energy dissipated by the fracture process or by the creation of newly cracked surface (fracture energy), which tends to resist crack propagation, on the other hand. When the applied load is sufficient to overcome dissipation, the crack propagates, otherwise it stops. The non trivial point is that the entity of fracture driving force available depends not only on the remotely applied load, but also on the geometry and scaling of the problem, in particular, the size of the preexisting crack.

Earthquake-stopping barriers were characterized, for example, in the Pioneer studies by Bouchon (1979) and by Aki (1974) for the Parkfiled, 1966 earthquake. Aki build his study on earlier theoretical work of Barenblat (1959) and Ida (1973) who described the fracture energy as a dissipation process taking place in a finite, cohesive zone around of the propagating fracture.

The size d of the cohesive zone controls the scaling of stress at the fracture tip, and, as a consequence the stress intensity factor K (in units of Pa $m^{1/2}$) and the fracture energy (usually named G, in units of J m^{-2}). Another length, the characteristic slip weakening distance δ_c, allows scaling of the dissipated energy G, as the work done by friction against slip inside the cohesive zone.

I will first recall the main relationships that allow to define energy flow during the propagation of simple fracture models, and argue that the distinction should be made between two independent estimates of the fracture energy. On the one hand we may define G_e as the energy flow into the fracture tip. G_e equates to the finite amount of elastic energy stored in the vicinity if the fracture tip, which is absorbed when the fracture advances of a unit length. Though G_e usually concerns a small region around the fracture tip, it depends on the load provided by the whole previous slip history on the crack faces, thus it cannot be defined a priori based on local fault properties. On the other hand, the dissipated energy G_w may be defined as the work against the friction excess during the initial, weakening part of slip, depending, in principle, on local friction parameters only. Obviously, the dynamic fracture process satisfies the energy balance so that at the available energy G_e and the dissipated energy G_w should coïncide during fracture propagation.

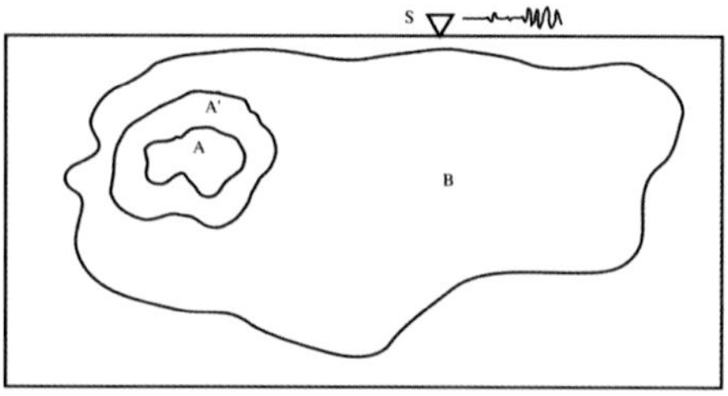

Fig. 2.1 Schematic representation of earthquake faulting process and formulation of the causality problem. Assume that in the source area, fracture has expanded into an area A at time t_a and continues to expand. The radiation produced by A at t_a reaches the receiver S at time $t_b > t_a$. At t_b the source has reached a larger area A' and continues to propagate. However, the signal arriving at S at time t_b only contains information on the early rupture patch A; it is not affected by A', all the less by the much larger area B, whose radiation will reach S at later times. The question is then: do the source properties in the initial area A affect the chances that rupture continues to grow and reaches a size B? If so, can such properties identified in the early radiated field reaching S at time t_b?

2.4 Defining and Quantifying Fracture Energy

Let's first define the dissipated energy G_w based the work against the friction excess during the initial, weakening part of slip, depending, in principle, on local friction parameters only:

$$G_w(\mathbf{x}) = \int_0^{D_c} \Big(\tau_f(\mathbf{x},\delta) - \tau_r(\mathbf{x})\Big)\, \mathrm{d}\delta = \int_0^t \dot{\delta}(\mathbf{x},\mathbf{t}') \Big(\tau_f(\mathbf{x},\mathbf{t}') - \tau_r(\mathbf{x})\Big)\, \mathrm{d}t' \tag{2.1}$$

where δ and $\dot{\delta}$ are slip and slip-rate, τ_w is the friction on the fault and τ_r is the relaxed (or minimum level) of the friction during slip. As usual, we can illustrate the G_w integral as the area below the initlal part of the frictional curve. For a simplified slip-weakening behavior as defined by Ida, where the friction drops linearly between the peak stress τ_y and the relaxed stress τ_r, the dissipation reduces to:

$$G_w = \tfrac{1}{2}\, D_c\, (\tau_y - \tau_r).$$

Let's now illustrate how the dissipated energy G_e may be estimated based on the fracture history instead. In certain cases knowledge of the previous slip history allows to compute the stress in the vicinity of the fracture tip and to derive the stress intensity factor K, which, in turn, allows to derive the the energy flow G per unit advancement of the fracture tip. If we consider only shear fracture (no opening), we may have anti-plane and in-plane motion (mode III and mode II, resp.), and the two intensity factors are defined as:

$$K_{III} = \lim_{r \to 0} \sqrt{2\pi r}\, \tau_{\perp}(r)\,, \tag{2.2}$$

$$K_{II} = \lim_{r \to 0} \sqrt{2\pi r}\, \tau_{//}(r)\,, \tag{2.3}$$

where r is the distance ahead of the crack tip and τ is the shear traction in the fracture plane, either parallel (//) or perpendicular ($\perp$) to the slip direction. Then the energy flow can be written, according to Irwin (1957), for a quasi-static crack:

$$G_e = (1-\nu)\,\frac{K_{II}^2}{2\mu} + \frac{K_{III}^2}{2\mu}$$

where μ is the shear stiffness and ν the poisson ratio. When the velocity of fracture propagation is not small, additional functions should be introduced to account for the dynamic propagation. For fractures propagating at a constant velocity, the additional dynamic terms can be evaluated analytically and we may write

$$G_e = (1-\nu)\,\frac{K_{II}^2}{2\mu}\,Y_{II}(v_r/\alpha) + \frac{K_{III}^2}{2\mu}\,Y_{III}(v_r/\beta) \tag{2.4}$$

Broberg (1999) called Y_{II} and Y_{III} he Yoffe functions in memory of the pioneer studies where Y_I was defined (Yoffe 1951). A slightly modified form of the Yoffe functions called $F(v_r)$ is also found in Freund (1979) and Rice (2005). The Yoffe functions only depend on the dimensionless ratios of fracture velocity to wave velocity, and hence remain the same whether fracture takes place as an expanding crack, a steady-state self-healing pulse, a self-healing self-similar pulse or a rupture with more complex history:

$$Y_{II} = \frac{2B\,(1-B^2)\,\gamma_\alpha^2\sqrt{B^2-\gamma_\alpha^2}}{4B^3\sqrt{1-\gamma_\alpha^2}\sqrt{B^2-\gamma_\alpha^2}-\left(2B^2-\gamma_\alpha^2\right)^2} \tag{2.5}$$

for the subsonic case ($v_r < v_{Rayleigh}$), and for the intersonic case ($\beta < v_r < \alpha$)

$$Y_{II}^{*} = \frac{2B\,(1-B^2)\,\gamma_\alpha^2\sqrt{B^2-\gamma_\alpha^2}}{\left(\gamma_\alpha^2-2B^2\right)^2\sqrt{1+\frac{16(1-\gamma_\alpha^2)(B^2-\gamma_\alpha^2)B^6}{(\gamma_\alpha^2-2B^2)^4}}} \tag{2.6}$$

With $B=\beta/\alpha$ and $\gamma_\alpha = v_r/\alpha$.The anti-plane function yields the much simpler expression:

$$Y_{III} = 1/\sqrt{1 - v_r^2/\beta^2} \tag{2.7}$$

Though *Y(.)* only depend on fracture velocity, the stress intensity factors *K(.)* vary greatly depending on the type and history of fracture, with high consequences on the energy flow.

For example, we an compare the stress intensity function for a mode III fracture, which can be derived according to expression (2.2), for some particular cases where an analytical expression is known. for the steady state pulse of length Λ:

$$K_{III} = (\tau_0 - \tau_r)\,\sqrt{2\,\pi\,\Lambda}$$

for an expanding, self-similar crack that has reached a radius of Λ (i.e., $\Lambda = t\, v_r$):

$$K_{III} = \frac{\sqrt{1 - v_r^2/\beta^2}}{\mathbf{E}(1 - v_r^2/\beta^2)}\,(\tau_0 - \tau_r)\,\sqrt{2\,\pi\,\Lambda}$$

where **E**(.) is the complete elliptic integral of the second kind, and finally, in the case of an expanding, self-similar pulse, that has reached a length Λ (i.e., $\Lambda = t\,(v_r - v_h)$):

$$K_{III} = \frac{\sqrt{\phi}\,\sqrt{\beta/v_r}\,\;(\tau_0 - \tau_r)\,\sqrt{2\,\pi\,\Lambda}}{4\left(1 - \frac{v_h}{v_r}\right)\left(\mathbf{F}\left(\frac{\lambda}{\phi}\right) - \mathbf{\Pi}\left(\frac{1+\frac{v_r}{\beta}}{1-\frac{v_r}{\beta}}, \frac{\lambda}{\phi}\right) + 2\sqrt{\frac{\beta\phi}{v_r}}\mathbf{E}\left(\frac{\lambda}{\phi}\right)\right)}$$

where, for clarity, the notations

$$\phi = \left(1 + \frac{\beta}{v_h}\right)\left(\frac{\beta}{v_r} - 1\right)$$

$$\lambda = \left(\frac{\beta}{v_h} - 1\right)\left(1 + \frac{\beta}{v_r}\right)$$

have been introduced. As seen above, the stress intensity factor depends on fracture velocity and healing front velocity if a healing is present (in the case of the steady state pulse, rupture velocity does not appear explicitly in the above *K* expression, but only if *K* is written in terms of final slip instead of stress drop, see for example in Freund 1979).

Though the above examples concern a *limited* set of rupture modes, their properties apply to all fracture types; In all cases, the stress intensity is proportional to the square root of Λ, so that the energy flow *G* will be proportional to Λ, the size of the actively slipping fracture, and we can write:

$$G_e = \psi\left\{\frac{vr}{\beta}, \frac{vr}{\alpha}, \frac{vh}{\beta}\right\} \frac{\pi\left(\tau_0 - \tau_r\right)^2}{\mu}\Lambda \tag{2.8}$$

though the functional form of φ varies depending on the fracture modality, the dependance of G_e on stress drop square $(\tau_0\text{-}\tau_r)^2$ and active fracture length Λ remains the same, even for complex fracture histories with no analytical conterpart. It is essential, however, that Λ in (2.8) describes the length of actively slipping rupture (rupture pulse), and not the length of fracture propagation (note, however, that in the case of crack-like rupture both lengths are the same).

As shown by Nielsen and Madariaga (2003), the presence of a propagating healing front in the fracture trail modifies the relationship between energy flow and propagation velocity, providing a self-locking mechanism for stopping fracture when the rupture front slows down. The Nielsen Madariaga (2003) solution correspond to a self-similar, expanding pulse, a fracture modality spontaneously developping under conditions where a healing front is triggered. Such conditions include the presence of a mild rate-weakening behavior in the friction law.

Steady-state fracture pulse of constant length were illustrated by Yoffe (1951), Freund (1979), Rice (2005) and Dunham and Archuleta (2005). The relevance of self-healing fracture pulses for earthquake faulting was also discussed in Heaton (1990) showing that kinematic inversions for several large earthquakes infer a systematically short rise-time T_r. Except for the case of intersonic fracture velocity illustrated in Dunham and Archuleta (2005), the nature of the steady-state solution is such that the kinetic wavefield around the fracture remains unchanged: this implies that no kinetic energy is radiated. As a consequence, the energy balance is greatly simplified energy flow reduces to a *locally* satisfied balance, where the energy dissipation is simply the product of slip by the dynamic stress drop. Indeed, the global energy balance for propagating fracture may be written:

$$\int_T \int_\Gamma \tau_0(\mathbf{x})\ \dot{\delta}(\mathbf{x},t)\mathrm{d}\mathbf{x}\,\mathrm{d}t = \int_T \int_\Gamma \tau_f(\mathbf{x},t)\ \dot{\delta}(\mathbf{x},t)\ \mathrm{d}\mathbf{x}\,\mathrm{d}t + \int_T \int_V \omega_{ksg}(\mathbf{x})\ \mathrm{d}v\,\mathrm{d}t$$

where Γ is the fault surface, ω_{ksg} groups the sum of kinetic, strain and gravity energy variations due to fracture, affecting volume V around the fault. Moreover, $\delta(\mathbf{x})$ is the slip at point $\mathbf{x}$, while τ_0 is the initial shear traction (prestress) and τ_f the frictional traction on the fault. The term on the left hand side is the work W done on fracture by the initial load τ_0, while first term on the right hand side describes the frictional work. If we introduce the relaxed frictional traction τ_r, the different terms can be re-grouped in order to make the fracture energy appear explicitly:

$$\int_T\int_\Gamma \tau_0(\mathbf{x})\,\dot{\delta}(\mathbf{x},t)\mathrm{d}\mathbf{x}\,\mathrm{d}t = \int_T\int_\Gamma \dot{\delta}(\mathbf{x},t)\left(\tau_f(\mathbf{x},\mathbf{t}) - \tau_r(\mathbf{x})\right)\mathrm{d}\mathbf{x}\,\mathrm{d}t + \int_T\int_\Gamma \dot{\delta}(\mathbf{x},t)\,\tau_r(\mathbf{x})\,\mathrm{d}\mathbf{x}\,\mathrm{d}t + \int_T\int_V \omega(\mathbf{x},\mathrm{t})\,\mathrm{d}v\,\mathrm{d}t \tag{2.9}$$

we recognize on the left hand of the equation, the energy dissipated in fracture advancement G_w (fracture energy) as defined in equation (2.1). On the right hand, the first term is the work W done by the prestress τ_0, while the second term corresponds to the energy dissipated into frictional heat Q. Integrating over time we can then write:

$$\int_\Gamma G_w(\mathbf{x})\,\mathrm{d}\mathbf{x} = \int_\Gamma \left(W(\mathbf{x}) - Q(\mathbf{x})\right)\mathrm{d}\mathbf{x} - \int_V \Omega(\mathbf{x})\,\mathrm{d}v \tag{2.10}$$

In the case of a sub-sonic, steady-state pulse the variation of volume energy Ω cancels (no variations of kinetic, strain or gravity energy around fracture, but a steady field). In this case, the equation is verified on any arbitrary small subset of Γ, so we can get rid of the surface integral. The requirement G_w=G_e is imposed by steady propagation, so that that (2.10) reduces to

$$G_w(\mathbf{x}) = G_e(\mathbf{x}) = \delta_{final}(\mathbf{x})\,(\tau_0(\mathbf{x}) - \tau_f(\mathbf{x})) \tag{2.11}$$

It is more troublesome to investigate the energy balance (2.10) for cases where of non-steady propagation. Indeed, the last term of the integral is essentially non-local and reflects the energy redistribution through elastic stress radiation. It can be argued, however, that the Ω term is negligible with respect to the remainder, and use (2.11) to evaluate seismic fracture energy from observation of earthquake parameters, an approach followed by Rice et al. (2005).

In any case, equating G_w and G_e provides a constraint on the possible fracture velocity and the viability of a stable fracture propagation (for example by equating (2.4) and (2.1) when K is known). If G_e becomes larger than G_w, the balance will be restored by an acceleration of fracture propagation; in the opposite case, the fracture would slow down and eventually stop. If no real solution to the equation is found, the fracture will not propagate.

2.5 Energy Flow, Moment Rate and Dominant Period

According to (2.8), the probability that an earthquake fracture will continue to propagate past a possible barrier, does not only depend on the relative local strength of the fault, but also on the dynamic stress drop and on the active slipping length inherited from the previous rupture history (both parameters which may somehow reflect in the early phases of the seismogram).

For an average fracture velocity v_r, the duration of active rupture will be of the order of $T_r = \Lambda / v_r$, as a consequence, the rise-time T_r reflects the pulse length Λ and we can rewrite (2.8) as

$$G_e = \psi \left\{ \frac{vr}{\beta}, \frac{vr}{\alpha}, \frac{vh}{\beta} \right\} \frac{\pi \left(\tau_0 - \tau_r \right)^2}{\mu} T_r \, v_r \quad . \tag{2.12}$$

Fractures with an initially larger rise-time will generate lower frequencies and the early phases of the seismogram should reveal a larger dominant period T_p, indicative of a larger available energy flow G_e for the advancement of fracture.

In addition, seismograms indicating a relatively large initial moment rate of the earthquake source, also should be diagnostic of a larger energy flow. Indeed, as seen from (2.8) G_e and the probability of fracture propagation increases as the square of the dynamic stress drop; the latter affects slip-rate on the fault, increasing the moment rate of the earthquake source. However, scaling laws show no evidence of stress drop increase with earthquake magnitude, so that may not be a realistic physical mechanism. On the other hand, if stress drop does not vary significantly, a larger Λ (or T_r) results in a larger actively slipping area, which also reflects in an increased moment rate of the earthquake source. Indeed, if we write the moment rate as

$$\dot{M}_o = \int_\Sigma \mu \, \dot{\delta} \, d\Sigma \approx \mu \, \dot{\bar{\delta}} \, A \quad . \tag{2.13}$$

For example, in the case of a steady-state pulse of dimension Λ spanning a fault of width W (Haskell-type faulting) yields:

$$\dot{M}_o \approx \mu \, \dot{\bar{\delta}} \, \Lambda \, W,$$

whereas for a circular fracture front of radius $R=v_r\, t$, where the active area is a ring-like pulse of constant width Λ:

$$\dot{M}_o \approx \mu \bar{\dot{\delta}}\, 2\,\pi\,(2\,t\,v_r - \Lambda)\,\Lambda.$$

Indeed, we note that the area A in (2.13), is not the final earthquake rupture surface but the actively slipping area at a given time (only a fraction of the total earthquake surface in the case of a pulse); since $\dot{\delta}=0$ outside the fracture pulse, the integral cancels on the inactive area (healed or not yet fractured fault portion). As a consequence we may write, for the steady-state pulse of Λ over a fault of width W:

$$A = W\,\Lambda$$

whereas for the ring-shaped pulse:

$$A = 2\,\pi\,(2\,t\,v_r\,\Lambda - \Lambda^2).$$

In the case of a crack,

$$\dot{M}_o \approx \mu \bar{\dot{\delta}}\, 2\,\pi\,t\,v_r.$$

According to classical elastodynamic results, the ascimptotic slip rate directly scales with the dynamic stress drop (τ_0- τ_r):

$$\dot{\delta} = C\,h(\frac{v_r}{\beta})\,\beta\,\frac{\tau_0 - \tau_r}{\mu}$$

(where μ is the shear rigidity, β the shear wave velocity, $h(v_r)$ a dimensionless function of fracture velocity v_r which tends to 1 as $v_r \rightarrow \beta$, and C a geometrical factor of order 1).

Finally, the far field displacement u can be schematically written, taking the example of a P wave and for a point source (for the sake of simplicity all complexity such as directivity effects are neglected) as a function of moment rate:

$$u(t,r) = \frac{A^{FP}}{4\,\pi\,\alpha^2\,r}\dot{M}_o(t - \frac{r}{\alpha}),$$

where A^{FP} is the radiation pattern, α the P wave velocity and r the source-reciever distance (Aki and Richards 2002).

A direct combination of the four above equations shows that the far field displacement u scales with the active rupture length Λ and with the stress drop (τ_0- τ_r). As mentioned earlier, the fracture energy G also scales with Λ and $(\tau_0 - \tau_r)^2$.

From the above discussion it appears that if the average slip rate is fixed (i.e., fixed dynamic stress drop), the moment rate should increase at least as Λ, in syntony with the available energy flow G_e. Large amplitude in the first seismogram phases thus implies either large Λ or large stress drop (τ_0-τ_r), or both, and an increased probability of propagation into a larger earthquake. Moment rate being relate to the radiated energy, this should reflect in the far field amplitude and/or period of the signal.

An increased moment rate implies a larger available energy flow G_e and thus increased probability of rupture continued propagation.

2.6 Predictive Statement on Final Rupture Size

According to (2.12), earthquake sources that initially generate large dominant periods and large moment rate functions indicate that rupture propagation is favoured by a large energy flow; large energy flow indicates that fracture is more likely to continue propagating and to grow into a large magnitude earthquake. Note that the above statement is predictive only in a probabilistic sense, because the strength distribution ($G_w(\mathbf{x})$) on the fault is unknown.

A stochastic distribution $\rho_w(G)$ for fault strength (in terms of G_w) can be assumed a priori, and a probabilistic statement can be issued, that the starting fracture with an energy flow G_e (estimated from moment rate and dominant period), encounters a barrier of strength $G_w \geq G_e$ and of size $l \geq \Lambda$ within a propagation distance L, For example, if G_w obeys a fractal distribution on earthquake faults,

the probability $\Pi s(G_w \geq G_e, l > \Lambda, L)$ of encountering a ``strong" cluster of size larger than Λ within a region L can be estimated with standard tools of percolation theory (Feder 1992). The probability that the intial rupture grows into an earthquake source of size L_0 or more is:

$$\Pi(L \geq L_0) = 1 - \Pi_s(G_w \geq G_e, l > \Lambda, L_0).$$

The above estimates and distributions can be adjusted by testing predictive statements on regional catalogs of past earthquakes.

References

Aki K (1979) Characterization of barriers on an earthquake fault. J Geophys Res 84:6140-6148

Aki K, Richards P (2002) Quantitative Seismology. University Science Books

Barenblat GI (1959) The formation of equilibrium cracks during brittle fracture: General ideas and hypothesis, axially symmetric cracks. J Appl Math Mech 23:434-444

Bouchon M (1979) Predictability of ground displacement and velocity near an earthquake fault: the Parkfield earthquake of 1966. J Geophys Res 84:6149-6156

Broberg KB (1999) Cracks and Fracture. Academic Press, London

Dunham EM, Archuleta RJ (2005) Near-source ground motion from steady state dynamic rupture pulses. Geophys Res Lett 32:L03302, doi:10.1029/2004GL021793

Feder (1992) An introduction to fractal geometry. University Press

Freund LB (1979) The mechanics of dynamic shear crack propagation. J Geophys Res 84:2199-2209

Griffith AA (1921) The phenomena of rupture and flow of solids. Phil Trans Roy Soc London 221:163-198

Heaton T (1990) Evidence for and implications of self-healing pulses of slip in earthquake rupture. Phys Earth and Planet Int 64:1-20

Ida Y (1973) Cohesive force across he tip of a longitudinal shear crack and Griffith's specific surface energy. J Geophys Res 77:3796-3805

Irwin (1957) Analysis of stresses and strains near the end of a crack traversing a plate. Jour Appl Mech Trans ASME 79 :361-364

Nielsen, Madariaga (2003) On the self-healing fracture pulse. Bull Seismol Soc Am

Olson EL, Allen RM (2005) The deterministic nature of earthquake rupture. Nature 438:212-215, doi:10.1038/nature04214

Rice JR, Sammis CG, Parsons R (2005) Off-Fault Secondary Failure Induced by a Dynamic Slip Pulse. Bulletin of the Seismological Society of America 95(1):109134, doi: 10.1785/0120030166

Allen RM, Kanamori H (2003) The Potential for Earthquake Early Warning in Southern California. Science 300:786-789

Yoffe EH (1951) The moving Griffith crack. Phil Mag 42:739-750

3 The ElarmS Earthquake Early Warning Methodology and Application across California

Richard M. Allen

University of California Berkeley

Abstract

Earthquake Alarms Systems, ElarmS, is a methodology for providing warning of forthcoming ground shaking during earthquakes. The approach uses a network of seismic instruments to detect the first-arriving energy at the surface, the P-waves, and translate the information contained in these low amplitude waves into a prediction of the peak ground shaking that follows. The instruments closest to the epicenter are the first to detect the seismic energy, and by using a seismic network this information can be integrated to produce a map of future ground shaking everywhere. The ElarmS methodology uses the frequency content of the P-wave arrival to estimate earthquake magnitude, arrival times to determine location, and then predicts the ground shaking using a radial attenuation relation. All data is gathered continuously and the hazard map updated every second. As observations of peak ground shaking are also made close to the epicenter they are integrated into the hazard assessment. Here, the methodology is applied to a set of 32 earthquakes in southern California to assess the accuracy and timeliness of warning if such a system was implemented using the existing seismic network. If there was no data telemetry delays the first warning would be available before the S-wave arrival at the epicenter for 56% of the earthquakes. The average absolute magnitude error at this time is 0.44 units and the error in the average absolute peak ground acceleration [ln(PGApredicted) – ln(PGAobserved)] is 1.08. Within 5 sec warning are available for 97% of the events, the average magnitude error is 0.33 units, and the average PGA error is 1.00. To further assess the utility of ElarmS implementation in California, probabilistic warning time distribution functions are determined for cities in northern California. Using the set of future likely earthquakes provided by the Working Group on California Earthquake Probabilities (2003) the warning times that the ElarmS methodology could provide (if implemented) can bc estimated, and a probability of occurrence associated. The alarm time is defined as the time when 4 sec of P-wave data is available at 4 seismic stations. At this point in time the average magnitude error is 0.5 units. The warning times range from zero seconds to over a minute, the most likely warning times range from seconds to a few tens of seconds depending on location. The largest magnitude earthquakes are also associated with the greatest warning times and it is more likely than not, that San Francisco would receive more than 20 sec warning for earthquakes generating the most damaging ground shaking.

3.1 Introduction

Current earthquake mitigation in the United States focuses on long-term characterization of the likely levels of ground shaking and the frequency of occurrence (e.g. Frankel et al. 1996). These estimates are the basis for building codes which aim to prevent collapse during earthquakes. The approach is highly effective at reducing deaths but not necessarily at reducing the cost of earthquakes. While buildings may not collapse during an earthquake, they may still suffer structural damage requiring them to be demolished. In other countries, including Mexico, Japan, Taiwan and Turkey, earthquake warning systems (EWS) are used in addition to building codes to further reduce the impact of earthquakes (Espinosa Aranda et al. 1995; Wu et al. 1998; Wu and Teng 2002; Erdik et al. 2003; Odaka et al. 2003; Boese et al. 2004; Kamigaichi 2004; Nakamura 2004; Horiuchi et al. 2005; Wu and Kanamori 2005). Short-term mitigation actions are taken in these countries to reduce both financial losses and casualties.

Earthquake warning systems (EWS) rapidly detect the initiation of earthquakes and warn of the forthcoming ground shaking. For a specific city, such as San Francisco, the warning time could be tens of seconds for some earthquakes, while zero seconds for others. However, in situations when San Francisco gets zero seconds warning, surrounding cities such as Oakland would likely get a few seconds and San Jose would get ~15 sec warning. Thus, for any earthquake scenario in a densely populated region, such as the San Francisco Bay Area (SFBA) or the Los Angeles Metropolitan Area (LAMA), an EWS could provide warning to at least some of the affected population in a damaging earthquake.

Here we present one methodology for an EWS that could be implemented in California and other regions around the world. The Earthquake Alarm System, "ElarmS," is designed to predict the distribution of peak ground shaking across the region affected by an earthquake before the beginning of significant ground motion (see http://www.ElarmS.org). ElarmS uses the first few seconds of P-wave arrivals at the closest stations to the epicenter to locate an earthquake and estimate its magnitude. A map—AlertMap—of predicted ground shaking is then generated and updated as more information becomes available. We apply the methodology to the specific problem of earthquake warning in southern and northern California using datasets of past and likely future earthquakes. In southern California, we use a set of past earthquakes and apply the methodology to determine the accuracy of the warnings generated. In northern California, we estimate the warning times that would be available for locations across the SFBA for all likely future earthquakes identified by the Working Group on California Earthquake Probabilities (2003).

3.2 The ElarmS Methodology

The ElarmS methodology was designed with the goal of predicting the distribution of peak ground shaking across the region affected by an earthquake before the beginning of significant ground motion at the epicenter. The first few seconds of the P-wave at the station and stations closest to the epicenter is used to estimate the magnitude of the earthquake and attenuation relations provide the predicted distribution of ground shaking as a function of distance from the epicenter. The complete ElarmS system is designed to generate a map of predicted peak ground shaking, a predicted-ShakeMap that we call "AlertMap." The first AlertMap is available 1 sec after the first P-wave trigger and is updated every second as additional data is gathered from stations farther from the epicenter. Below, we describe the three components of ElarmS.

3.2.1 Earthquake Location and Warning Time Estimation

Earthquakes are located using the arrival times of P-waves. When the first station triggers, an event is located at that station with a depth typical of events in the region. The earthquake is then located between the first two, and then the first three, stations to trigger. Once four stations have triggered a grid search method is used to locate the event, minimizing the misfit between predicted and observed arrival times.

The warning time is defined as the remaining time until the onset of peak ground shaking and can be estimated given the origin time and location of the earthquake using S-wave arrival time curves. The use of the predicted S-arrival time provides a conservative estimate of the remaining warning time. In larger magnitude earthquakes, such as Northridge and

Loma Prieta, peak ground shaking occurred 5-10 sec after the S-arrival at stations tens of kilometers from the epicenter.

3.2.2 Rapid Earthquake Magnitude Estimation

The magnitude of an earthquake is rapidly estimated using the frequency content of the first four seconds of the P-wave arrival. The predominant period, τ_p, of the vertical component waveform is calculated using the method first described by Nakamura (1988), and the maximum value within 4 sec, $\tau_p^{\max}$, is found to scale with event magnitude (Allen and Kanamori 2003; Lockman and Allen 2005; Olson and Allen 2005; Lockman and Allen 2007). Before calculation of τ_p, accelerometer recordings are converted to velocity and all processing is done recursively in a causal fashion. τ_p is determined continually in realtime from the vertical component velocity waveform using the relation

$$\tau_i^p = 2\pi\sqrt{X_i / D_i} \tag{3.1}$$

where

$$X_i = \alpha X_{i-1} + x_i^2 \tag{3.2}$$

$$D_i = \alpha D_{i-1} + \left({}^{dx}\!/_{dt} \right)_i^2 \tag{3.3}$$

x_i is the ground motion recorded at time i and α is a 1 sec smoothing constant (for 100 sps data α=0.99, for 20 sps data α=0.95). The higher frequency content of smaller magnitude earthquakes is measurable within a shorter time period after the P-wave arrival than the low frequency energy of larger events. Correspondingly, the magnitude of smaller events can be determined more rapidly than that of larger events. This also means that the magnitude estimate after 1 s is a minimum estimate, and once 2, 3 and 4 s of data are available, the magnitude estimate may increase.

Two linear relations between $\tau_p^{\max}$ and magnitude are used (Allen and Kanamori 2003). For smaller earthquakes (magnitudes 3.0 to 5.0), broadband data low-pass filtered at 10 Hz is used and a good magnitude estimate is possible given just 1 s of data. With 2 s of data the magnitude error reduces slightly, but additional data does not improve the estimate. Using our τ_p^{max} observations from the broadband waveforms after 2 s and minimizing the average absolute deviation we determine the relation

$$m_l = 6.3\log\left(\tau_p^{max}\right) + 7.1 \tag{3.4}$$

to estimate the magnitude of low-magnitude earthquakes. For larger magnitude events (magnitude > 4.5), better estimates are possible with the application of a 3 Hz low-pass filter and the best estimates of magnitude require 4 s of data, although minimum magnitude estimates can be made as soon as 1, 2 and 3 s after the P-arrival. The best-fit high-magnitude relation is

$$m_h = 7.0\log\left(\tau_p^{max}\right) + 5.9. \tag{3.5}$$

Both m_l and m_h are used by ElarmS to produce the best estimate of magnitude. Initially, 1 sec after a station triggers, m_l is calculated from τ_p^{max}, and when 2 s of data are available the estimate is updated. Station-magnitude estimates (one from each triggered station) are averaged to provide an event-magnitude estimate. If the event-magnitude estimate becomes greater than 4.0, then m_h is also calculated and the event-magnitude estimate is the average of both m_l and m_h from each triggered station.

τ_p^{max} has been calculated for earthquakes with magnitudes ranging from 3.0 to 8.3 from various regions around the world (Fig. 3.1). Datasets with a wide magnitude range from southern California and Japan show a similar scaling relation between the value of τ_p^{max} and magnitude (Allen and Kanamori 2003; Lockman and Allen 2007), and a global dataset including waveforms from southern California, Japan, Taiwan and the Denali earthquake in Alaska suggest that the scaling relation does not break down for even the largest magnitude earthquakes (Olson and Allen, 2005).

The accuracy of magnitude estimates is a function of the number of stations providing P-wave data. Figure 3.2 shows how the average error of magnitude estimates decreases as τ_p^{max} observations at multiple stations are combined to provide an average magnitude estimate. Datasets from southern California and Japan show a similar relation. Using just the closest station to the epicenter the average magnitude error is ~0.75 magnitude units; once data from the closest 2 stations is available the error drops to ~0.6, and to ~0.5 magnitude once 4 stations provide data.

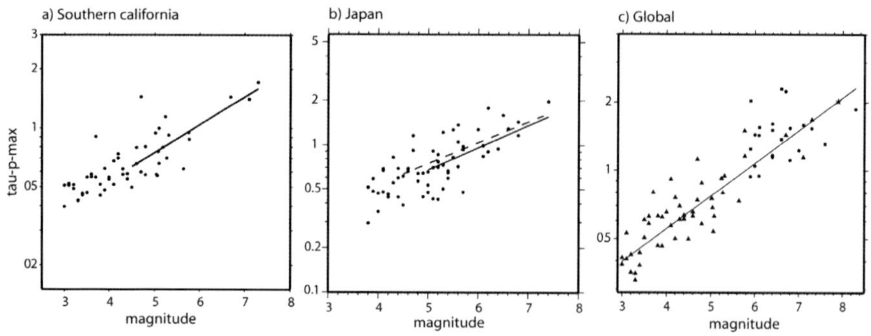

Fig. 3.1 Scaling relation between event-averaged $\tau_p^{\max}$ and magnitude. All data has been processed using the same recursive algorithms. **A)** Southern California earthquakes and best fit relation (solid line). **B)** Earthquakes in Japan and best fit relation (solid line). The dashed line is the best fit relation for California shown in A, which is nearly identical. **C)** Global compilation of earthquakes including southern California, Japan, Taiwan and the Denali earthquake. Waveforms are a mixture of accelerometers and broadband velocity instruments.

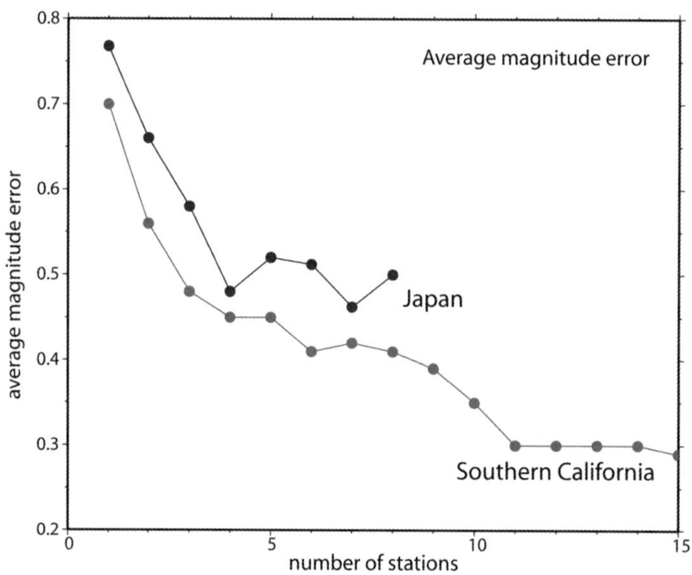

Fig. 3.2 Average absolute error in magnitude estimates as a function of the number of stations providing P-wave data for all events studied in southern California (green) and Japan (red). Using 1 station, the average error is ~0.75 magnitude units, and drops to ~0.6 with 2 stations and ~0.5 once 4 stations provide data.

3.2.3 Predicting the Distribution of Ground Shaking

Given the location and magnitude of an earthquake, the spatial distribution of peak ground shaking can be estimated using attenuation relations. Most existing relations use only ground motion observations for earthquakes with magnitudes greater than 5.0. ElarmS uses its own attenuation relations developed from regional observations for events with magnitudes greater than 3.0. Designing ElarmS to be operational during the frequent low magnitude events as well as large events is desirable in order to continually test the system.

Many different functional forms have been used for different types of earthquakes in different regions (e.g. Campbell 1981; Joyner and Boore 1981; Fukushima and Irikura 1982; Abrahamson and Silva 1997; Boore et al. 1997; Campbell 1997; Sadigh et al. 1997; Field 2000), however most are based on the functional form

$$A = A_0 r^n e^{-kr} \tag{3.6}$$

where A is the peak ground acceleration (PGA) at a distance r, and A_0, n and k are constants to be determined. This functional form has a term for geometric spreading, r^n, and one for intrinsic attenuation, e^{-kr}. Using a dataset of local earthquakes ranging in magnitude from 3.0 to 7.3 from southern California, best-fit attenuation relations were determined for the region. The effect of intrinsic attenuation was not significant within 200 km of an event, so k was set to zero to reduce the unknowns in the regression. n was determined as a function of magnitude by grouping PGA observations by magnitude and calculating the best fitting n. Having determined n, A_0 was calculated for each event and the best fitting linear relation between A_0 and magnitude was obtained. Figure 3.3 shows how A_0 and n vary as a function magnitude.

ElarmS uses the attenuation relations in a two-stage process. One second after the first P-wave trigger the first estimate of magnitude is available. Given the magnitude, A_0 and n are determined from the relations shown in Fig. 3.3, and estimated PGA is calculated as a function of distance. As time progresses during the event sequence, the stations closest to the epicenter measure their PGA. Once this information is available from a few stations, it is used to adjust the attenuation relation by keeping n fixed but allowing A_0 to change in order to best-fit the attenuation relation to PGA observations. Figure 3.4 shows examples of the attenuation relations for several earthquakes. Note the discrepancy between the observations and predictions of thc Field (2000) attenuation relations. This discrepancy is a common problem when using attenuation relations determined from larger-magnitude events only. The attenuation relations described here do not account for near-surface amplification effects, such as rock versus soil, which are responsible for much of the scatter in the acceleration observations shown in Fig. 3.4. Although site corrections are not currently part of ElarmS, they can easily be included when known (e.g. Wald et al. 1999; Wald et al. 1999).

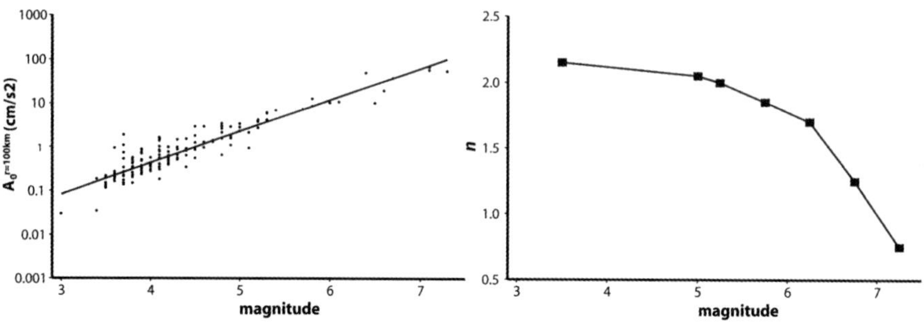

Fig. 3.3 Empirically determined values of n and A_0 as a function of earthquake magnitude. PGA observations were initially grouped by magnitude and n determined for each group by regression. Having determined n, the best fitting A_0 (defined as the amplitude at r=100 km) was calculated for each event. Linear regression provides A_0 as a function of magnitude.

3.3 Accuracy and Timeliness of Warnings

To test the accuracy and timeliness of warning information we gather a dataset of 32 earthquakes from southern California. The events were selected to give as wide a range of magnitude as possible and to include the events occurring beneath the denser portions of the existing broadband network. Earthquakes with magnitude ranging from 3.0 to 5.4 were included as shown in Fig. 3.5. All events with magnitudes larger than 5.4 either occurred before the existing seismic network was in place (e.g. Landers and Northridge), or were in locations where the network is sparse (e.g. Hector Mine).

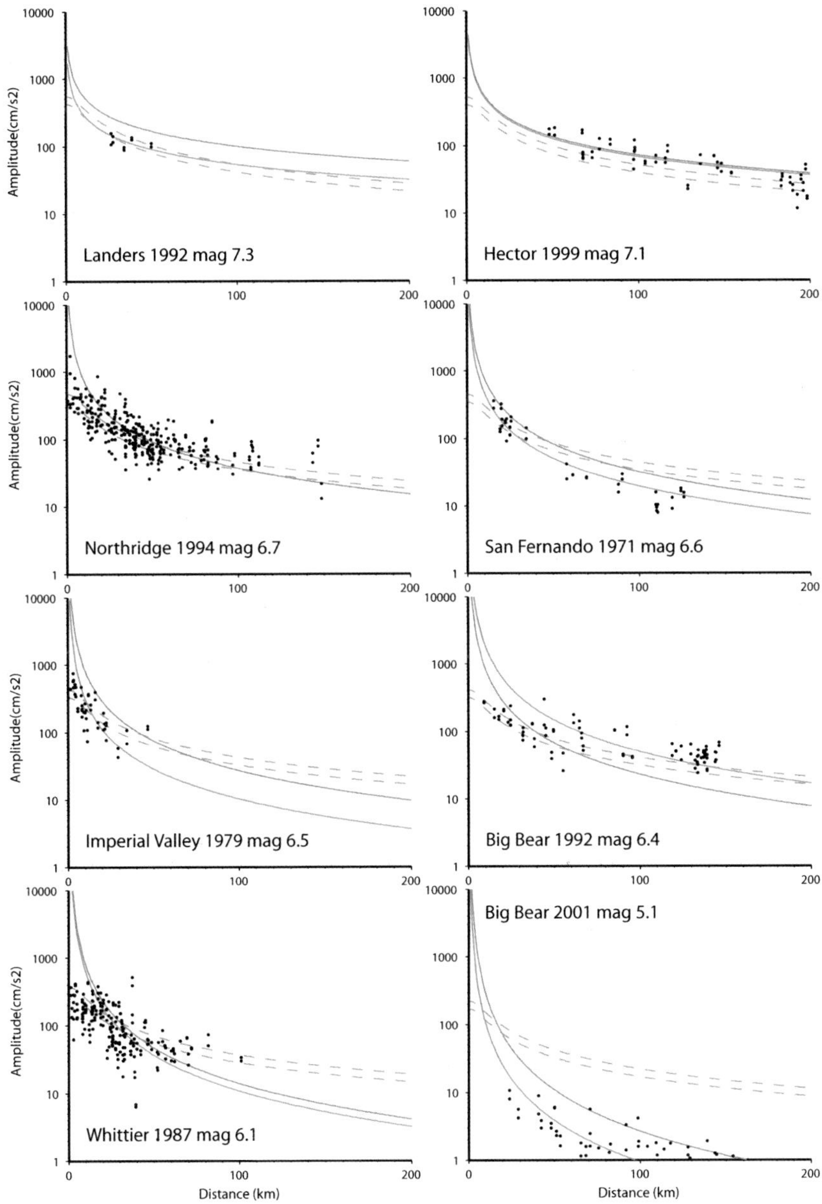

Fig. 3.4 Examples of attenuation relations (lines) and PGA observation (dots) for eight southern California earthquakes with magnitudes ranging from 5.1 to 7.3. Grey lines show the ElarmS attenuation relations determined given just earthquake magnitude and the green lines are the result of adjusting the relation based on PGA observations. Dashed lines show the Field (2000) attenuation relations for rock and soil for comparison.

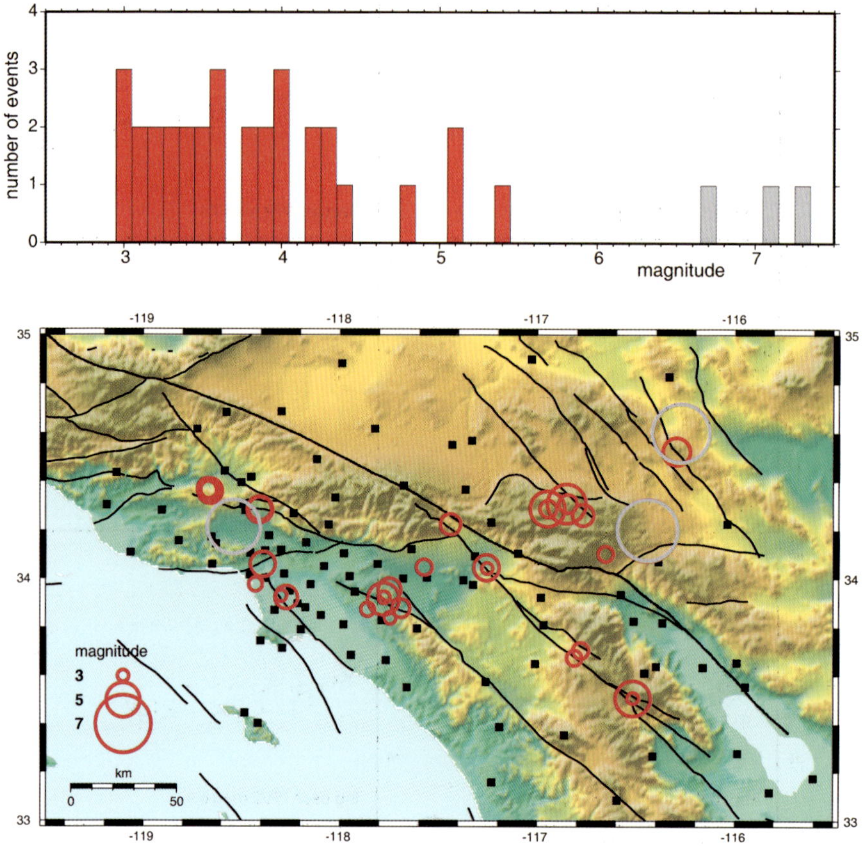

Fig. 3.5 Map of southern California showing the 32 earthquakes used to assess the accuracy and timeliness of ground shaking warnings. The 32 events occurred beneath the denser portions of the seismic network in the regions which are also the most densely populated. The histogram shows the magnitude distribution of the events included (red). The three largest magnitude events (grey on map and histogram) were not included as they did not occur beneath the current dense array.

The waveform datasets from these events are processed using the ElarmS methodology to determine the magnitude and the predicted ground shaking (PGA) as a function of time. We find that initial magnitude and PGA estimates are available for 56% of the earthquakes by the time the S-wave arrived at the surface, Fig. 3.6. We use the S-arrival at the epicenter as the zero time, because this is the earliest possible time of peak ground shaking at the surface, although in large magnitude earthquakes the peak ground shaking typically occurs 5 to 10 sec after the S-wave arrival at local sites. This test does not include any delays in data transmission, which would delay warnings by 1 or 2 sec, depending on how the early warning algorithms are implemented. Note, however, that likely telemetry delays are less than peak ground shaking delays for large magnitude earthquakes. With no telemetry delay and peak ground shaking at the time of the S-wave arrival, warnings would be available for more than 50% of earthquakes at the epicenter. If there is a 2 sec telemetry delay, then warning would be available for more than 50% of earthquakes at locations greater than ~8 km from the epicenter. This "blind zone" close to the epicenter, where warning may not be available using ElarmS, is also the region where some of the most severe damage would likely occur. Single station approaches to onsite early warning can offer timely hazard information in these regions (e.g. Nakamura 1996, 2004; Lockman and Allen 2005; Wu and Kanamori 2005; Wu and Kanamori 2005; Wu et al. in review). It should also be noted that, although the intensity of ground shaking may be lower outside the blind-zone than within, the total hazard exposure outside the blind-zone may be greater than within it. For example, buildings were red-tagged as structurally unsafe and scheduled for demolition as far as 60 km from the epicenter in the 1994 Northridge earthquake in LAMA. In that event, an 8 km radius blind-zone represented less than 2% of the total area severely affected by the earthquake.

While the first hazard prediction is available 1 sec after the first P-wave arrival, the majority of the initial predictions in Fig. 3.6 are based on trigger times and magnitude estimates from more than one seismic station. The offline algorithms used in this test gather all available information and update hazard estimates once per second. The density of seismic stations (typically 20 km spacing in the populated regions) means that, within a 1 sec time interval, usually two, and often three, stations trigger. The first event location, hazard, and warning time estimates, therefore, are based on information from multiple stations, providing a more accurate location and magnitude estimate than using a single station.

The test shows that magnitude estimates are available for 56% of earthquakes at the time of the S-arrival with an average magnitude error of 0.44 magnitude units, Fig. 3.6A. Within 5 sec, magnitude estimates are available for 97% of events and the average error is down to 0.33 magnitude units. Figure 3.6B shows the error in the PGA estimates as a function of time. PGA is estimated at each station within 100 km of the event using the available ElarmS magnitude and location and the attenuations relations described above. The error in the PGA estimate is calculated in the usual way: it is the natural logarithm of the predicted PGA minus the natural logarithm of the observed PGA for the event. At the time of the S-arrival, the average absolute error is 1.08. It drops to 1.00 within 5 sec, 0.98 within 10 sec, and reaches 0.95 at 15 sec. When the correct magnitude is used in the attenuation relations (thus removing the error in the ElarmS magnitude estimate), the error is only slightly lower: 0.89. An error in PGA of 1 is equivalent to the difference between a Modified Mercalli Intensity (MMI) of IV to V or alternatively of VIII to X.

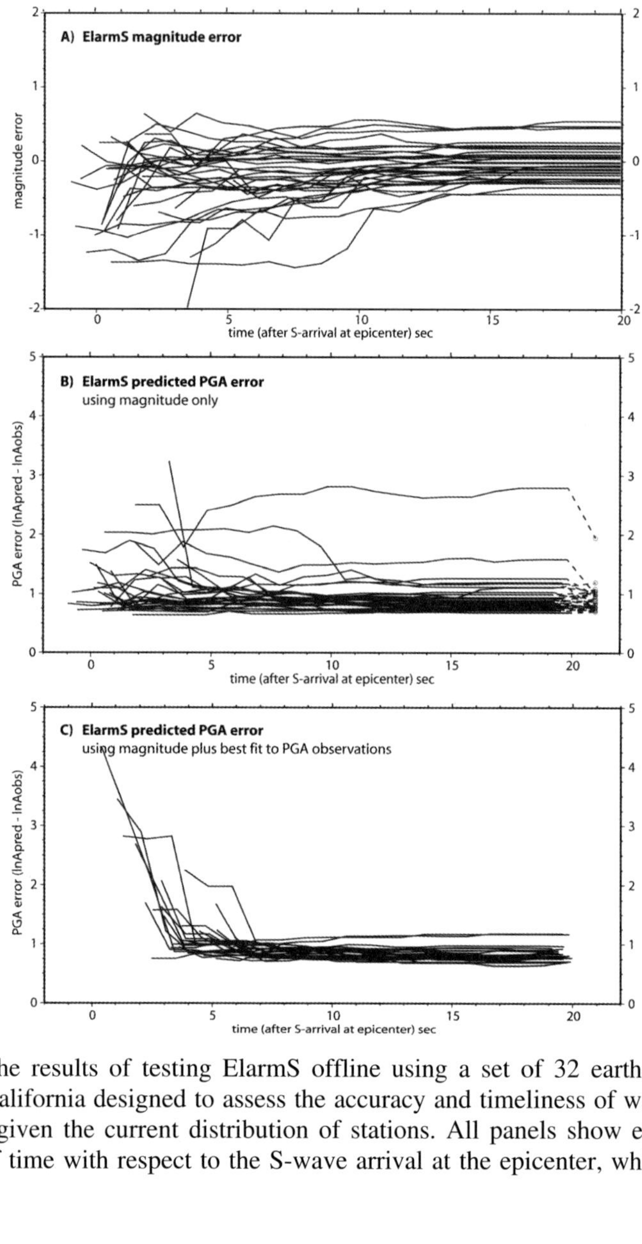

Fig. 3.6 The results of testing ElarmS offline using a set of 32 earthquakes in southern California designed to assess the accuracy and timeliness of warning information given the current distribution of stations. All panels show errors as a function of time with respect to the S-wave arrival at the epicenter, which repre-

sents the earliest time of peak ground shaking during an earthquake. **A)** The error in the magnitude estimate. **B)** Average absolute error in PGA estimates at all stations using available magnitude and location estimates and the ElarmS attenuation relations. The open circles at the far right are the errors when the true magnitude is used. **C)** Average error in PGA once available PGA observations are incorporated. The error in the PGA estimates is calculated in the usual way: the error is the natural logarithm of the predicted PGA minus the natural logarithm of the observed PGA for the event.

As time progresses during an earthquake, the closest stations record their PGA and this information is included in the prediction for stations at greater distances from the epicenter. The error in the PGA prediction once PGA observations from near stations are incorporated, is shown in Fig. 3.6C. At 5 sec, the average error is 1.02, similar to when PGA observations are not included, but it drops to 0.85 at 10 sec and 0.82 at 15 sec, which is slightly better than just using magnitude estimates alone. The most important use of PGA observations is to remove outliers, that is, cases when the magnitude-based estimate is very high or low.

3.4 Warning Time Distributions for Northern California

Having assessed the timeliness and accuracy of warnings in southern California, we look to the likely distribution of warning times in northern California should the ElarmS methodology be implemented using the existing seismic network. We use the set of likely earthquake scenarios for northern California identified by the Working Group on California Earthquake Probabilities (2003). Each earthquake scenario has an associated probability of occurrence by the year 2032, allowing determination of probabilistic warning time distributions for any location in the region.

To calculate the warning times, we define an "alert time" intended to represent the time when sufficient information about an earthquake is available for users to take action. During any earthquake the accuracy of warning information will increase with time and specific users will define the certainty level required for their own mitigation action (Grasso and Allen in review). Here, we choose a single threshold based on the accuracy of the warning and use the point in time when 4 sec of P-wave data are available at four seismic stations. This is defined as the alert time and represents the time when the average error in the magnitude estimate will be ~0.5 magnitude units based on tests in southern California and Japan (see Fig. 3.2). The warning time is the difference between the alert time and the estimated time of peak ground shaking for a given location. For the arrival-time of peak ground shaking as a function of epicentral distance we use the S-wave arrival-time curve out to a distance of 150 km and then a constant moveout of 3.55 km/s based on the observed moveout of peak ground shaking in California.

Warning times are calculated for a total of 4070 earthquake epicenters. These epicenters were distributed at 1 km intervals along the faults identified as those most likely to cause damaging earthquakes in northern California by the Working Group on California Earthquake Probabilities (2003). The study identified seven fault systems, each of which has one or more rupture segments, as shown in Fig. 3.7, that can rupture on their own or with adjacent segments. In all, 35 earthquake rupture scenarios were identified and a probability of occurrence within 30 years was estimated for each. The total probability of one or more of these earthquake scenarios (with magnitudes ranging from 5.8 to 7.9) occurring before 2032 is 84%. Within the SFBA, the faults that are most likely to rupture are the San Andreas Fault and the Hayward-Rodgers Creek Fault with probabilities of producing a magnitude 6.7 or greater earthquake of 21% and 27% respectively. The aggregate probability of one or more magnitude 6.7 or greater earthquakes within the next 30 years (from 2003 to 2032) in the SFBA is 62%.

Each of these earthquake scenarios involves rupture across a finite fault plane. The warning time in a given earthquake is dependent on the epicentral location where the rupture initiates. We do not know the likely point of initiation for the 35 scenarios; therefore, we accommodate the uncertainty in epicentral location by distributing epicenters at 1 km intervals along each fault. The probability of an earthquake with each epicentral location within one rupture scenario is set equal, and the aggregate probability of all the epicenters is equal to the scenario probability.

Given the epicenter of an earthquake, the alert time is dependent on the relative locations of seismic stations to detect the P-wave arrivals. Several thousand seismic stations are operated in northern California by the California Integrated Seismic Network (CISN), which consists of multiple, complementary seismic networks (see http://www.cisn.org). The ElarmS methodology requires continuous seismic waveforms recorded by instruments with a broad frequency sensitivity, i.e. continuous broadband stations. Such instruments are operated by the University of California Berkeley, which contributes a network of 24 stations, each with a broadband velocity seismometers and an accelerometer, and the U.S. Geological Survey, which operates approximately 100 accelerometers, located mostly in the SFBA, and 15 broadband velocity seismometers. In total, there are approximately 140 seismic stations across northern California which could be used in an EWS, Fig. 3.7.

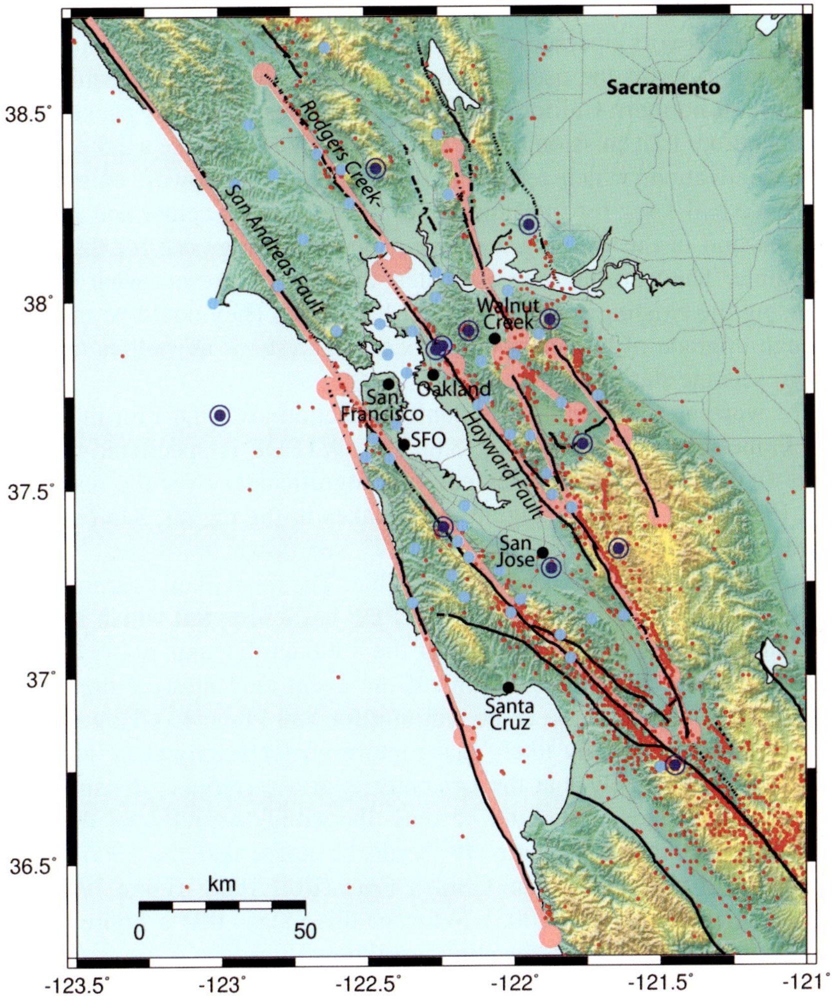

Fig. 3.7 Map of the San Francisco Bay Area (SFBA) showing mapped faults (bold black lines) and the location of earthquakes with magnitude greater than 3 since recording began (red dots). Existing continuous broadband stations operated by UC Berkeley (dark blue) and the U.S. Geological Survey (light blue) are shown with circles for broadband velocity seismometers and dots for accelerometers. The fault segments identified by the Working Group on California Earthquake Probabilities (2003) are shown with pink dots at the ends of segments joined by broad pink lines. The six "warning points" included in Figs. 3.8 and 3.9 are shown as black dots.

The alert time for each earthquake epicenter is calculated as the time at which 4 sec of P-wave data are available at the 4 closest continuous broadband stations plus a fixed telemetry and processing delay of 4.5 sec. A 4.5 sec delay accounts for transmission of waveform data from each station to one of the network operation centers, processing time, and transmission of the warning out to the user community. Given the current seismic infrastructure in northern California, the most significant delay is packetization of data before transmission from each station. We introduce a 2.5 sec delay for packetization, which represents the delay at the slowest existing stations. We add 1 sec for transmission to the processing center and 1 sec for transmission of the warning message. The processing time for the data is negligible. The warning time estimates, therefore, represent what is possible using the existing seismic network hardware. They could be improved through upgrade of telemetry and processing systems as well as the addition of seismic stations.

The warning time probability density function (WTPDF) for the city of San Francisco is shown in Fig. 3.8. This WTPDF is specifically for the Civic Center; however, it does not vary significantly over the rest of the city. For all the likely damaging earthquakes in the region, San Francisco could receive warnings varying from 77 sec down to -8 sec. Negative warning times mean no warning is possible. The most likely warning times are less than 25 sec; however, the WTPDF has a long tail which is due to the San Andreas Fault. In a repeat of the 1906 earthquake, a 450 km long segment of the fault could rupture. If the event nucleates off the Golden Gate, there would be little or no warning for San Francisco. However, assuming that it is equally likely that rupture nucleates anywhere along the fault, it is more likely that the epicenter is at a significant distance from San Francisco and there could be tens of seconds warning for this most damaging earthquake scenario. It should be noted that the 1906 rupture probably did nucleate off the Golden Gate (Bolt 1968; Boore 1977; Zoback et al. 1999; Lomax 2005). Whether this means that a future rupture would nucleate in the same location is unknown.

In addition to the warning times for each earthquake, we also estimate the likely intensity of ground shaking at the warning point, i.e. the Civic Center in the case of Fig. 3.8. These intensities are derived from ShakeMap scenario calculations (Working Group on California Earthquake Probabilities 2003). The grey regions in Fig. 3.8 represent earthquakes for which shaking intensity at the Civic Center is less than V on the MMI scale (Richter, 1958) and there is unlikely to be damage. Above a MMI V, the likely damage increases with the severity of shaking from light (V: unstable objects displaced), to strong (VII: broken furniture and damage to masonry), to violent (IX: masonry seriously damaged or destroyed, frames displaced from foundations).

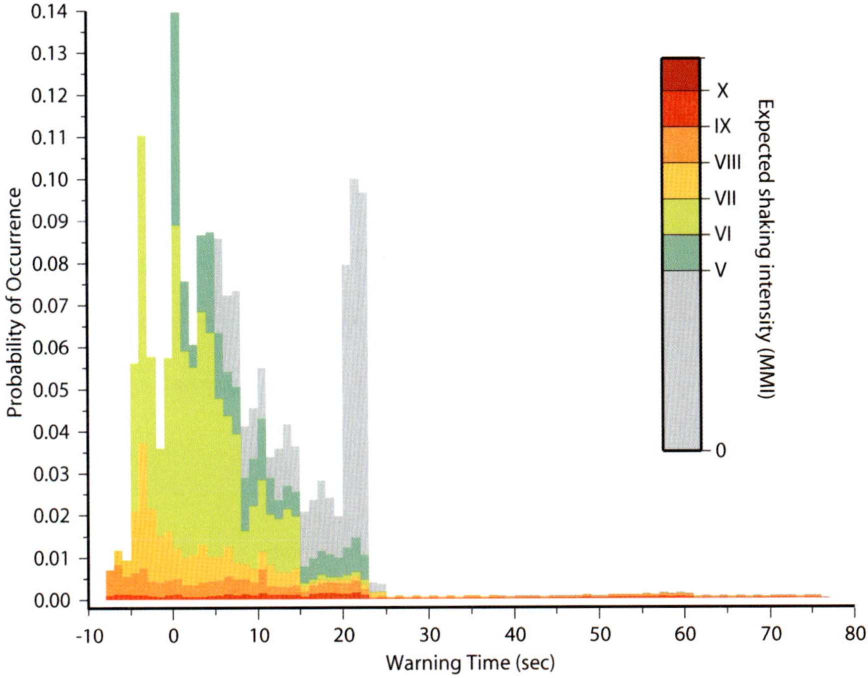

Fig. 3.8 Warning time probability density function (WTPDF) for the Civic Center of San Francisco (37.78°N, 122.42°W). The warning times for all likely earthquakes range from -8 sec to 77 sec, negative warning times mean no warning is possible. Earthquakes are in 1 sec bins and the vertical axis shows the total probability of one or more earthquakes occurring before 2032 with a given warning time. The color represents the estimated intensity of ground shaking for each event. Damage is unlikely for MMI < V (grey); MMI > IX means violent shaking likely to cause serious damage to buildings (red).

In the case of the WTPDF for San Francisco, Fig. 3.8, the long tail of large warning times includes a large portion of the earthquake scenarios which will cause violent (MMI > IX) ground shaking. This is because the intensity of ground shaking in a given earthquake is dependent on the closest distance to the fault rupture, while the warning time is dependent on the distance to the epicenter. Our warning time estimates are conservative in that they represent the traveltime of shear energy directly from the epicenter to the warning point. The true time of peak ground shaking may not occur until the rupture has propagated along the fault to the closest point, which is typically at less than the shear-wave speed, and then from the fault to the warning point at the shear-wave speed.

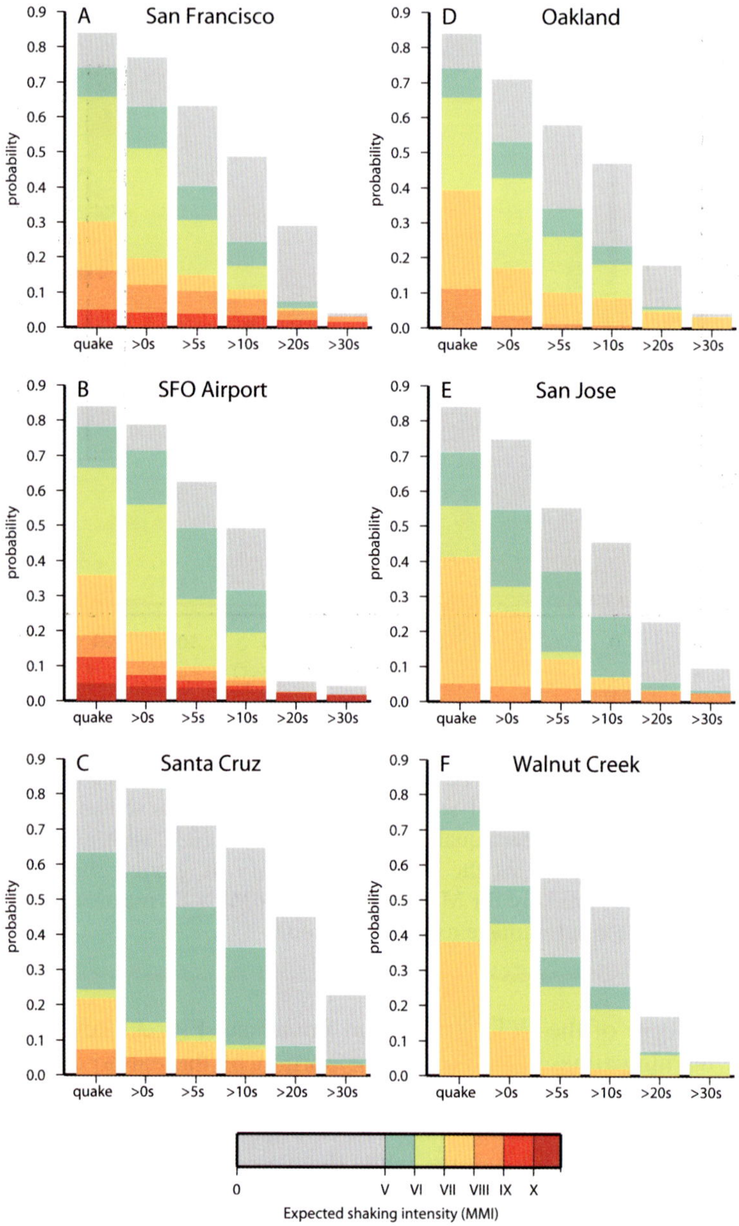

Fig. 3.9 Simplified warning time probability density functions (WTPDF) for six locations around the SFBA. In each panel, the first column shows the aggregate

probability of all likely earthquakes in the region before 2032 (84%) and the expected intensity of ground shaking. The remaining columns show the probability of an earthquake occurring for which more than 0, 5, 10, 20 and 30 sec warning could be available, and the distribution of ground shaking intensities for those events. The six locations are shown on Fig. 3.7. **A)** The city of San Francisco (37.78°N, 122.42°W). **B)** San Francisco International Airport, SFO (37.62°N, 122.37°W). **C)** The city of Santa Cruz (36.97°N, 122.03°W). **D)** The city of Oakland (37.805°N, 122.270°W). **E)** The city of San Jose (37.33°N, 121.90°W). **F)** The city of Walnut Creek (37.90°N, 122.06°W).

Simplified representations of the WTPDF for six locations around the SFBA are shown in Fig. 3.9. The full WTPDF for these locations and other cities and sites of engineering interest are available at http://www.ElarmS.org. Figure 3.9A shows that there is a 74% probability of one or more earthquakes that will cause some damage (MMI ≥ V) in San Francisco by 2032, and a 63% probability of a damaging event for which a warning could be available. There is a 5% probability of an earthquake that causes violent ground shaking (MMI ≥ IX), and a 3% chance of one for which greater than 10 sec of warning could be available. It is therefore more likely than not that more than 10 sec warning would be available before violent ground shaking in the city. The WTPDF for the San Francisco International Airport (Fig. 3.9B) is similar to that for the city, except that the intensity of ground shaking could be greater given the closer proximity to the San Andreas Fault.

The most severe earthquakes for East Bay Cities occur on the Hayward-Rodgers Creek Fault. Its close proximity to cities such as Oakland (Fig. 3.7) make for reduced warning times, but also lower intensities due to the shorter length of the fault. It is still more likely than not that a warning will be available for a damaging earthquake, Fig. 3.9D. Most of the hazard for San Jose comes from the San Andreas Fault. As with San Francisco, this means there is a high probability of large warning times for the most damaging earthquakes. While there is a 5% probability of an event causing MMI VIII in San Jose, there is a 3% chance of an event for which there could be greater than 20 sec warning (Fig. 3.9E). In the October 17, 1989, Loma Prieta earthquake (M_w 6.9), the closest city to the epicenter, Santa Cruz, experienced MMI VIII. There is a 7% probability of a similar intensity of ground shaking by 2032, and a 3% chance of similar ground shaking for which greater than 30 sec warning could be available (Fig. 3.9C). Finally, the rapidly growing urban areas east of the Berkeley Hills, such as Walnut Creek, are as likely to experience damaging ground shaking as San Francisco, although the most severe events have a lower intensity (Fig.

3.9F). As is the case for all locations in the SFBA, Walnut Creek could receive a warning before ground shaking starts for the majority of damaging earthquakes.

3.5 Earthquake Warning Outlook

An EWS for San Francisco was first suggested by Cooper (1868), who proposed that the telegraph cables radiating from the city could transmit warning ahead of ground shaking. He also noted that this would not work if the center of the "shock" was close to the city, but estimated such a scenario to occur less than 1% of the time. His estimate was not far from our current estimates today. A more recent study by Heaton (1985), using a theoretical distribution of earthquakes in southern California, concluded that there could be more than a minute of warning for the larger, most damaging earthquakes. Here, we come to a similar conclusion using the set of past earthquakes in southern California and future likely earthquakes and existing seismic stations in northern California.

Active early warning systems are now operational in Mexico, Japan, Taiwan and Turkey (Espinosa Aranda et al. 1995; Wu et al. 1998; Wu and Teng 2002; Erdik et al. 2003; Odaka et al. 2003; Boese et al. 2004; Kamigaichi 2004; Nakamura 2004; Horiuchi et al. 2005; Wu and Kanamori 2005). Their warning messages are currently used by transportation systems such as rail and metro systems, as well as private industries, including construction, manufacturing and chemical plants. They are also used by utility companies to shut down generation plants and dams, and emergency response personnel to initiate action before ground shaking. In addition, schools receive the warnings allowing children to take cover beneath desks, housing units automatically switch off gas and open doors and windows, and entire complexes evacuate. Many of these applications would also be appropriate in California. WTPDF for the specific location of any user can be calculated and used to determine the cost-benefit of implementing an automated response to warning messages.

EWS are no panacea for the mitigation of seismic hazard. While EWS cannot warn everyone prior to all ground shaking events, they can offer warning to many affected people most of the time. No approach to natural hazard mitigation is perfect. Building codes are intended to prevent collapse of most structures in most earthquakes. If the mitigation of natural hazards is our intent, it is important to ensure that we continually ask what more could be done, what new technologies can be applied. As the December 26, 2004, tsunami disaster demonstrated most clearly, complacency is not an option.

3.6 Acknowledgments

This work benefited from discussions with Hiroo Kanamori, Yih-Min Wu, David Wald, Michael Brudzinski, and the seismic network operators of the CISN, in particular David Oppenheimer, Lind Gee, Douglas Neuhauser and Egill Hauksson. Andrew Lockman, Erik Olson and Gilead Wurman contributed to the analysis. Support for this work was provided by the USGS National Earthquake Hazard Reduction Program and National Science Foundation.

References

Abrahamson NA, Silva WJ (1997) Empirical response spectral attenuation relations for shallow crustal earthquakes. Seismo Res Lett 68, 94-127

Allen RM, Kanamori H (2003) The potential for earthquake early warning in Southern California. Science 300:786-789

Boese M, Erdik M, Wenzel F (2004) Real-time prediction of ground motion from p-wave records. Eos Trans AGU Fall Meet Suppl 85, Abstract S21A.0251

Bolt BA (1968) The focus of the 1906 California earthquake. Bull Seismol Soc Am 50:457-471

Boore DM (1977) Strong-motion recordings of California earthquake of April 18, 1906. Bull Seismol Soc Am 67:561-577

Boore DM, Joyner WB, Fumal TE (1997) Equations for estimating horizontal response spectra and peak acceleration from western North American earthquakes; a summary of recent work. Seismo Res Lett 68:128-153

Campbell KW (1981) Near-source attenuation of peak horizontal acceleration. Bulletin of the Seismological Society of America 71:2039-2070

Campbell KW (1997), Empirical near-source attenuation relationships for horizontal and vertical components of peak ground acceleration, peak ground velocity, and pseudo-absolute acceleration response spectra. Seismo Res Lett 68:154-179

Cooper JD (1868) Earthquake indicator. In: Evening Bulletin (ed) San Francisco

Erdik MO, Fahjan Y, Ozel O, Alcik H, Aydin M, Gul M (2003) Istanbul earthquake early warning and rapid response system. Eos Trans AGU Fall Meet Suppl 84, Abstract S42B.0153

Espinosa Aranda JM, Jimenez A, Ibarrola G, Alcantar F, Aguilar A, Inostroza M, Maldonado S (1995) Mexico city seismic alert system. Seismo Res Lett 66:42-52

Field EH (2000) A modified ground-motion attenuation relationship for southern california that accounts for detailed site classification and a basin-depth effect. Bull Seismol Soc Am 90:S209-S221

Frankel A, Mueller C, Barnhard T, Perkins D, Leyendecker EV, Dickman N, Hanson S, Hopper M (1996) National seismic hazard maps. U.S. Geological Survey, Open-File Report 96-532, Denver

Fukushima Y, Irikura K (1982) Attenuation characteristics of peak ground motions in the 1995 Hyogo-Ken. J Phys Earth 45:135-146

Grasso VF, Allen RM (in review) Uncertainty in real-time earthquake hazard predictions

Heaton TH (1985) A model for a seismic computerized alert network. Science 228;987-990

Horiuchi S, Negishi H, Abe K, Kamimura A, Fujinawa Y (2005) An automatic processing system for broadcasting earthquake alarms. Bull Seismol Soc Am 95:708-718

Joyner WB, Boore DM (1981) Peak horizontal acceleration and velocity from strong-motion records including records from the 1979 Imperial Valley, California, earthquake. Bulletin of the Seismological Society of America 71:2011-2038

Kamigaichi O (2004) Jma earthquake early warning. J Japan Assoc Earthquake Eng 4

Lockman A, Allen RM (2005) Single station earthquake characterization for early warning. Bull Seism Soc Am 95:2029-2039

Lockman A, Allen RM (2007) Magnitude-period scaling relations for Japan and the Pacific Northwest: Implications for earthquake early warning. Bull Seismol Soc Am

Lomax A (2005) A reanalysis of the hypocentral location and related observations for the great 1906 California earthquake. Bull Seismol Soc Am 95:861-877

Nakamura Y (1988) On the urgent earthquake detection and alarm system (Uredas). Proc. 9th World Conf. Earthquake Eng., VII, 673-678

Nakamura Y (1996) Real-time information systems for hazards mitigation. Proceedings of Eleventh World Conference on Earthquake Engineering, Paper No. 2134

Nakamura Y (2004) Uredas, urgent earthquake detection and alarm system, now and future. Proc. 13th World Conf. Earthquake Eng., August 2004, Paper No. 908

Odaka T, Ashiya K, Tsukada S, Sato S, Ohtake K, Nozaka D (2003) A new method of quickly estimating epicentral distance and magnitude from a single seismic record. Bull Seismol Soc Am 93:526-532

Olson E, Allen RM (2005) The deterministic nature of earthquake rupture. Nature 438:212-215

Richter CF (1958) Elementary seismology. W. H. Freeman, 768 pp

Sadigh K, Chang CY, Egan JA, Makdisi F, Youngs RR (1997) Attenuation relationships for shallow crustal earthquakes based on California strong motion data. Seismo Res Lett 68:180-189

Wald DJ, Quitoriano V, Heaton TH, Kanamori H (1999) Relationships between peak ground acceleration, peak ground velocity, and modified Mercalli intensity in California. Earthquake Spectra 15:557-564

Wald DJ, Quitoriano V, Heaton TH, Kanamori H, Scrivner CW, Worden CB (1999) Trinet "shakemaps": Rapid generation of peak ground motion and intensity maps for earthquakes in Southern California. Earthquake Spectra 15:537-555

Working Group on California Earthquake Probabilities (2003) Earthquake probabilities in the San Francisco Bay region: 2003 to 2031. U.S. Geological Survey

Wu Y-M, Kanamori H (2005) Experiment on an onsite early warning method for the Taiwan early warning system. Bull Seismol Soc Am 95:347-353

Wu Y-M, Kanamori H, Allen RM, Hauksson E (in review) An onsite earthquake early warning method for the Southern California seismic network. Bull Seism Soc Am

Wu Y-M, Teng T-l (2002) A virtual subnetwork approach to earthquake early warning. Bull Seismol Soc Am 92:2008-2018

Wu YM, Kanamori H (2005) Rapid assessment of damage potential of earthquakes in Taiwan from the beginning of p waves. Bull Seism Soc Am 95:1181-1185

Wu YM, Shin TC, Tsai YB (1998) Quick and reliable determination of magnitude for seismic early warning. Bull Seismol Soc Am 88:1254-1259

Zoback ML, Jachens RC, Olson JA (1999) Abrupt along-strike change in tectonic style: San Andreas fault zone, San Francisco Peninsula. J Geophys Res 104:10719-10742

4 Real-time Estimation of Earthquake Magnitude for Seismic Early Warning

Aldo Zollo, Maria Lancieri

RISSC-Lab, Dipartimento di Scienze Fisiche, Università di Napoli Federico II, Napoli, Italy

Abstract

A prototype system for earthquake early warning and rapid shake map evaluation is being developed and tested in southern Italy based on a dense, dynamic seismic network (accelerometers + seismometers) under installation in the Apenninic belt region (Irpinia Seismic Network). It can be classified as a regional Earthquake Early Warning System consisting of a broad-based seismic sensor network covering a portion or the entire area which is threatened by the quake's strike.

The real time magnitude estimate will take advantage from the high spatial density of the network in the source region and the broad dynamic range of installed instruments. Based on the offline analysis of high quality strong-motion data bases recorded in Italy, several methods are envisaged, using different observed quantities (peak amplitude, dominant frequency, square velocity integral, …) to be measured on seismograms, as a function of time, both on P and early-S wave signals.

Results from the analysis of the Italian strong motion database point out the possibility of using low-pass filtered displacement and velocity peak amplitudes measured in time windows lasting less than 3-4 sec after the first P- or S-wave arrivals. These parameters show they are robustly correlated with moment magnitude.

The correlation found of 3Hz low-pass filtered PGV and PGD with magnitude is discussed and interpreted in terms of plausible dynamic models of the earthquake rupture process during its initial stage.

4.1 Introduction

Over the last few decades, there has been ongoing experimentation into seismic early-warning (SEW) systems in several active seismic areas of the world. Prototype SEW systems have been developed and implemented in Taiwan, Japan, USA and Mexico, where warning signals from dense seismograph networks in the earthquake source areas are sent to nearby urban settlements.

SEW systems that are based on real-time automated analysis of ground-motion measurements can have a role in reducing regional resiliency by promoting action for the protection of built environments and lifelines. The early information provided by SEW systems when the seismic waves are still propagating can be used to activate various security measures, such as the shutting down of critical systems, the stopping of transportations systems and the shutting-off of lifelines. Depending on the network geometry and configuration around the potential seismic source and/or target area, SEW systems can be classified according to (Kanamori 2005):

- regional (dense seismic network deployed in the potential earthquake source area)
- on-site (single instrument or array of instruments deployed at the target site, which is distant from the earthquake source area)

For regional SEW systems, the earthquake-warning window begins at the time of the first P-wave detection by the network deployed in the earthquake source area. This window can last from a few to several tens of seconds, depending on the distance between the source and the target area.

For on-site SEW systems, the lead time is given by the difference between the first recorded P-wave motion at the target site and the later arrival of the energy amplitudes (carried by the primary S-waves, secondary body waves or surface waves), which will also depend on the distance from the epicentral area.

In both cases, fully automated, robust and reliable real-time estimates of the main earthquake parameters (location and magnitude) must be obtained in an evolving, continually updated form, so that they can be used for warning purposes or to simulate realistic shake maps, and be helpful for emergency preparations and management.

The ca. 6 million inhabitants and large number of industrial plants in the Campania region are highly exposed to seismic risk, which is related to the moderate to large magnitude seismicity that originates from the active fault systems in the Apennine belt. The 1980, M=6.9, Irpinia earthquake was the most recent destructive earthquake to occur in the region: it caused more than 3,000 casualties and massive, widespread damage to buildings and infrastructure across the whole region. In the framework of an ongoing project financed by the Regional Department of Civil Protection, a prototype system for earthquake early-warning and rapid shake-map evaluation is being developed and tested in southern Italy (Weber et al. 2007, this issue). This is based on a highly dynamic and dense seismic network (accelerometers + seismometers) that is being installed in the Apennine belt region (ISNet) (Fig. 4.1).

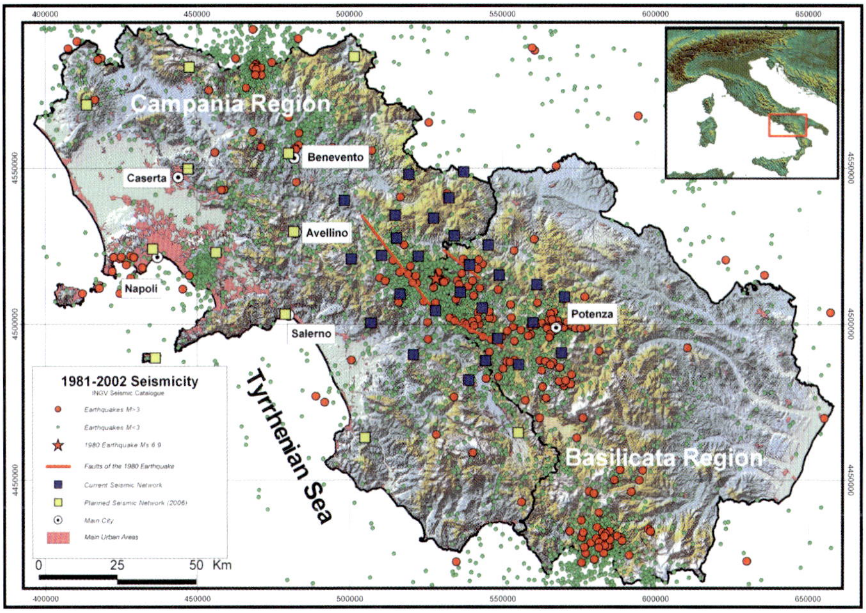

Fig. 4.1 Map of the ISNet network and seismicity in the Campania region.

This system can be classified as a regional SEW system as it will consist of a wide seismic sensor network that covers part or all of the area under threat from an earthquake strike. Application of such an SEW system in the Campania region that will be based on the ISnet network will need to consider an expected time delay to the first S-wave energy train that will vary between 14 and 20 sec at distances of 40 to 60 km, and between 26 to 30 sec at 80 to 100 km, from a shallow crustal earthquake occurring in the source region. The latter delay is the typical time window available for the mitigation of earthquake effects through early warning in the city of Naples (with ca. 2 million inhabitants, including the suburbs).

Considering such a warning window that will range from tens of seconds before to hundreds of seconds after an earthquake, several public infrastructures and buildings of strategic importance (e.g., hospitals, gas pipelines, railways, railroads) to Campania can be considered as potential test-sites for testing the innovative technologies for data acquisition, processing and transmission.

Specifically, for real-time magnitude estimations for SEW purposes, a number of methods have very recently been proposed that are based on the dominant frequency/period and/or peak ground-motion amplitudes measured over a narrow time window (between 3 and 4 seconds) extending from just after the first P-wave arrival (Allen and Kanamori 2003, Wu and Teng 2004) In particular, the methods based on real-time magnitude estimations from the predominant period parameter have been validated and calibrated for regional recordings of velocity ground motion that are acquired by seismographic stations equipped with vertical short-period seismometers.

Due to the great dynamic range and high density of the seismic network under installation in the seismogenetic area of the Campania region (Weber et al. 2007, this issue), in this article we seek to investigate the possibility of measuring different observed quantities in real-time signals acquired by this seismic network, including the dominant period parameter, that can be used as magnitude-moment estimators. Assuming a moderate to large potential event that occurs at shallow crustal depths (< 20 km) beneath the seismic network, the signals from the first P-waves and S-waves are expected to be detected within 1.5 to 3.5 sec and 2.6 to 6.0 sec, respectively, after the origin time. These rather short time windows will provide the opportunity to integrate the P-wave and early-S-wave information for both fast earthquake localization and magnitude estimation.

In the present study, we analyzed the Italian strong-motion database, the data type which is most closely related to that expected to be recorded by the Campania region SEW system, in both instrumental and seismotectonic terms. Based on this existing information about event locations and moment magnitudes, we investigated the correlations of peak strong-motion parameters and dominant frequency parameters as functions of magnitude for increasing time windows from the first P-wave and S-wave arrivals.

This study represents a basic strong-motion data analysis that should lead to the calibration, validation and testing of the algorithms that can be used for real-time estimations of magnitudes from the Campania region seismic network.

4.2 Strong-Motion Data Analysis

4.2.1 The Italian Strong Motion Data Base

Due to the dynamic range and density characteristics of the SEW system network under construction in the Campania region, peak-amplitude and dominant period information from unsaturated early P-wave and S-wave signals can be jointly used for magnitude estimations. With the aim of searching for correlations between the observed parameters and magnitudes, we analyzed the three-component records from the European Strong-Motion Database (ESD) (Ambraseys et al. 2000), using small-to-large earthquakes that have occurred in Italy over the last three decades.

The ESD was created as the result of a European project under the 5th Framework Program. It is an internet-searchable data-bank that spans the period from 1972 to 1999. It has thus collected and archived, and freely distributes, more than 3,000 acceleration time histories from earthquakes in Europe and adjacent areas. More than 2,000 acceleration time histories are now archived in the databank as uncorrected and corrected records, together with their corresponding elastic-response spectra. The main source parameters (location, moment magnitude) for each recorded earthquake are also available in the database following their validation, and if necessary, the re-calculation or re-estimation of their seismological, instrumental and site-specific parameters.

The strong-motion records of the Italian earthquakes that occurred between 1976 and 1998 represent a major part of the databank, and were acquired for the most by the ENEL-ENEA strong-motion network that consists of 300 accelerograph stations installed all over Italy. This network is now operated by the Italian Civil Protection Department (DPC) through the National Seismic Service (SSN). Data from other Italian local and regional networks are also contained in the databank. We refer to Ambraseys et al. (2000) for a complete description of the participating networks, instruments and methods for the building of the databank.

Most of the strong-motion records for past Italian earthquakes were acquired by using Kinemetrics SMA-1 analogue accelerographs. These are threshold-based instruments that record the ground motion in the form of either a photographic trace on film or paper, or a scratch trace on waxed paper. The threshold levels are usually set to 0.005 to 0.010g in the vertical direction, and thus very frequently they do not record the whole earthquake signal, but rather a portion of the signal starting after the first P-wave train that is able to trigger the strong-motion recording for optimal conditions of event distance and magnitude. The processing of SMA-1 data contained in the ESD involves digitization, sensitivity correction, linear base-line correction and filtering in the frequency band 0.25 to 25Hz using an eighth-order elliptical bandpass filter (Sunder and Connor 1982). After visual inspection of samples of displacement and velocity records obtained by double and single integration of acceleration time series, we decided to apply an additional high-pass, 2-pole, zero-phase-shift Butterworth filter with a corner frequency of 0.075Hz, to provide a more suitable base-line and long-period trend correction.

In order to investigate the correlations between real-time estimates of strong-motion quantities and magnitudes, we selected 116 three-component strong-motion records of Italian earthquakes occurring from 1976 to 1998 with moment magnitudes ranging from 3.5 to 7.0, and with epicentral distances less than 50 km. This maximum recording distance was chosen according to the general observation that high-frequency, direct body waves radiating from extended earthquake ruptures dominate in amplitude in the near-source distance range, i.e., at distances from the source comparable with the rupture length (Beroza 1996, Zeng et al. 1993, Emolo and Zollo 2005).

Figure 4.2 illustrates the locations of the stations and earthquakes used in this study, following this selection based on the maximum distance between source and receiver.

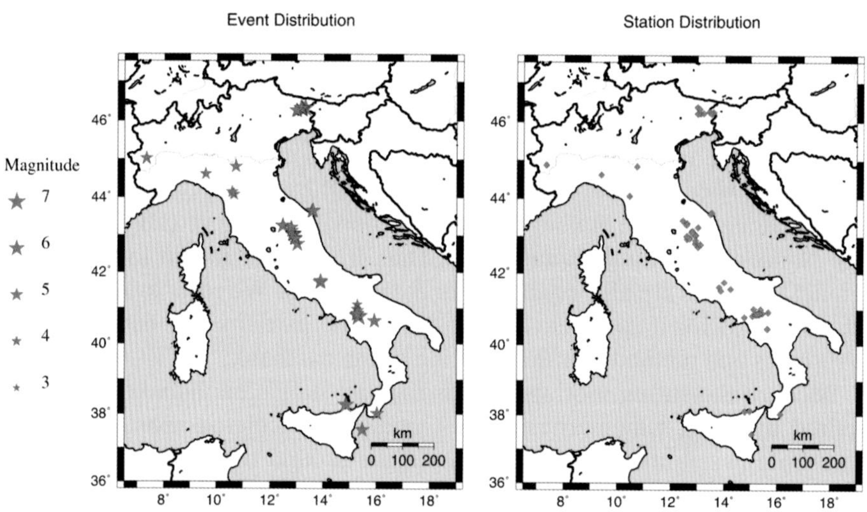

Fig. 4.2 Map of the strong-motion stations(left) and earthquakes(right) used in this study. The sizes of symbols are proportional to the event magnitudes.

A histogram demonstrating the number of records vs their magnitudes is given in Fig. 4.3. Most of the recorded events are in the range M 4.5-6.5, while the largest event in the catalogue is the M_w=7.0, 1980 Irpinia earthquake, for which 10 records are available within the considered distance range.

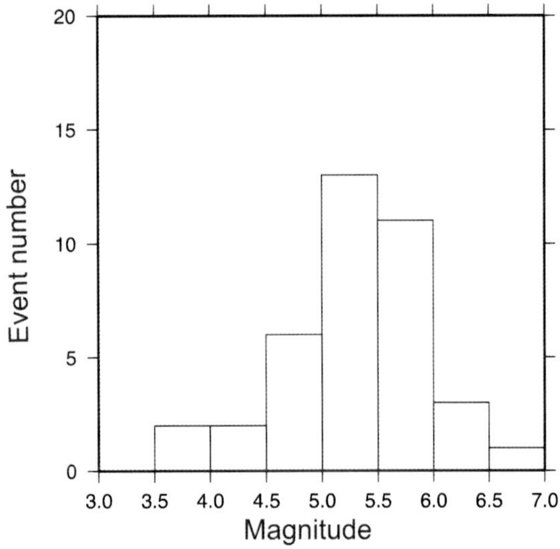

Fig. 4.3 Histograms showing the number of records as related to the magnitudes.

4.2.2 Measurements of Strong Ground-motion Quantities

The SEW network under construction in the Campania region is a highly dynamic and particularly dense seismic array, such that unsaturated P-wave and S-wave signals will be available a few seconds after the occurrence of an event where the epicenter is located within the area covered by the network.

The first analysis we performed was to identify and choose the first S-arrivals from all the selected strong-motion records. Using the availability of these chosen first S-waves we calculated the origin times of the earthquakes, the expected first P-arrivals and the triggering times of each of the records, i.e., the times to be associated to the first sample of the time series. A homogeneous crustal velocity model was assumed, Vp=5.5 and Vs=3.2 km/s. This procedure is particular relevant for SMA-1 records for which the absolute times of the traces are not available.

The S-signal detection is based on the analysis of variation of amplitude, frequency and horizontal polarization as functions of time along the low-pass-filtered accelerograms. The availability of the first S-wave arrival times allowed us to classify the records according to the estimated S-P times (Fig. 4.4a).

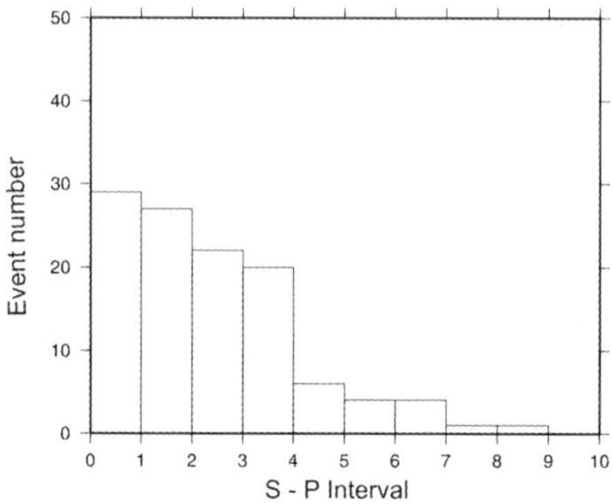

Fig. 4.4a Histogram for the number of records vs S-P times

Since SMA1 accelerographs generally trigger on P-wave arrivals or later, we also determined the records for which the triggering times were later than the estimated first P-arrivals (Fig. 4.4b). For the large majority of records analyzed, the S-P times are less than 4 sec, with less than 25 records showing a Tf_s –Tp that is greater than 4 sec, where Ts and Tp are the estimated times of the first sample and the first P-arrival, respectively.

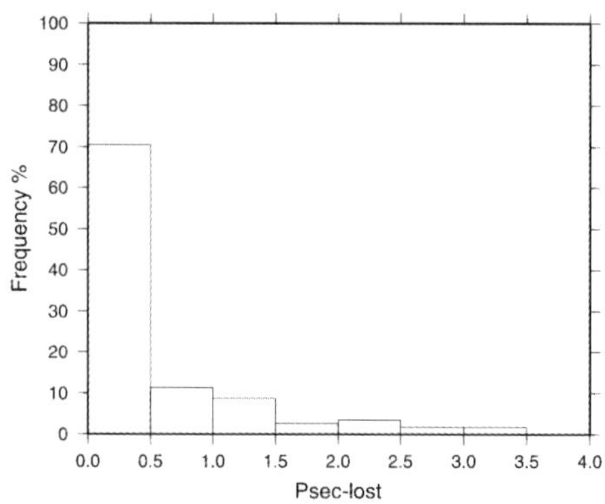

Fig. 4.4b Histogram for the number of events vs Tsf -Tp (Tf s and Tp (b), which are the estimated times of the first samples and first P-arrival times, respectively, relative to the selected strong-motion dataset.

This rather short S-P time interval for strong-motion stations located in the near-source window suggests that the information carried by both P-waves and early S-waves can be used for estimations of source magnitudes.

Starting from the estimated first P-wave and S-wave arrivals on each strong-motion record, we considered the increasing time windows of the low-pass-filtered records to determine the tau_c (Allen and Kanamori 2003), peak ground displacement, velocity and acceleration parameters (Fig. 4.5). A zero-phase-shift, 2-pole, Butterworth filter was used. After a series of trials using different low-pass corner frequencies, we chose a low-pass frequency of 3 Hz, which provided the best results in terms of the correlations between the observed ground-motion quantities and moment magnitudes. This value of 3 Hz was also used by Allen and Kanamori (2003) to retrieve the tau_c vs magnitude relationships (for M> 5.5) using the Californian earthquake database.

The selected strong-motion records from the Italian earthquake dataset were processed as described above, for the following strong-motion quantities:

- PGA_t (peak ground acceleration in a time window of duration t),
- PGV_t (peak ground velocity in a time window of duration t),
- PGD_t (peak ground displacement in a time window of duration t),
- τ_c [dominant period, according to the definition of Allen and Kanamori (2003)].

These were measured within increasing time windows, with an incremental time of 1 sec, starting from the estimated first P-wave and S-wave arrivals. The vertical and the "root-squared sum" of horizontal components are used for the P-wave and S-wave measurements, respectively.

Figure 4.5 illustrates an example of the measurements of ground-motion quantities with their relative time windows, performed along a whole seismogram.

Figures 4.6 and 4.7 show the plots of the logarithm of peak ground-motion quantities as a function of the moment magnitude (Mw) for the time windows after the first P- and S-arrivals, respectively. Figure 4.8 shows the similar plots for the dominant period parameter, calculated for both the P-wave and S-wave trains.

For each time slice, the unweighted, earthquake-averaged logarithmic values of the strong-motion quantities were used to estimate the regression lines and correlation coefficients, representing the relationships between the observed ground-motion quantities and moment-magnitudes.

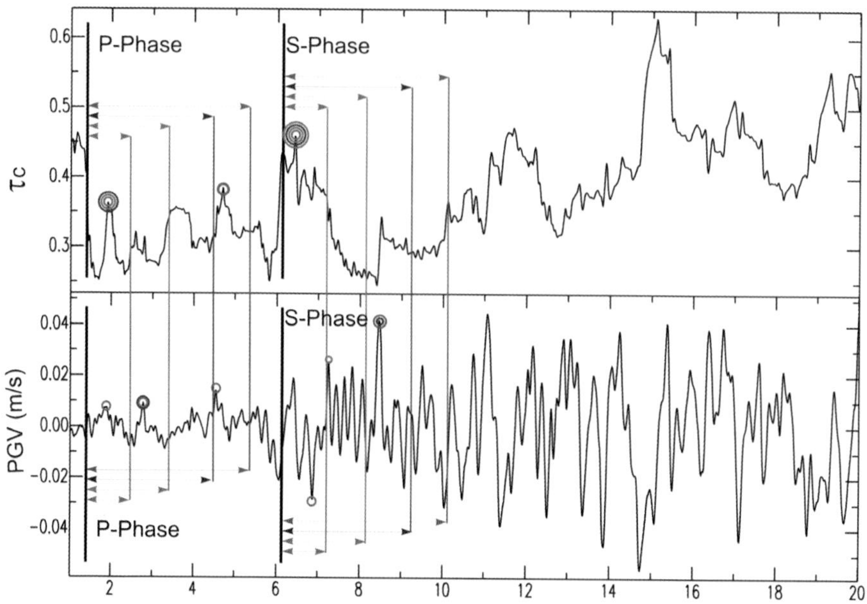

Fig. 4.5 Example of strong-motion record analysis. Top. Tc vs time. Bottom: Peak ground velocity vs time. The time windows increase from 1 to 5 sec after the initial P-wave and S-wave arrivals. The vertical and the root-squared sum of horizontal components are used for the P-wave and S-wave measurements, respectively.

The statistical significance of the log(parameter) vs magnitude correlations were tested by applying the "correlation test" (Taylor 1997). Given the amount of data and the expected significance level, the correlation test provides a threshold of the correlation coefficient above which the linear relationships obtained can be considered statistically significant, with an error of type I defined by the assigned significance level. For each selected time window after the first P- and S-arrivals, Fig. 4.9 shows the measured correlation coefficient for each measured ground-motion quantity, along with the correlation test threshold for the assigned significance level. Since the correlation coefficients are much higher for the S-wave measurements than for those of the P-waves, we chose two different significance levels of the test for the two data types: 0.5 % and 5.0 %, respectively. This means that with *alpha* denoting the significance level of the test, if the correlation coefficient is larger than *alpha*, the probability of making an error of type I (i.e., accepting the hypothesis of a linear relationship between the log(parameter) and magnitude) is smaller than *alpha*.

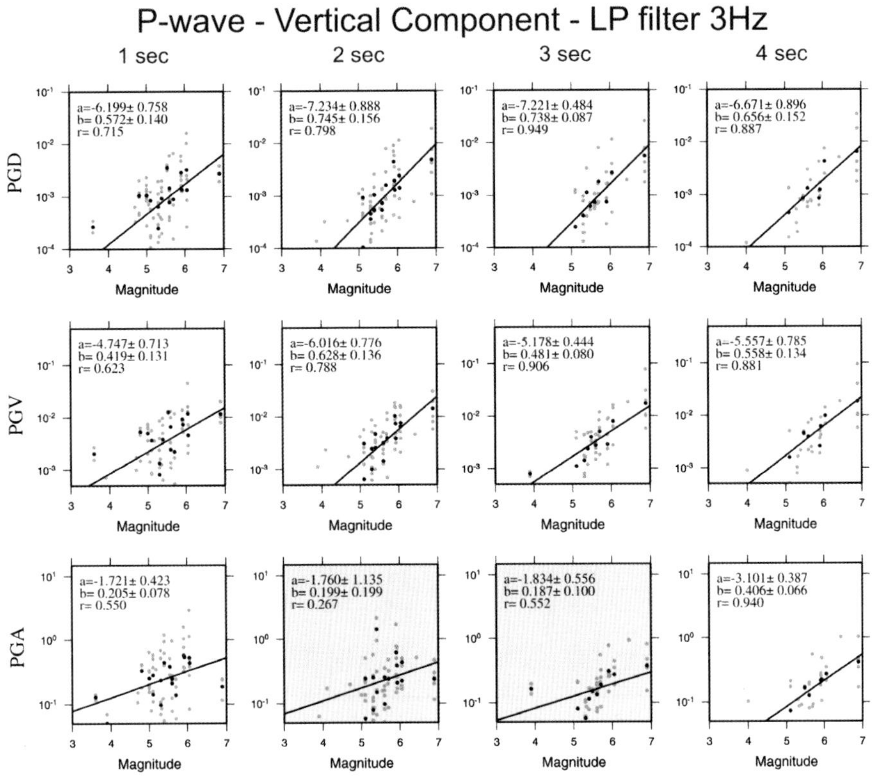

Fig. 4.6 Plot of PGA,PGV and PGD vs magnitude for each time window, and for the P-wave records. The black dots represent event-averaged measures and the shaded panels show low-correlation data according to the statistical test (see text and Fig. 4.9).

For the sake of clarity, the grey shaded panels in Figs. 4.6, 4.7 and 4.8 correspond to the data that did not pass this correlation test.

The results of the correlation analysis between the ground-motion quantities and the moment magnitudes show that 3-Hz, low-pass-filtered displacement and velocity peak quantities can be considered good magnitude estimators on strong-motion records, when measured with 2-3-sec-wide time windows after the initial P-wave and S-wave arrivals. The tau_c parameter appears to be particularly well-correlated with the magnitude on the S-wave records at $t > 3$ sec, while for the P-waves an acceptable correlation was seen only in the $t=3$ sec time window.

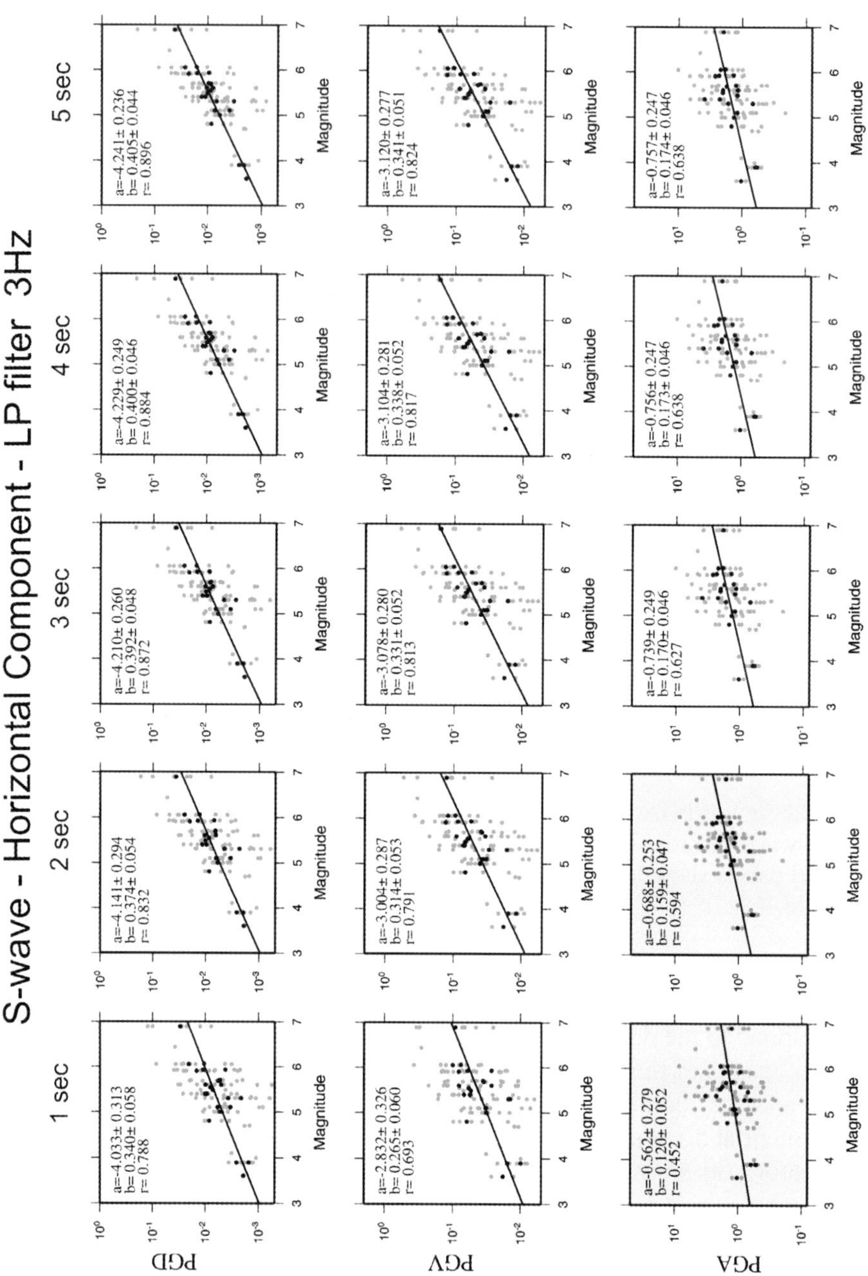

Fig. 4.7 As for Fig. 4.6 but for time windows after the estimated first S-arrivals.

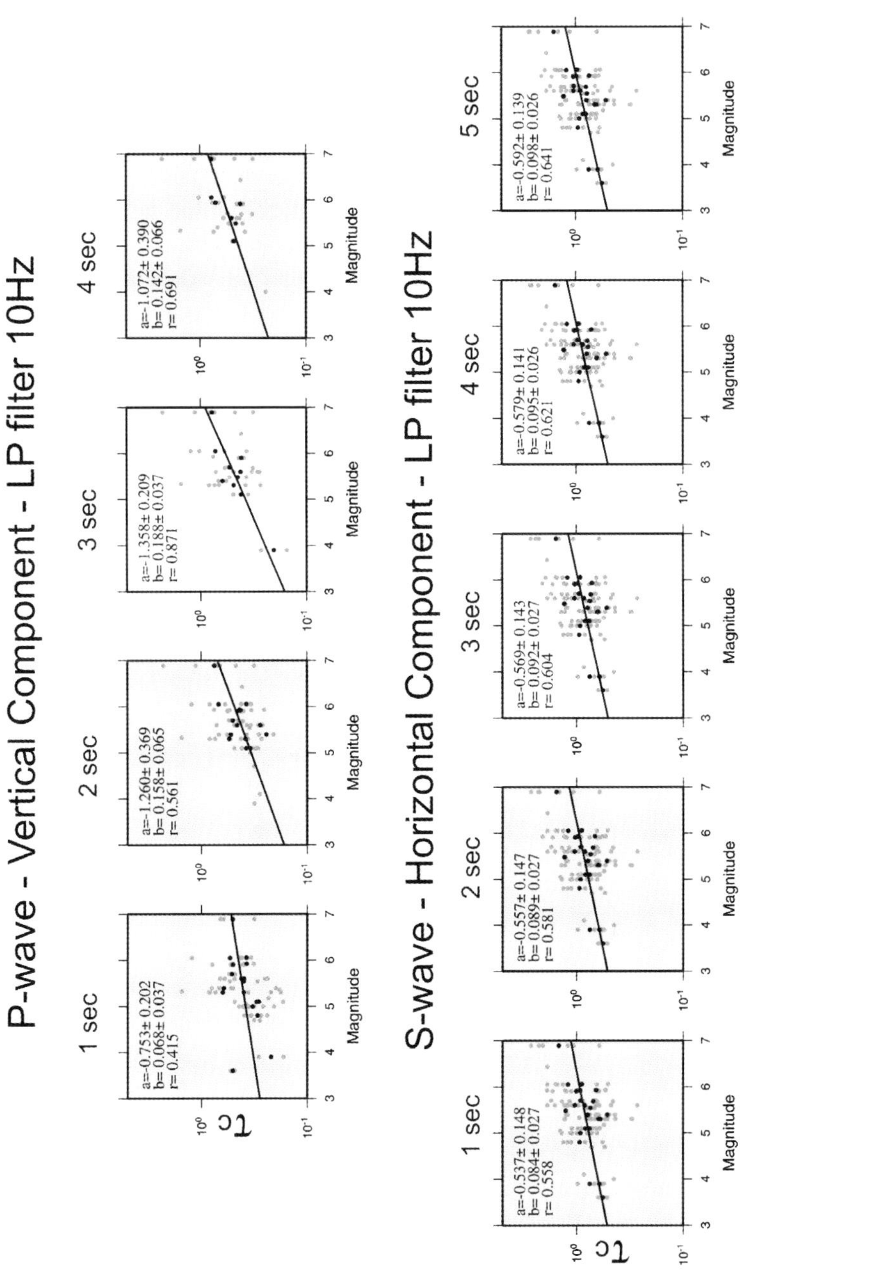

Fig. 4.8 Plot τ_c vs magnitude for each considered time window after the first P-wave and S-wave arrivals. The black dots are event-averaged measures and the shaded panels show low-correlation data according to the statistical test (see text and Fig. 4.9).

4.3 Discussion and Conclusions

Based on these 116 strong-motion records of Italian earthquakes that occurred from 1976 to 1998 and that were acquired at epicentral distances less than 50 km, we investigated the correlations between several observed ground-motion quantities and the moment magnitudes within an increasing time window starting from the first P- and S-arrivals. This was based on the need to implement a real-time procedure for magnitude estimations with the SEW system under construction for the Campania region in southern Italy.

Regression analysis of log (parameter) vs Mw indicates that 3-Hz, low-pass-filtered peak ground velocity (PGVt) and displacement parameters (PGDt) measured in time windows within $t \geq 2$sec after the initial P-wave and S-wave arrivals correlate well with moment magnitude. In particular, the parameters of the retrieved regression lines are stable for the different time windows, and are associated with statistically significant correlation coefficients. This was not the case for the peak ground-acceleration parameter (PGAt), which shows rather scattered and uncorrelated log-values with respect to moment magnitude.

Regression analysis applied to the predominant period parameter introduced by Allen and Kanamori (2003) shows less stable results than the PGVt and PGDt in the post-P-wave windows (an acceptable correlation was found only in the 3-sec time window), while it shows significant correlation with magnitude on S-wave strong-motion records for time windows greater than 3 sec.

This analysis indicates that a significant improvement in the reliability and robustness of earthquake size estimations in real-time procedures can be obtained by combining magnitude estimations obtained from different ground-motion quantities measured at different stations as functions of time from the first P-wave detections at a strong-motion dense network.

A highly dynamic, high-density network around the epicentral area of expected moderate to large earthquakes will also allow for real-time use of early S-wave information, in addition to that of the P-waves. Integration of the different wave-type information is expected to reduce uncertainties in magnitude estimation, as a function of time or of the number of triggered stations. The latter aspect is fundamental in the case of SEW applications that are to provide automatic activation of security measures, such as the shut-down of critical systems that are highly exposed to risk following earthquakes. The possibility of releasing upgraded estimations of source parameters (location and magnitude) along with their uncertainties as a function of time is a basic requirement for a reliable SEW system. Indeed, any control system interfaced with a SEW system can use this information for the step-by-step prediction of peak ground acceleration and intensity, and, in general, the required engineering demand parameter for the target structure. This can allow automatic evaluation of the possibility of activating security measures, based on a probabilistic estimation of false/missed alarms (Iervolino et al. 2007, this issue).

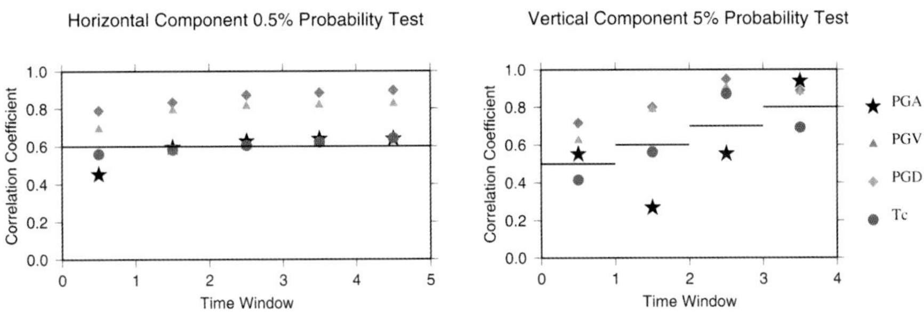

Fig. 4.9 The correlation test. The calculated correlation coefficients are plotted for each time window considered for both the P-wave and S-wave measurements. The horizontal segments give the correlation coefficient thresholds above which the measured values indicate a statistically significant correlation between parameter and magnitude.

The basic scientific question originating from this and other similar studies on the correlations between ground-motion quantities and magnitudes measured during the very early stages of the recording of earthquake signals relates to what earthquake physics and rupture mechanisms can be used to explain the paradox of being able to estimate the earthquake size using just a few seconds of signals from the first P-wave arrivals, i.e. while the rupture itself is still propagating and when the whole rupture process has not yet been achieved. This is especially true for shallow crustal earthquakes of M >7 for which the total rupture times are expected to be in the range of 10 to 15 sec, depending on the average rupture velocity.

In a recent review dedicated to the issues of real-time seismology, Kanamori (2005) discusses and debates different hypotheses and observations regarding the rupture nucleation problem and possible correlations of early P-wave dominant periods with magnitude.

Here, our analysis of near-source strong-motion records primarily shows that 3 Hz low-pass filtered peak velocity and displacement parameters measured in a rather short time window (2-4 sec) after the initial P-wave and S-wave arrivals correlate with earthquake moment.

Given the considered high frequency limit and receivers located at distances greater than 4-6 km from the rupturing fault, the far-field approximation of P-wave seismic radiation predicts a displacement or velocity amplitude proportional to the moment rate function $\dot{M} = \mu\langle\Delta\dot{u}\rangle\Sigma$, where μ is the rigidity, $\langle\Delta\dot{u}\rangle$ is the local slip rate averaged on the ruptured surface Σ.

Two alternative hypotheses can therefore be postulated to explain the observed correlation between peak displacement and velocity correlation with moment magnitude. Following the first hypothesis, in the early stage of the seismic ruptures the slip-rate should scale with earthquake magnitude, i.e., initial large slip-rates are premonitory of large earthquake ruptures and inversely for small magnitude events. According to theoretical models of rupture dynamics (Kostrov 1964, Scholz 1990), the slip-rate amplitude correlates with dynamic stress drop, which would therefore imply a scaling of this quantity with moment.

Several measurements of stress release, including the apparent, static and dynamic stress drop, as inferred from teleseismic, broadband or acceleration recordings of body waves, have been proposed in the literature [for a comprehensive review refer to Boatwright (1984)]. In particular, Boatwright (1984) pointed out that the measurement of Brune static stress drop (Brune 1970) from displacement spectra provides, rather, an estimate of the average dynamic stress drop, as Brune originally proposed.

The scaling of apparent stress with magnitude has been proposed by several authors (Kanamori and Heaton 2000, Wu 2001) but it is still a controversial issue (Ide and Beroza 2001), due to possible limited bandwidth effects on seismic energy measurements from teleseismic and broadband seismograms.

Specifically concerning stress release estimates performed on strong-motion records, De Natale et al. (1987) show a significant variation of Brune stress drop with moment, based on data sets from different world-wide earthquake sequences with $10^{11}<M_0<10^{18}$ $N\ m$. More recently, based on the analysis of earthquakes recorded at Canjon Pass borehole data $10^{10}<M_0<10^{15}$ $N\ m$ of Abercrombie (1995), Beeler et al. (2003) observe that apparent stress and (Brune) stress drop co-vary with magnitude. Kanamori and Rivera (2004), using a wider moment range dataset ($10^{10}<M_0<10^{19}$ $N\ m$), conclude that static stress drop and rupture velocity can scale differently for small and large earthquakes, and in particular stress drop may not necessarily be scale independent, although this scale independence is often implied.

Alternatively, one can assume that in the early stage of the rupture, dynamic stress drop is scale independent, and so the\ slip-rate rate function. The observed correlation of peak displacement and velocity should therefore be attributed to variation in the surface of active slip with magnitude. In other words, for a given time window after the rupture initiation, the surface where maximum slip occurs has a dimension which is magnitude dependent. Since the area of active slip on a fault at a given time is controlled by the kinematic rise-time (τ) parameter (i.e. the slip duration at a given point on the fault surface), this hypothesis would imply a dependence of rise-time with magnitude. Assuming a constant and scale independent slip-rate, since $\Delta\dot{u} \approx \Delta u / \tau = const$ an increase(decrease) in rise-time corresponds to a compensating increase/decrease in slip, which is therefore also dependent on magnitude. The final slip and rise-time parameters are generally obtained through the kinematic inversion of strong-motion and/or teleseismic waveforms. Unfortunately, the estimate of τ is quite a difficult task due to the bias in data between the slip duration, slip amplitude and rupture velocity, which all concur to modify the amplitude and shape of the observed seismic radiation. Based on an analysis of a large number of well-constrained kinematic source models retrieved from different large-size earthquakes, Heaton (1990) observed that the slip duration is generally much smaller than the total rupture time (about 1 sec for earthquake magnitude around M=7).

To our knowledge, no incontrovertible evidence for rise-time scaling with moment actually exists, while the average slip dependence with magnitude is a rather well-known scaling law of earthquakes that has been inferred from a large number of geological and seismological observations (Wells and Coppersmith 1994, Madariaga and Perrier 1998).

On the other hand, if we assume that the predominant period parameter τ_c proposed by Allen and Kanamori (2003) is correlated to the slip duration in the early stages of the rupture, the observed correlation of τ_c with magnitude, partially confirmed also by the present study, indicate that rise-time scales with earthquake size. This hypothesis has been recently advanced by Nielsen (2007, this issue) and Allen (2005).

In conclusion, we consider the observations and ideas developed in this work as a preliminary but important step towards developing a robust, reliable procedure for real-time estimations of magnitudes for SEW applications. Nevertheless, a more refined study is planned for the near future, which will provide the integration of data uniformly covering the magnitude range under investigation, and which will further explore the implications for earthquake-source physics that will arise from confirmation of the results of the present study.

4.4 Acknowledgements

We are grateful to Stefan Nielsen and Gaetano Festa for many useful discussions on the implication of our results on earthquake physics processes and their valuable comments on the preliminary version of this article. We also acknowledge the useful suggestions and remarks of Vincenzo Convertito, Claudio Satriano, Raffaella De Matteis and Giovanni Iannaccone on data analysis. This research was carried out in the framework of the AMRA Center of Competence. This work was also partially funded by MIUR through the PON-Tecsas.

References

Abercrombie R (1995) Earthquake source scaling relationships from 1 to 5 ML using seismograms recorded at 2.5 km depth. J Geophys Res 100(24):24015-24036

Allen RM, Olson EL (2005) The relation between rupture initiation and earthquake magnitude. In: AGU (ed) Eos Trans. AGU Fall Meet Suppl, Abstract S33C-06, 2005

Allen RM, Kanamori H (2003) The Potential for Earthquake Early Warning in Southern California. Science 300:786-789

Beeler NM, Wong TF, Hickman SH (2003) On the expected relationships among apparent stress, static stress drop, effective shear fracture energy, and efficiency. Bull Seism Soc Am 93(3):1381-1389

Beroza GC (1996) Rupture history of the earthquake from high frequency strong motion data. In: Spudich P (ed) The Loma Prieta, California, Earthquake of October 17, 1989: Main Shock Characteristics, p. 9-32. USGS-Prof. Pap. 1550-A

Boatwright J (1984). Seismic estimates of stress release. J Geophys Res 89:6961-6968

Brune JN (1970) Tectonic stress and the spectra of seismic shear waves from earthquakes. J Geophys Res 75:4997-5009

De Natale G, Madariaga R, Scarpa R, Zollo A (1987) Source parameter analysis from strong motion records of the Friuli, Italy, earthquake sequence (1976-1977). Bull Seism Soc Am 77(4):1127-1146

Emolo A, Zollo A (2005) Kinematic source parameters for the 1989 Loma Prieta earthquake from the nonlinear inversion of accelerograms. Bull Seism Soc Am 95:981-994

Ide S, Beroza GC (2001) Does apparent stress vary with earthquake size? Geophys Res Lett 28:3349-3352

Iervolino I, Convertito V, Giorgio M, Manfredi G, Zollo A (2007) The crywolf issue in earthquake early warning applications for the Campania region. In:

Gasparini P, Manfredi G, Zschau J (eds) Earthquake Early Warning Systems. Springer

Kanamori (2005) Real-time seismology and earthquake damage mitigation. Annu Rev Earth Planet Sci 33:195-214

Kanamori H, Heaton T (2000) Microscopic and macroscopic physics of earthquakes. In: Rundle JB, Turcotte DL, Klein W (eds) GeoComplexity and the Physics of Earthquakes, pp. 147-163, AGU

Kanamori H, Rivera L (2004) Static and dynamic scaling relations for earthquakes and their implications for rupture speed and stress drop. Bull Seism Soc Am 94(1):314-319

Kostrov BV (1964) Selfsimilar problems of propagation of shear cracks. Applied Mathem and Mech 28(5)

Nielsen S (2007) Can earthquake size be controlled by the initial seconds of rupture? In: Gasparini P, Manfredi G, Zschau J (eds) Earthquake Early Warning Systems. Springer

Perrier G, Madariaga R (1998) Les Tremblements de Terre. CNRS France, pp. 216

Scholz CH (1990) The Mechanics of Earthquakes and Faulting. Cambridge University Press, see Chapter 4.2, equation 4.23

Sunder S, Connor J (1982) A new procedure for processing strong-motion earthquake signals. Bull Seism Soc Am 72:643-6621

Taylor JR (1997) An Introduction to Error Analysis. University Science Book

Weber E, Iannaccone G, Zollo A, Bobbio A, Cantore L, Corciulo M, Convertito V, Di Crosta M, Elia L, Emolo A, Martino C, Romeo A, Satriano C (2007) Development and testing of an advanced monitoring infrastructure (ISNet) for seismic early-warning applications in the Campania region of southern Italy. In: Gasparini P, Manfredi G, Zschau J (eds) Earthquake Early Warning Systems. Springer

Wells DL, Coppersmith KL (1994) New empirical relationships among magnitude, rupture width, rupture area, and surface displacement. Bull Seism Soc Am 84:974-1002

Wu YM, Kanamori H (2005) Rapid Assessment of Damage Potential of Earthquakes in Taiwan from the Beginning of P Waves. Bull Seism Soc Am 95(3):1181-1185

Wu YM, Teng T (2004) Near real-time magnitude determination for large crustal earthquakes. Tectonophys 309:205-216

Wu ZL (2001) Scaling of apparent stress from broadband radiated energy catalogue and seismic moment catalogue and its focal mechanism dependence. Earth Planets Space, 53:943-948

Zeng Y, Aki K, Teng T (1992) Mapping of the high-frequency source radiation for the Loma Prieta earthquake, California. J Geophys Res 98:11981-11993

5 A New Approach to Earthquake Early Warning

Maren Böse[1], Mustafa Erdik[2], Friedemann Wenzel[1]

[1] Karlsruhe University, Geophysical Institute, Karlsruhe, Germany
[2] Bogazici University, Kandilli Observatory, Istanbul, Turkey

Abstract

Earthquake early warning systems utilize the capability of modern real-time systems to process and transmit information faster than seismic waves propagate (3-6 km/sec); they provide first information on forthcoming ground shaking prior to the arrival of seismic waves at potential user sites. The possible warning time ranges up to 70 seconds, depending on the distances between the seismic source, sensor and user sites. The main effort in the design and implementation of early warning systems in the last few years aims at increased warning times and thus at an expansion of areas that can be effectively warned. This requires new strategies and methods for the rapid characterization of earthquakes: on-site warning systems are based on observations at single stations for the quick estimate of source parameters. The on-site warning approach is significantly faster than network based regional strategies for early warning even though often at the expense of accuracy. We have developed an alternative method for earthquake early warning – called PreSEIS (pre-seismic shaking) – that is as fast as the on-site warning approach but combines information from several sensors within small seismic networks with apertures of about 100 kilometers to estimate source parameters from the first few seconds of seismic recordings. Therefore, it is not necessary for seismic waves to have arrived at all sensors before estimates can be issued because even a non-triggered station provides valuable information about the possible location of the seismic source and confines the space of possible solutions to the problem of a magnitude-distance trade-off. With on-going time, longer time series at more stations become available, allowing improved information to be inferred about the earthquake. PreSEIS is based on artificial neural networks. The algorithm will be demonstrated using the example of the mega-city Istanbul, Turkey.

5.1 Introduction

Substantial progress in seismic real-time acquisition and communication technologies, aside from enhancements of seismic processing software, has been made over the past few years, paving the way for the design and implementation of earthquake early warning systems world-wide. Systems are now in operation in Japan, Taiwan and Mexico (Nakamura 1989; Wu and Teng 2002; Wu and Kanamori 2005; Espinosa-Aranda et al. 1995). In other countries, such as Romania or Turkey, systems are well under way (Wenzel et al. 1999; Erdik et al. 2003b).

The maximum warning time that can be theoretically achieved by an earthquake early warning system is defined by the time difference between P-wave detection by one or more seismic sensors close to the earthquake source and the arrival of high-amplitude S- or surface waves at potential user sites. The available time is generally less than 70 seconds. Earthquake early warning systems are effective tools for disaster mitigation if used for the triggering and execution of automatisms to prepare vulnerable systems and dangerous processes for the imminent danger. For example, seismic warnings can be used to slow down high-speed trains to avoid derailments, to shutdown pipelines and gas lines to minimize fire hazards, to shutdown manufacturing operations to decrease potential damage to equipment, or to save vital computer information to inhibit losses of data. A compilation of effective measures in response to warnings is given by Goltz (2002).

While hitherto existing (traditional) regional warning systems have mostly focused on potential users in epicentral distances of about 100 kilometers, on-site warning systems are supposed to serve broader user communities on a local scale. On-site systems save warning times by decentralized processing and restriction on information derived from single stations instead of station networks; generally, only the first few seconds of P-wave arrivals are analyzed and used to rapidly estimate the magnitude and location of the triggering earthquake. A promising approach to early warning is the usage of frequency content of seismic motion as an indicator of earthquake magnitude (Nakamura 1985; Kanamori 2005). Due to large scatter in observations at single stations, magnitude predictions are generally obtained by averaging over estimates at several sites which are considered independently of each other. Wu and Kanamori (2005), for example, suggest averaging over the first 8 stations in epicentral distances of less than 21 kilometers to obtain reliable predictions of magnitudes in Taiwan, which is strictly speaking in contradiction with the primary definition of on-site early warning.

We have developed a new approach to earthquake early warning that takes advantage of both – regional and on-site – warning concepts. PreSEIS combines observations of seismic ground motion at several sites to solve the problem of the magnitude-distance trade-off. PreSEIS does not require the arrival of seismic waves at all sites before estimates are issued since even non-triggered stations provide valuable information on likely source locations. Like the on-site warning approach PreSEIS requires the triggering of only one sensor to come into action. Considering observations of seismic ground motion at several stations of a local seismic network simultaneously (even if they are not triggered) provides more details on the earthquake – for example with respect to rupture expansion and radiation patterns – than using single stations. PreSEIS provides estimates as rapidly as the on-site warning method and is therefore suitable for earthquake early warning for users in epicentral distances of less than 100 kilometers. With on-going time and increasing availability of information, estimates are automatically updated.

In this paper we will demonstrate PreSEIS using the example of the mega-city Istanbul, Turkey. Istanbul faces a huge seismic risk caused by the Main Marmara Fault as the western continuation of the North Anatolian Fault Zone. Segments of the Main Marmara Fault partly run only about 20 kilometers from the coast of the Sea of Marmara and Istanbul (Fig. 5.1) and pose a high seismic threat to more than 11 million people. A historic earthquake catalogue for the Marmara region over the past 2,000 years (Ambraseys 2002) reveals that on average at least one medium intensity (VII-VIII) earthquake has hit Istanbul every 50 years; the average return period for high intensity (VIII-IX) events is 300 years. From previous earthquakes in the Sea of Marmara and the stress transfer of the 17 August 1999 Izmit earthquake (M_w = 7.4) Parsons (2004) determine a 53 ± 18% probability of strong shaking in Istanbul metropolitan area during the next 30 years. Erdik et al. (2003a) estimate that in case of a M_w = 7.5 earthquake in the Sea of Marmara destructions might amount to about USD 11 billion losses, with 40,000 to 50,000 deaths and between 430,000 and 600,000 destroyed households.

The devastating earthquakes of Kocaeli and Düzce in 1999 opened a window of opportunity for seismic risk reduction in the Istanbul metropolitan area and the Marmara region. One component of seismic risk mitigation is a real-time earthquake information system, the Istanbul Earthquake Rapid Response and Early Warning System (IERREWS). Details on IERREWS are given by Erdik et al. (2003b). The earthquake early warning system which is part of IERREWS comprises ten tri-axial strong motion instruments deployed along the coast of the Sea of Marmara (Fig. 5.1), a real-time communication radio- and satellite-link, and a central processing facility in the Kandilli-Observatory of the Bogazici University in Istanbul. The proximity of seismogenic faults to Istanbul poses a great challenge for the design of an early warning system since the time window between seismic detection and high amplitude wave arrivals in Istanbul is extremely short.

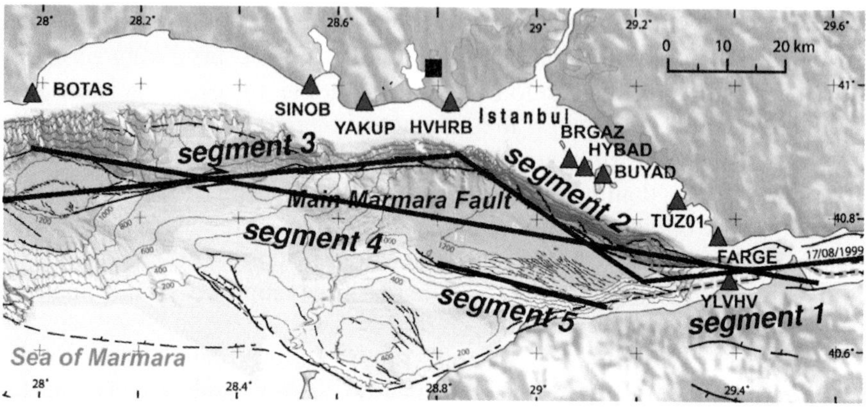

Fig. 5.1 Locations of fault segments in the Sea of Marmara, close to Istanbul, as input for the simulation of ground motion scenarios in this work (map after Armijo et al. 2002; own modifications). Segment 4 stands for joint ruptures of segments 1 to 3. Triangles mark locations of the ten tri-axial accelerometers with real-time communication link to the Kandilli-Observatory as part of the earthquake early warning system in Istanbul (Erdik et al. 2003b).

5.2 Method

As soon as the first sensor of the employed seismic network (here: the ten stations of the Istanbul earthquake early warning system) detects the P-wave, PreSEIS provides estimates on magnitude and location of the triggering earthquake. PreSEIS utilizes artificial neural networks (ANNs) for rapid source characterization. ANNs provide statistical models of systems and processes whose design has to be based on empirical experience rather than explicit rules (e.g., Swingler 1996). They are built up by simple processing units - called *neurons* - which are highly interconnected with each other. The importance of each link within the network is governed by its assigned weight.

PreSEIS uses ANNs for rapid estimation of source parameters from the initial parts of seismograms; these ANNs approximate the propagation and attenuation of seismic waves without requiring specifications of crustal properties such as velocity distributions or attenuation models. Short processing times and a high tolerance of noise turn ANNs into interesting tools for seismic early warning.

PreSEIS uses so-called Two-Layer Feed-Forward neural networks with neurons arranged in a *hidden* and an *output* layer. The mapping function of such a network can be written as

$$y_k^{net} = \tilde{g}\left(\sum_{j=0}^{M} w_{2kj}^{net}\, g\left(\sum_{i=0}^{d} w_{1ji}^{net}\, x_i^{net}\right)\right), \tag{5.1}$$

where y_k^{net} is the output of the kth and x_i^{net} the input of the ith neuron in network *net*; $g(\ )$ and $\tilde{g}(\ )$ are non-linear activation functions; w_{1ji}^{net} describes the weight of the link between input neuron i to neuron j in the hidden layer; w_{2kj}^{net} describes the weight of the link between neuron j in the hidden layer to neuron k in the output layer.

The availability of information about the earthquake is clearly time-dependent: with on-going time t approximations of source parameters within PreSEIS are continuously updated. Using all available information at each time step – even the information that certain sensors are not yet triggered - helps to confine the space of possible solutions. The problem of magnitude-distance trade-off is clearly underdetermined using data from a single station. A favorable design of early warning algorithms, however, supports the ability to consider all available information – including also *a priori* information about fault locations etc. – with a rapid and reliable convergence towards the correct magnitude and location solution.

The two tasks of rapid source localization and magnitude estimation in PreSEIS are addressed by distinct neural network designs, abbreviated by *net hypo* and *net* M_w. Localization is based on (scaled) time delays Δt_j^{trigg} between P-wave detections at sensors j=1,..., 10, t_j^{trigg}, and the first triggered station i, t_i^{trigg}, whereby $t_i^{trigg} \le t_j^{trigg}$. If station j is not triggered at time t, PreSEIS assumes $t_j^{trigg} = t$. Hence, the inputs to *net hypo* can be written as

$$x_j^{hypo}(t) = \Delta t_j^{trigg}(t),\ j\text{=}1,\ldots,\ 10, \tag{5.2}$$

with

$$\Delta t_j^{trigg}(t) \equiv \left(t - t_i^{trigg}\right),\ \text{if } t < t_j^{trigg},\ \text{and} \tag{5.2a}$$

$$\Delta t_j^{trigg}(t) \equiv \left(t_j^{trigg} - t_i^{trigg}\right),\ \text{if } t \ge t_j^{trigg}. \tag{5.2b}$$

For the estimation of earthquake magnitude, *net* M_w uses the observed ground motions at the different sensors of the seismic network $a_j(t)$, j=1,…,10, that are available at time t. More strictly speaking, PreSEIS uses the (bracketed) cumulative absolute velocity (CAV) (Benjamin et al. 1988), i.e. the integrated velocity measured in the time interval between t_j^{trigg} and t. One of the obvious advantages of CAV relative to other ground motion parameters is its integrative character that makes the CAV independent of details of rupture histories, e.g. caused by inhomogeneous slip distributions. PreSEIS uses the logarithmic values of the (scaled) CAV as input to *net* M_w:

$$x_j^{M_w}(t) = \log_{10}\left(CAV_j(t)+1\right), j=1,\ldots, 10, \tag{5.3}$$

whereby the CAV at station j and time t is defined by

$$CAV_j(t) \equiv 0\text{, if } t < t_j^{trigg}\text{, and} \tag{5.3a}$$

$$CAV_j(t) \equiv \int_{t_j^{trigg}}^{t} \left|a_j(t)\right| dt\text{, if } t \geq t_j^{trigg}. \tag{5.3b}$$

As further input PreSEIS includes the estimated source locations provided by *net hypo*.

Let us assume that the true source location of earthquake l is given by $\overline{X}_l = \left(X_l^{hypo}, Y_l^{hypo}, Z_l^{hypo}\right)$, its magnitude by $M_{w,l}$; the corresponding estimates by PreSEIS are given by $\hat{\overline{X}}_l = \left(\hat{X}_l^{hypo}, \hat{Y}_l^{hypo}, \hat{Z}_l^{hypo}\right)$ and $\hat{M}_{w,l}$. Model parameters of the ANNs are determined from example patterns by minimizing the errors of the n pairs of target t_k and output values y_k in the training dataset. For example, choosing the mean-absolute-error function E^{net} with

$$E^{net} = \frac{1}{n}\sum_{l=1}^{n}\left(\sum_{k=1}^{c}\left|y_{k,l}^{net} - t_{k,l}^{net}\right|\right), \tag{5.4}$$

we obtain for *net hypo*

$$\widetilde{E}^{hypo} = \frac{1}{n}\sum_{l=1}^{n}\left(\left|\hat{\widetilde{X}}_l^{hypo} - \widetilde{X}_l^{hypo}\right| + \left|\hat{\widetilde{Y}}_l^{hypo} - \widetilde{Y}_l^{hypo}\right| + \left|\hat{\widetilde{Z}}_l^{hypo} - \widetilde{Z}_l^{hypo}\right|\right) \tag{5.4a}$$

and for *net* M_w

$$\tilde{E}^{M_w} = \frac{1}{n}\sum_{l=1}^{n}\left|\hat{\tilde{M}}_{w,l} - \tilde{M}_{w,l}\right|, \tag{5.4b}$$

where ˜ indicates scaled and ^ estimated parameters (e.g., Swingler 1996).

In addition to the described magnitude and location estimates, PreSEIS is designed to predict the full expansion of the evolving rupture. Even though these estimates are generally associated with significant errors, this information is essential - if combined with empirical attenuation laws – to calculate real-time *alert maps*. Alert maps show the likely distribution of peak ground acceleration (PGA), seismic intensity (I) or any other ground motion parameter of interest, and are therefore strongly related to shake maps (Wald et al., 1999) with the distinction of being based on predicted rather than observed ground motion data. Alert maps aim to rapidly identify areas likely to be affected by strong ground shaking. Relative to the simplified assumption of the earthquake source being in a single point, the estimation of rupture expansions helps to significantly improve alert maps.

5.3 Database

The North Anatolian Fault shows high seismic activity. Most earthquakes recorded during recent decades, however, have occurred along the middle and eastern part of the 1,500 km long fault. Seismicity in the Marmara region, on the other hand, is relatively low – a circumstance that significantly aggravates the design and verification of a seismic early warning system for the region. Simulations of seismic ground shaking help to overcome the lack of appropriate data.

While long-period motions of seismic ground shaking are in principle predictable, short-period motions are hard to assess because of the increasing incoherence of seismic waves at high frequencies caused by small-scale heterogeneities in the source process and crustal properties. Short-period motions can be treated as stochastic phenomena: at large distances from the earthquake source seismic ground motions can be well described as band-limited Gaussian noise of finite duration with a characteristic ω^n Fourier amplitude spectrum (FAS):

$$|S(x,\omega)| = \frac{R^{\Theta\Phi} M_0}{4\pi\rho\beta^3}\left[1+\left(\frac{\omega}{\omega_c}\right)^2\right]^{-(n+1)/2}, \tag{5.5}$$

where $R^{\Theta\Phi}$ is the radiation pattern, M_0 the seismic moment, ρ the density and β the seismic velocity of the medium through which the seismic waves propagate; ω is the circular frequency and ω_c the corner frequency of the FAS.

By the specification of parameters that describe the shape of the earthquake source spectrum and simple models that account for propagation and site effects, the stochastic method allows for simulating ground motion time series at arbitrary sites (Boore 1983). At smaller distances, however, finite source dimensions have to be considered. In this work we make use of an extension of the point source method to finite faults as proposed by Beresnev and Atkinson (1997). The basic idea of stochastic modeling for finite faults is to treat the rupture as a system composed of several point sources: each of these subfault elements is assigned a ω^n Fourier amplitude spectrum as described by equation (5.5). The relation between target spectrum and diverse subfault parameters is studied by Atkinson and Beresnev (1997, 1998).

To provide a realistic database for subsequent studies we assign variability to certain source parameters, including stress drops, slip velocities, rupture dimensions and average asperity sizes on the fault (Böse 2006). Diverse slip distributions are simulated and combined with different directions of rupture propagation, including bi- and unidirectional cases. Site effects as a combination of high-frequency diminution and site amplification - for example by soft sediments - are included as frequency-dependent factors based on the NEHRP soil classification scheme (Building Seismic Safety Council 1995) as determined by Boore and Joyner (1997). Hence we assume that our synthetic database includes a realistic degree of inter- and intra-event variability.

Using the stochastic modeling procedure for finite faults we simulate ground motion records for 250 earthquakes in the magnitude range of M_w = 4.5 to 7.5 at the ten stations of the Istanbul earthquake early warning system. In addition, we model ground motion time series at an industrial settlement close to Istanbul (*UserX*) as a potential user of warnings. Earthquake epicenters are located along the five fault segments in the Sea of Marmara as shown in Figs. 5.1 and 5.2. The introduction of segment 4 was necessary for the simulation of earthquakes associated with joint ruptures of fault segments 1 to 3; the simulation procedure allows only linear rupture propagation. Source depths of simulated earthquakes vary between 5 and 20 kilometers. A detailed description of the synthetic database is given by Böse (2006).

For neural network learning in PreSEIS we subdivide the database of simulated earthquake records randomly into a training and a test dataset of 200 and 50 patterns, respectively. Earthquakes epicenters of the synthetic database are plotted in Fig. 5.2: black stars indicate locations of training, white stars of test events. By applying the Resilient Backpropagation algorithm (Riedmiller and Braun 1993) we determine the combination of network weights at which $\widetilde{E}^{net}$ is sufficiently small at simultaneous high generalization capability of the ANNs which is quantified with help of the independent test dataset.

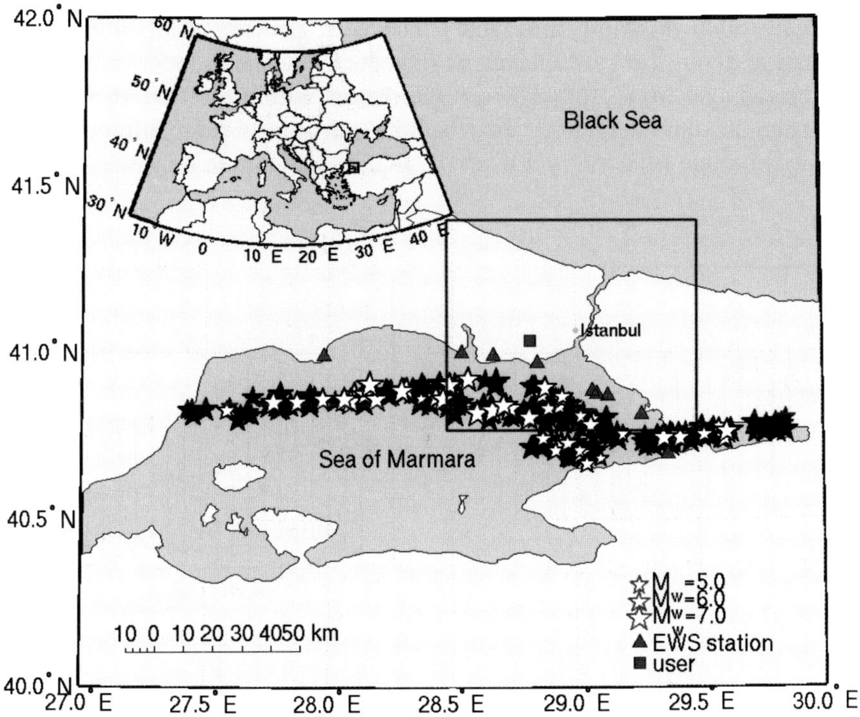

Fig. 5.2 Distribution of epicenters of 250 stochastically simulated earthquakes in the Sea of Marmara (4.5 ≤ M_w ≤ 7.5); black stars mark locations of events of the training, white stars of the test dataset for the artificial neural networks in PreSEIS. The earthquakes occur along the 5 fault segments shown in Fig. 5.1 in depths between 5 to 20 kilometers. Source parameters are randomly varied; different slip distributions and slip directions are modeled, including uni- and bidirectional ruptures.

5.4 Results

The absolute prediction errors of (unscaled) locations E^{hypo} and magnitudes E^{M_w} at time steps between 0.5 to 10.0 seconds after triggering of the first sensor are plotted in Figs. 5.3a and 5.3b. For this error analysis we consider all n = 250 earthquakes in the database, i.e. all training and test events. The reliability of predictions of both source parameters increases gradually with on-going time and hence with growing availability of information about the earthquakes as expected: the mean localization error can be reduced from almost 9 to 6 kilometers within the first 10 seconds, the mean absolute magnitude error and its unit standard deviation decrease during the same time interval from 0.5 to 0.25 magnitude units.

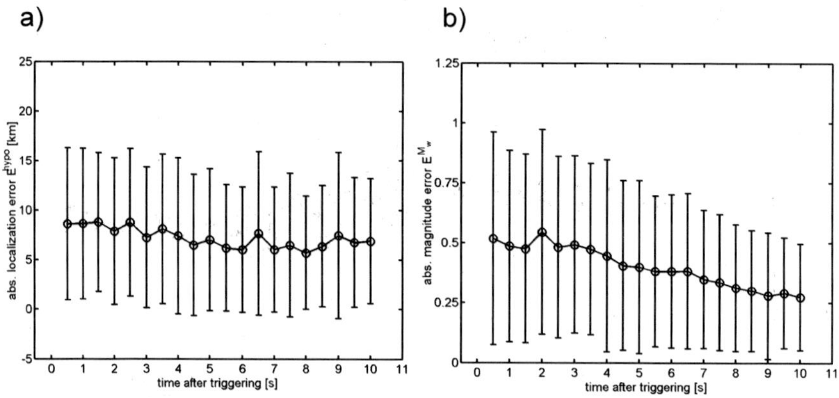

Fig. 5.3 Absolute errors of estimated a) hypocenter locations and b) magnitudes by PreSEIS for the entire training and test datasets as a function of time after triggering of the first station. Shown are the mean errors and unit standard deviations. The mean localization error can be reduced from 9 to 6 kilometers in the first 10 seconds after triggering of the first sensor. The mean magnitude error can be almost halved during the same time interval from 0.5 to 0.25 units.

In the following we will demonstrate PreSEIS using two earthquake scenarios from the test database. The first scenario is a M_w = 6.7 earthquake on fault segment 3 (*Scenario 1*), the second a M_w = 5.4 earthquake on fault segment 5 (*Scenario 2*). Epicenter locations are indicated in Figs. 5.4b and 5.5b by stars; the associated fault ruptures are visualized in Figs. 5.4c and 5.4c. Note that the density of early warning stations in the western Sea of Marmara is relatively low. For *Scenario 1* potential users of PreSEIS located at *USERX* (indicated in Fig. 5.4b by a square) would have

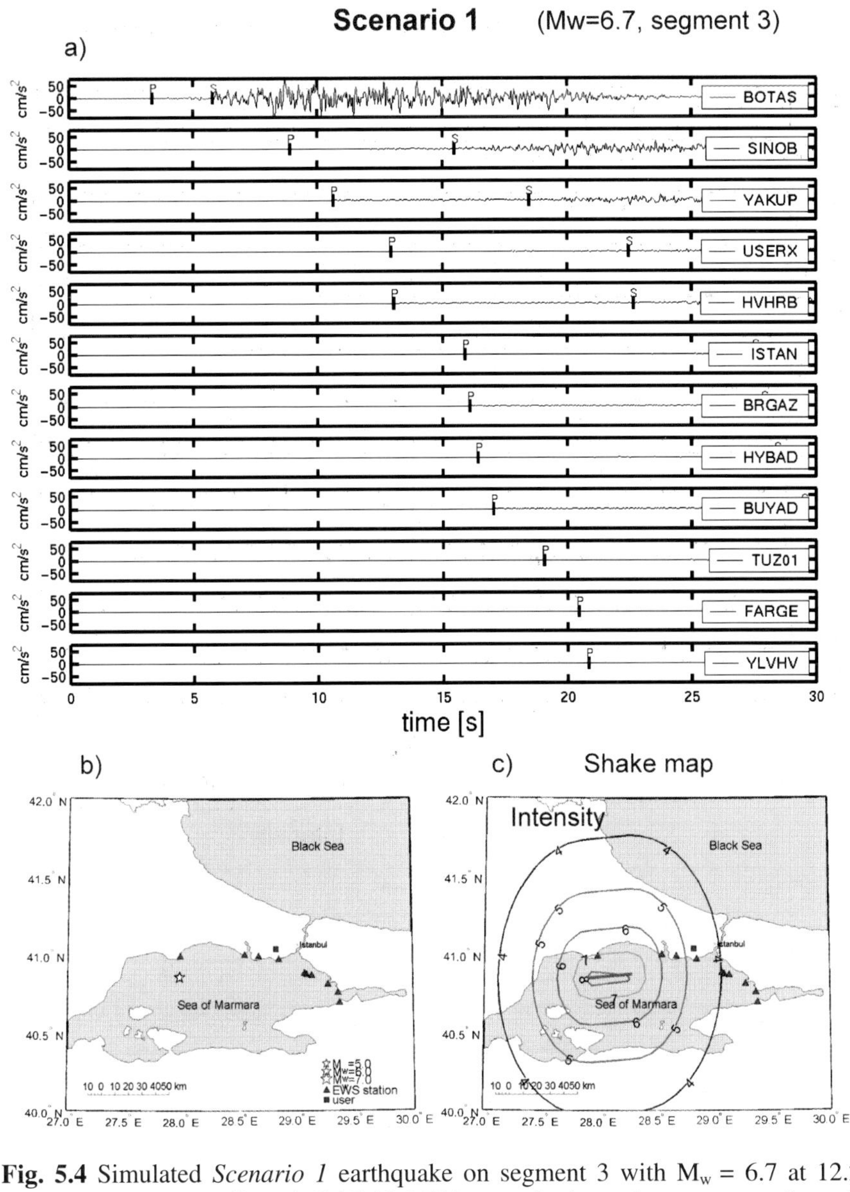

Fig. 5.4 Simulated *Scenario 1* earthquake on segment 3 with M_w = 6.7 at 12.2 kilometers depth. a) Filtered (0.05-12.0 Hz) mean horizontal component of acceleration simulated at the early warning stations and the industrial settlement (*UserX*). The records are sorted according to P-wave arrivals. b) Location of the earthquake epicenter indicated by a star. c) Fault rupture and shake map of seismic intensity for rock condition.

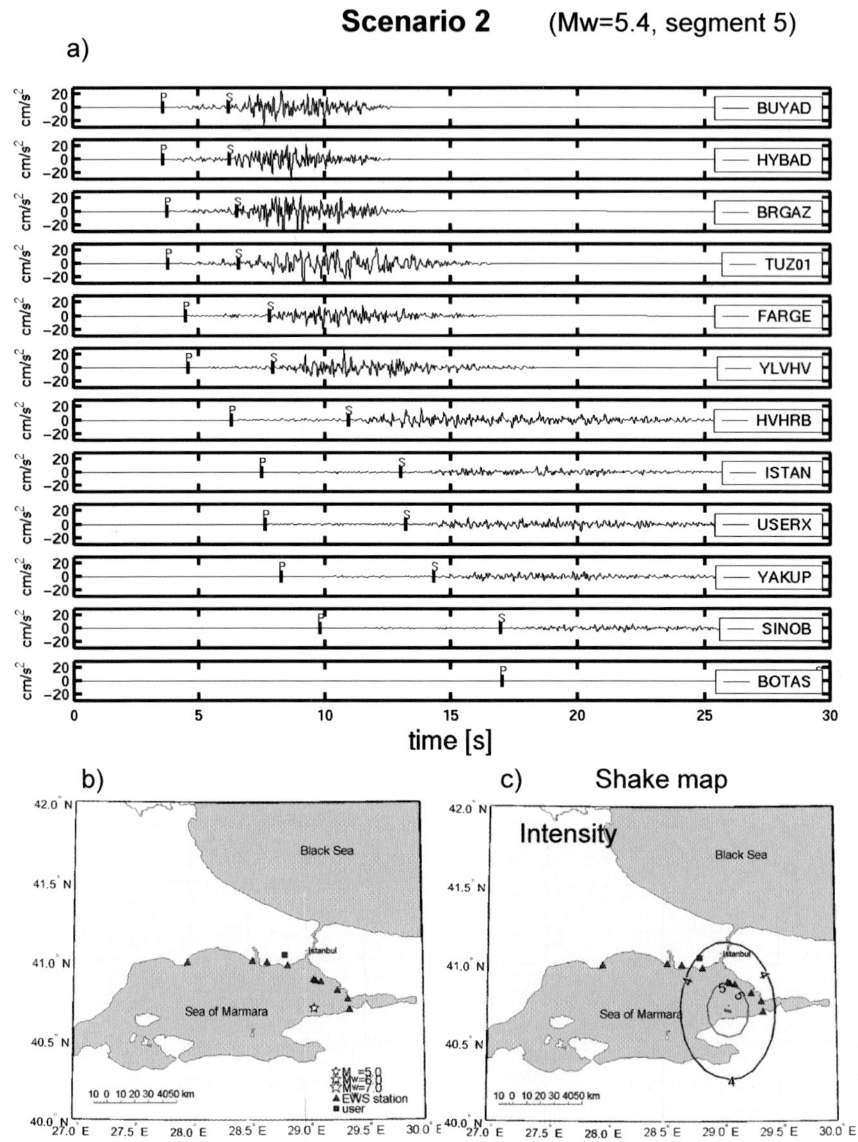

Fig. 5.5 Simulated *Scenario 2* earthquake on segment 5 with $M_w = 5.4$ at 7.2 kilometers depth. For further explanations see caption of Fig. 5.4.

a maximum warning time of almost 20 seconds available. The *Scenario 2* earthquake, on the other hand, occurs in the eastern Sea of Marmara where the station density is significantly higher; warning times are only 10 seconds for users located at *USERX*. The simulated mean horizontal components of ground acceleration at the early warning stations and the industrial settlement (*UserX*) are shown in Figs. 5.4a and 5.5a. The records are sorted according to P-wave arrivals.

To calculate shake and alert maps of seismic intensity we apply the following attenuation law derived from observational data and simulated ground motion records in Northwestern Turkey (Böse 2006) where we make use of an empirical relationship between spectral amplitudes and seismic intensity (Sokolov 2002):

$$\ln(I) = 0.8089 + 0.2317M_w - 0.1073\ln(r_{jb} + 0.6M_w) - 0.0052r_{jb} + C(site, M_w). \qquad (5.6)$$

The Joyner-Boore distance r_{jb} defines the closest horizontal distance in [km] between the point of observation and the vertical projection of the rupture onto the surface; correction term $C(site, M_w)$ accounts for magnitude-dependent site effects on ground motion (Böse, 2006).

Figures 5.4c and 5.5c show shake maps of seismic intensity for *Scenario 1* and *Scenario 2* assuming rock condition. The corresponding predictions by PreSEIS at four time steps between 0.5 to 2.0 seconds after triggering of the first sensor are illustrated in Figs. 5.6 and 5.7: maps in the left columns show the estimated locations of the epicenters; figures in the right columns visualize alert maps of seismic intensity determined from the estimated values of magnitudes and rupture expansions by PreSEIS using equation (5.6).

Comparisons of predicted and target values of epicenter locations, magnitudes and rupture expansions are shown for *Scenario 1* and *Scenario 2* in Figs. 5.6 and 5.7 and Figs. 5.4 and 5.5, respectively. The importance of *a priori* information on likely source locations by the training patterns is nicely demonstrated in case of *Scenario 1* (Fig. 5.8, left): even though only one station provides data in the first two seconds after the triggering of station *BOTAS*, PreSEIS maps the epicenter close to the true location (Fig. 5.4b). Like estimates on magnitude and rupture expansion the predicted hypocenter location is very stable; seismic intensity is underestimated by 0.5 units. Estimates of magnitude vary in the first 2 seconds between $\hat{M}_w = 6.3$ and $\hat{M}_w = 6.4$. Predictions of source parameters for *Scenario 2* within the first two seconds after triggering of station *BUYAD* are based on data from 4 to 6 sites. Source location and earthquake size are quickly recognized (Fig. 5.7). PreSEIS issues magnitude estimates between $\hat{M}_w = 5.3$ and $\hat{M}_w = 5.7$.

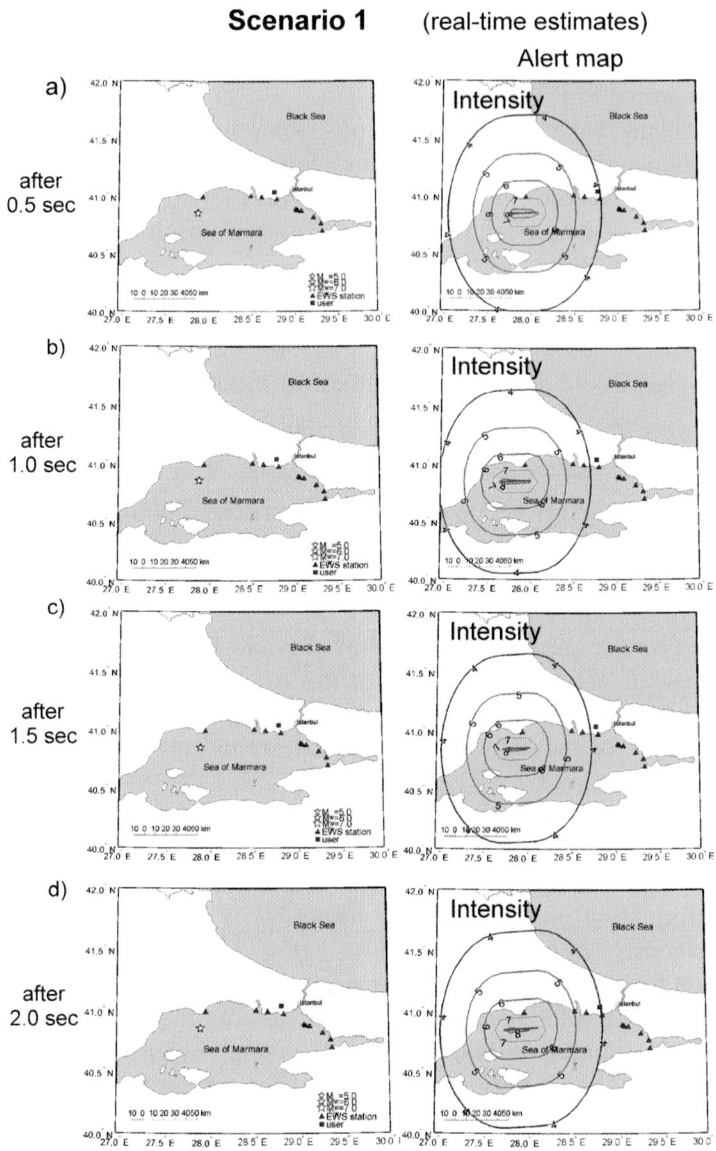

Fig. 5.6 Real-time estimates of epicenter locations (left) and alert maps of seismic intensity (right) for *Scenario 1* at a) 0.5, b) 1.0, c) 1.5 and d) 2.0 seconds after triggering of station *BOTAS*. Simulated acceleration time series and target values are shown in Fig. 5.4. For the calculation of alert maps PreSEIS estimates the earthquake magnitude, start and end points of fault rupture and uses attenuation laws such as given by equation (5.6). Predictions are very stable with on-going time and give good approximations of the true source parameters (Fig. 5.4).

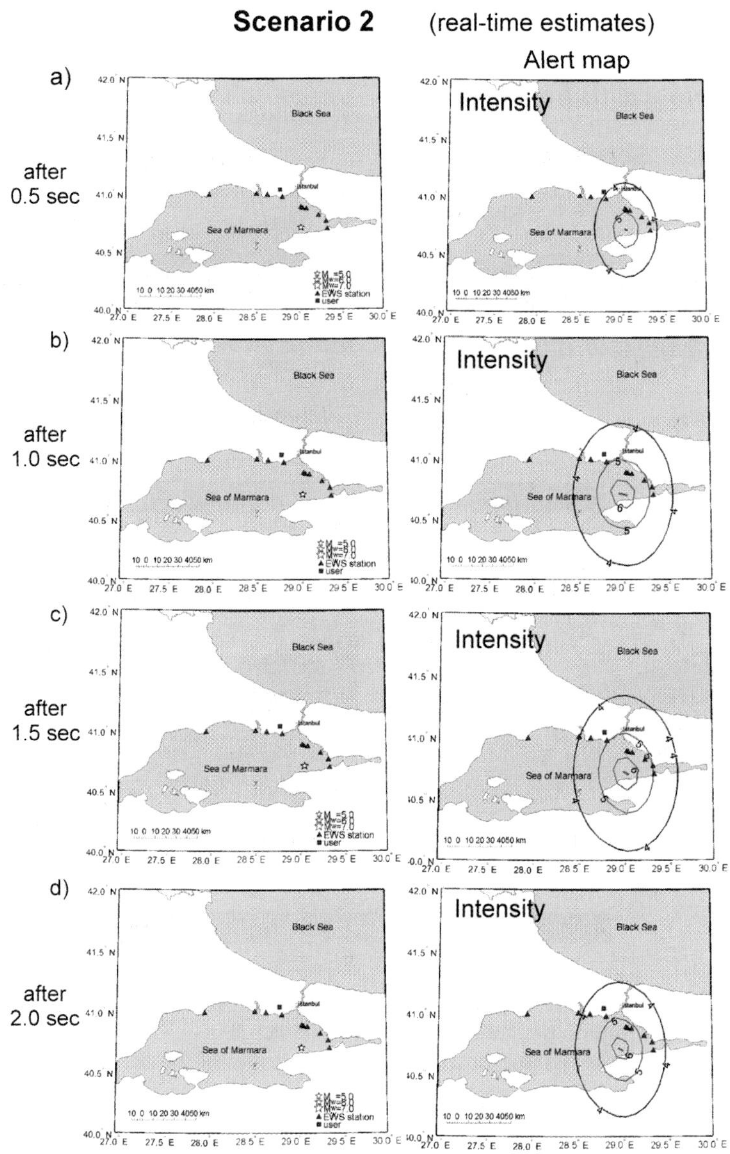

Fig. 5.7 Real-time estimates of epicenter locations (left) and alert maps of seismic intensity (right) for *Scenario 2* at a) 0.5, b) 1.0, c) 1.5 and d) 2.0 seconds after triggering of station *BUYAD*. Simulated acceleration time series and target values are shown in Fig. 5.5. For explanation see Fig. 5.6.

Figure 5.8 compares shake maps and alert maps of seismic intensity for *Scenario 1* and *Scenario 2* including magnitude-dependent site effects for the Istanbul area (Böse 2006).

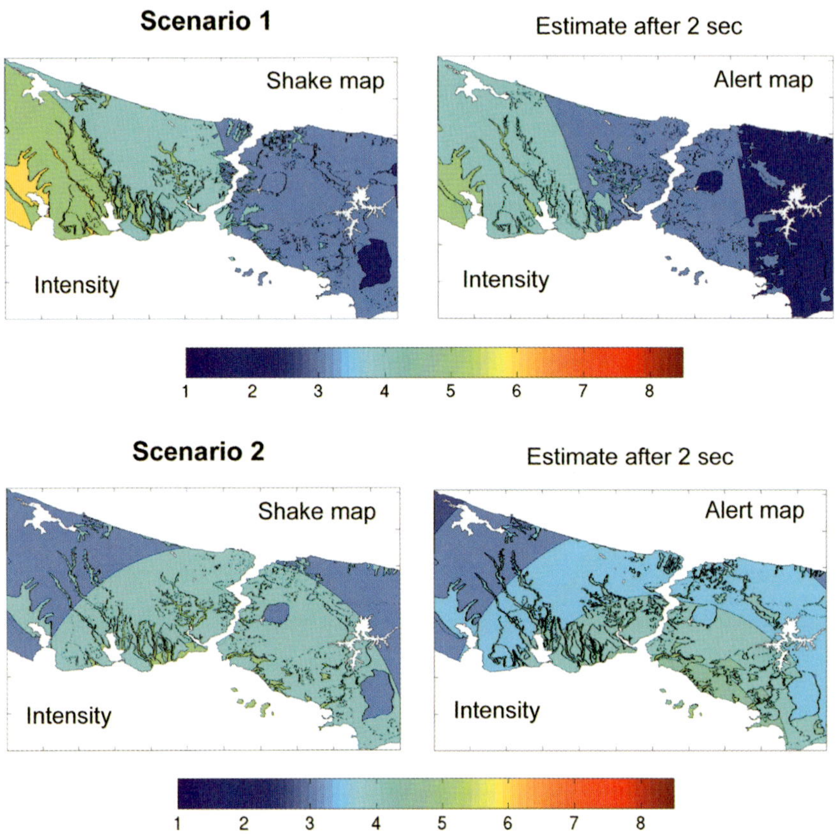

Fig. 5.8 Shake and alert maps of seismic intensity for *Scenario 1* and *Scenario 2* considering magnitude dependent site effects (Böse 2006). The shake map is calculated from the true magnitude and rupture expansion of the earthquake, the alert map from estimated source parameters provided by PreSEIS, here 2 seconds after triggering of station *BOTAS* or station BUYAD, respectively.

5.5 Conclusions

We have developed an approach for earthquake early warning that takes advantage of regional and on-site warning concepts: PreSEIS combines measurements of seismic ground motions at different sites without requiring seismic waves to have already arrived at all sensors before warnings are issued. PreSEIS is as fast as on-site single station based approaches for early warning, with the advantage of more information about the earthquake entering the estimates of source parameters.

PreSEIS utilizes artificial neural networks (ANNs) for the rapid approximation of earthquake magnitude, source and rupture locations. The estimation of rupture expansion produces closer approximations of the distribution of ground shaking in the form of alert maps than if the earthquake source is assumed to be in a single point.

From a training dataset ANNs learn essential background knowledge about fault locations, the relative frequency of earthquake magnitudes and others, similar to the Virtual Seismologist method for earthquake early warning (Cua and Heaton 2004). ANNs, however, integrate *a priori* information indirectly through the training patterns, i.e. it is not necessary that background knowledge is explicitly available.

PreSEIS uses the (bracketed) cumulative absolute velocity (CAV) for parameterization of seismic ground motion as ANN input. We expect other parameters describing, for example, the signal envelope or the frequency content of seismic ground motion to be suitable as well.

Using the example of the mega-city Istanbul, which faces a huge seismic risk, we have demonstrated in this paper the applicability of PreSEIS to earthquake early warning in areas with very short warning times.

5.6 Acknowledgments

This study was supported by the Collaborative Research Center (CRC) 461: Strong earthquakes – A Challenge for Geosciences and Civil Engineering, funded by the Deutsche Forschungsgemeinschaft (DFG).

References

Ambraseys N (2002) The seismic activity of the Marmara Sea region over the last 2000 years. Bull Seism Soc Am 92(1):1-18

Armijo R, Meyer B, Navarro S, King G, Barka A (2002) Asymmetric slip partitioning in the Sea of Marmara pull-apart: a clue to propagation processes of the North Anatolian Fault? Terra Nova 14(2):80-86

Benjamin JR and associates (1998) A criterion for determining exceedance of the operating basis earthquake. EPRI-report NP-5930, Electric Power Research Institute, Palo Alto, California

Beresnev IA, Atkinson GM (1997) Modeling finite-fault radiation from the ω^n spectrum. Bull Seism Soc Am 87(1):67-84

Beresnev I, Atkinson G (1998) FINSIM - a FORTRAN program for simulating stochastic acceleration time histories from finite faults. Seism Res Lett 69(1):27-32

Böse M (2006) Earthquake early warning for Istanbul using artificial neural networks. PhD thesis, Karlsruhe University, Germany, in prep.

Boore DM (1983) Stochastic simulation of high-frequency ground motions based on seismological models of the radiated spectra. Bull Seism Soc Am 73:1865-1894

Boore DM, Joyner WB (1997) Site amplifications for generic rock studies. Bull Seism Soc Am 87(2):327-341

Building Seismic Safety Council (BSSC) (1995) NEHRP Recommended Provisions for Seismic Regulations for New Buildings, FEMA 222A/223A. Technical report, Federal Emergency Management Agency, Washington D.C.

Erdik M, Aydinoglu N, Fahjan Y, Sesetyan K, Demircioglu M, Siyahi B, Durukal E, Özbey C, Biro Y, Akman H, Yuzugullu O (2003a) Earthquake risk assessment for Istanbul metropolitan area. Earthquake Engineering and Engineering Vibration 2(1):1-23

Erdik M, Fahjan Y, Özel O, Alcik H, Mert A, Gul M (2003b) Istanbul Earthquake Rapid Response and the Early Warning System. Bulletin of Earthquake Engineering 1;157-163, Kluwer Academic Publishers, Netherlands

Espinosa-Aranda J, Jimenez A, Ibarrola G, Alcantar F, Aguilar A, Inostroza M, Maldonado S (1995) Mexico City Seismic Alert System. Seism Res Lett 66(6):42-53

Goltz JD (2002) Introducing earthquake early warning in California: A summary of social science and public policy issues. Caltech Seismological Laboratory, Disaster Assistance Division, A report to OES and the Operational Areas

Kanamori H (2005) Real-time seismology and earthquake damage mitigation. Annual Reviews of Earth and Planetary Sciences 33:5.1-5.20

Nakamura Y (1985) A concept of one point detection system and its example using personal computer for earthquake warning. In Proceedings of 18th Earthquake Engineering Symposium of Japan

Nakamura Y (1989) Earthquake alarm system for Japan Railways. Japanese Railway Engineering 8(4):3-7

Parsons T (2004) Recalculated probability of M $\geq$ 7 earthquakes beneath the Sea of Marmara, Turkey. J Geophys Res 109:B05304, doi:10.1029/2003JB002667

Riedmiller M, Braun H (1993) A direct adaptive method for faster backpropagation learning: The RPROP algorithm. In Proceedings of the IEEE International Conference on Neural Networks

Sokolov V (2002) Seismic intensity and Fourier acceleration spectra: Revised relationship. Earthquake Spectra 18(1):161-187

Swingler K (1996) Applying neural networks - A practical guide. Academic Press Inc., San Diego

Wald D, Quitoriano V, Heaton T, Kanamori H, Scrivner C, Worden C (1999) TriNet ShakeMaps: Rapid generation of instrumental ground motion and intensity maps for earthquakes in southern California. Earthquake Spectra 15:537-556

Wenzel F, Oncescu M, Baur M, Fiedrich F (1999) An early warning system for Bucharest. Seism Res Lett 70(2):161-169

Wu Y-M, Kanamori H (2005) Experiment on an onsite early warning method for the Taiwan early warning system. Bull Seism Soc Am 95:347-353

Wu Y-M, Teng T-l (2002) A virtual subnetwork approach to earthquake early warning. Bull Seism Soc Am 92(5):2008-2018

6 Optimal, Real-time Earthquake Location for Early Warning

Claudio Satriano[1], Anthony Lomax[2], Aldo Zollo[1]

[1] RISSC-Lab, Dipartimento di Scienze Fisiche, Università di Napoli Federico II, Napoli, Italy
[2] Anthony Lomax Scientific Software, Mouans-Sartoux, France

Abstract

An effective early warning system must be capable of estimating the location and size of a potentially destructive earthquake within a few seconds after the event is first detected.

In this work we present an evolutionary, real-time location technique, based on the equal differential time (EDT) formulation and on a probabilistic approach for hypocenter estimation. The algorithm, at each time step, relies on the information coming from triggered arrivals and not yet triggered station. With just one recorded arrival, hypocentral position can be constrained by the Voronoi cell associated to the first triggering station. As time passes and more triggers become available, the evolutionary location converges to a standard EDT location.

We performed synthetic location tests using the actual geometry of the ISNet (Irpinia Seismic Network, Southern Italy) in order to evaluate the accuracy of the algorithm and its robustness in the presence of outliers.

6.1 Introduction

Destructive S and surface waves from a large earthquake can take several tens of seconds to travel from the earthquake source region to distant populated areas and sensitive infrastructure. If there is a seismological monitoring network in the source region, modern seismological analysis methods and communications systems allow characterization of the event and the issuing of alarm messages within seconds, leaving tens of seconds for mitigating actions to be taken. This procedure is known as early warning. For example, for an earthquake in the Irpinia region of Southern Italy there is a delay of about 25-30 sec before the first energetic S wave trains arrive at Naples at about 80-100 km distance. With an early warning system, alarm messages could be sent to critical sites in Naples 20 or more seconds before strong shaking commences.

The characterization of an earthquake includes, most importantly, estimates of its location and size (Zollo et al. 2007, this issue). Here we are concerned with obtaining the most constraint possible on the location of the event hypocenter as time passes after event detection. This constraint is expressed as a probability density function (*pdf*) for the hypocenter location in 3D space. This time-evolving, probabilistic, optimal location information will form a critical part of early warning messages, allowing actions to be taken based on the range of likely source distances and directions as estimated at the time of each message (Iervolino et al. 2007 this issue).

There are many approaches to standard earthquake location, which is performed when all the phase arrival times for an event are available. Our optimal, real-time location methodology is based on the equal differential-time formulation (EDT) of Font et al. (2004) and Lomax (2005) for standard earthquake location. EDT is a generalization of the master-station method (Zhou 1994) and the "method of hyperbolas" cited by Milne (1886). The EDT location is given by the maximum of a stack over quasi-hyperbolic surfaces, on each of which the difference in calculated travel-time to a pair of stations is equal to the difference in observed arrival times for the two stations. The EDT location determination is independent of origin time and reduces to a 3D search over latitude, longitude and depth. Because it uses a stack, EDT is highly robust in the presence of outliers in the data (Lomax 2005). This robustness is critical for the present problem, since we will often work with small numbers of data and may have outlier data such as false triggers and misidentified picks due to energetic, later phases.

Previous work on earthquake location for early warning includes several novel approaches to gain constraint on the location at an earlier time and with fewer observations than for standard earthquake location. Horiuchi et al. (2005) combine standard L_2–norm event location, EDT location on quasi-hyperbolic surfaces, and the information from not yet arrived data to constrain the event location beginning when there are triggered arrivals from two stations. The two arrivals define a hyperbolic surface on which the event can be located. The largest volume which may contain the hypocenter is bounded by EDT surfaces constructed using the current time (t_{now}) as a substitute for future, unknown arrival times at the stations which have not yet recorded arrivals. This volume shrinks as T_{now} progresses, even if no further stations record an arrival. Rydelek and Pujol (2004) apply the approach of Horiuchi et al. (2005) for the case of only two stations triggered.

6.2 Method

We assume that a seismic network has known sets of operational and non-operational stations, that when an earthquake occurs P wave arrival picks will become available from some of the operational stations, and that there may be outlier picks which are not due to P arrival. Our methodology is related to that of Horiuchi et al. (2005), which we extend and generalize by a) starting the location procedure after only one station has triggered, b) using the EDT approach throughout to incorporate the triggered arrivals and the not yet triggered stations, c) estimating the hypocenter probabilistically as a *pdf* instead of as a point, and d) applying a full, global search for each update of the location estimate.

When a first station S_n triggers with an arrival at $t_n = t_{\mathrm{now}}$, we can already place useful limits on a *pdf* volume that is likely to contain the hypocenter. These limits are given by conditional EDT surfaces on which the P travel time to the first triggering station $tt_n(\mathbf{x})$ is equal to the travel-time to each of the operational but not yet triggered stations, $tt_l(\mathbf{x})$, $l \neq n$. In the case of a homogeneous medium, the hypocentral *pdf* volume is the Voronoi cell around the first recording station defined by the perpendicular bisector surfaces with each of the immediate neighbor stations (Fig. 6.1b).

As the current time t_{now} progresses we gain the additional information that the not yet triggered stations can only trigger with $t_l > t_{\mathrm{now}}$. Thus the *pdf* volume is bounded by conditional EDT surfaces that satisfy the inequality $tt_l(\mathbf{x})$ - $tt_n(\mathbf{x}) < t_{\mathrm{now}} - t_n$, $l \neq n$. This hypocentral *pdf* volume will be smaller than the previous *pdf* volume estimate since the updated, conditional EDT surfaces tend to fold towards and around the first triggered station (Fig. 6.1c).

When the second and later stations trigger, we construct standard, true EDT surfaces between each pair S_l, S_m of the triggered stations using the equality $tt_l(\mathbf{x})$ - $tt_m(\mathbf{x}) = t_l - t_m$, $l \neq m$. These EDT surfaces are stacked with the volume defined by the not yet triggered stations, as described above, to form the current hypocentral *pdf* volume (Figs. 6.1d-f). In practice, all EDT surfaces are given a finite width by including errors in the arrival time picking and the travel-time calculation.

As more stations trigger, the number of not yet triggered stations becomes small, and the stacked true EDT surfaces and volumes bounded by conditional EDT surfaces converge towards the hypocentral *pdf* volume that is obtained with standard EDT location using the full set of data from all operational stations.

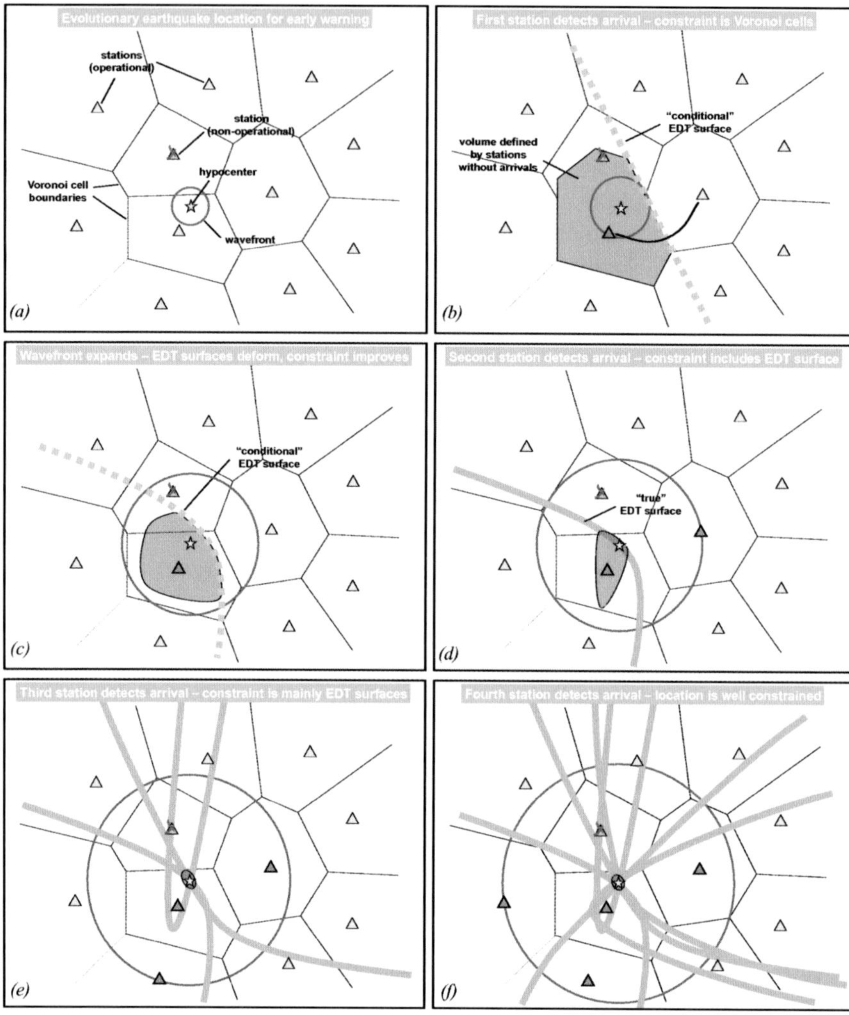

Fig. 6.1 Evolutionary earthquake location algorithm **(a)** Given a seismic network with known sets of operational and non-operational stations, we can *a priori* define the Voronoi cell associated to each station. **(b)** When the first station triggers, we can define a volume that is likely to contain the hypocenter limited by the "conditional" EDT surfaces on which the P travel time to the first triggering station is equal to the travel time to each of the operational but not yet triggered stations. **(c)** As time progresses, we gain additional information from the stations not yet triggered: the EDT surfaces move toward and bend around the first triggering station, and the hypocenter volume shrinks. **(d)** When the second station triggers,

we can define a "true" EDT surface and the actual hypocenter is likely to be at the intersection between this surface and the previous defined volume (which keeps shrinking). **(e)** When a third station triggers, we can define two more "true" EDT surfaces, increasing the constraint on hypocenter position. **(f)** As more stations trigger, the location converges to the standard EDT location.

If there are few outlier data, the final hypocentral *pdf* volume will usually give an unbiased estimate of the hypocentral location, as with standard EDT location. However, if one or more of the first arrival times is an outlier, the early estimates of the hypocentral *pdf* volume may be biased. If N_{out} is the number of outlier data, the bias should be significantly reduced after about 4+N_{out} arrivals have been obtained, and should be further reduced as the solution converges towards a standard EDT location.

6.2.1 Algorithm

We consider a network of N stations (S_0, …, S_N), a gridded search volume V containing the network, and the travel times from each station to each grid point in V computed for a given velocity model. If S_n is the first station to trigger, we search for grid points (i, j, k) in V where the following system of differential time inequalities is satisfied:

$$(tt_l - tt_n)_{i,j,k} \geq \delta t_{n,l} \; ; l \neq n \, , \tag{6.1}$$

where tt_i is the travel time from the grid point (i, j, k) to the station S_i and δt is the time interval between the arrival time at station Sn and the latest time for which we have information from station S_l:

$$\delta t_{n,l} = t_{\text{now}} - d_k - t_n \, , \tag{6.2}$$

where t_{now} is the current clock time and d_l is the delay time for receiving information from station S_l.

The system (1) defines the volume where the hypocenter may be located given that, at current time t_{now}, only the station S_n has triggered. In the case of a homogeneous medium and all $\delta t_{n,l} = 0$ (*i.e.*, $t_{\text{now}} = t_n$ and $d_l = 0$), (1) defines the *Voronoi cell* for the station S_n relative to the positions of the other operational stations. For each inequality in (1), we define a value $p_{n,l}$ which is 1 if the inequality is satisfied and 0 if not. Then we sum the $p_{n,l}$ for each station l at each grid point, obtaining a non-normalized probability density $P(i, j, k)$, where $P(i, j, k)=N$-1 for grid points where all the inequalities are satisfied, and a value less than N-1 elsewhere.

When an additional station triggers, we re-evaluate the system (1) for all pairs of triggered stations S_n and all not yet triggered stations S_l. Next, we search for grid points where the following equation is satisfied:

$$\left|(tt_l - tt_m) - (t_l - t_m)\right|_{i,j,k} \leq \sigma; l \neq m, \tag{6.3}$$

where S_l and S_m are triggered stations and σ gives the uncertainty in the arrival time picking and the travel-time calculation. This is the standard EDT equation.

We define a value $q_{l,m}$ which is 1 if the inequality (3) is satisfied and 0 otherwise. We sum the $q_{l,m}$ with the $p_{n,l}$ obtained from the re-evaluation of (1) to obtain a new $P(i, j, k)$. The maximum value of P is

$$P_{\max} = (N - n_T)n_T + n_T(n_T - 1)/2 \tag{6.4}$$

where nT is the number of stations that have triggered. The first term in (4) counts the number of inequalities from (1) and the second term the number from (3).

Starting from P, we define a value:

$$Q(i, j, k) = \left(\frac{P(i, j, k)}{P_{\max}}\right)^N, \tag{6.5}$$

which can be taken as the relative probability density (with value between 0 and 1) for the given grid cell to contain the hypocenter.

We calculate an updated value for $Q(i, j, k)$ when a new station triggers or after a predetermined time interval, whichever is earlier. Then an alarm message can be sent including information on the current constraint on the hypocentral location. This information may include, for example, the grid point where $Q(i, j, k)$ is greatest, or an uncertainty on the hypocentral location given by the largest horizontal and vertical distances between cells where $Q(i, j, k) > \alpha Q$ and α is a constant < 1. For message recipients at specific localities, the hypocentral location and uncertainty message might be provided as the likely epicentral distance range to the locality.

6.3 Location tests

In order to evaluate the accuracy and robustness of the location technique, we conducted several synthetic tests using the geometry of the ISNet (Irpinia Seismic Network, Southern Italy) (Weber et al. 2007, this issue) and a 1D Vp model for the region (Table 6.1) with a constant Vp/Vs of 1.68.

Table 6.1.

Depth (km)	Vp (km /s)
0.0	2.0
1.0	3.2
2.5	4.5
15.0	6.2
35.0	7.4
40.0	8.0

Our first test considers a shallow earthquake, occurring at the center of the network at a depth of 1 km. The event is located using only P triggers. Each panel in Fig. 6.2 represents the projection along three orthogonal planes passing through the true hypocenter of the earthquake location probability density *Q(i,j,k)*. The first snapshot is taken when the first station, ST24, triggers (T=0); the constraint on the earthquake location is given by the volume defined by equation (1), there is no constraint on depth. After 1 second, station ST25 triggers; the location is now constrained by the previously defined volume (which has been collapsing around station ST24) and the EDT surface defined by equation (3). After 2 seconds, 4 stations have triggered and the location is already well constrained for early warning purposes.

Figure 6.3 shows a location performed using only P triggers for an earthquake occurring outside the network at a depth of 10 km. At T=0 the maximum probability volume is bounded only towards the network. After 1 second, two more stations trigger and the volume is bounded in all directions. As time evolves, the constraint on the location volume improves, but it retains an elongated shape because, for events outside the network, the event distance is poorly determined. The depth is only constrained by a lower limit, but this depth bounds includes the true value.

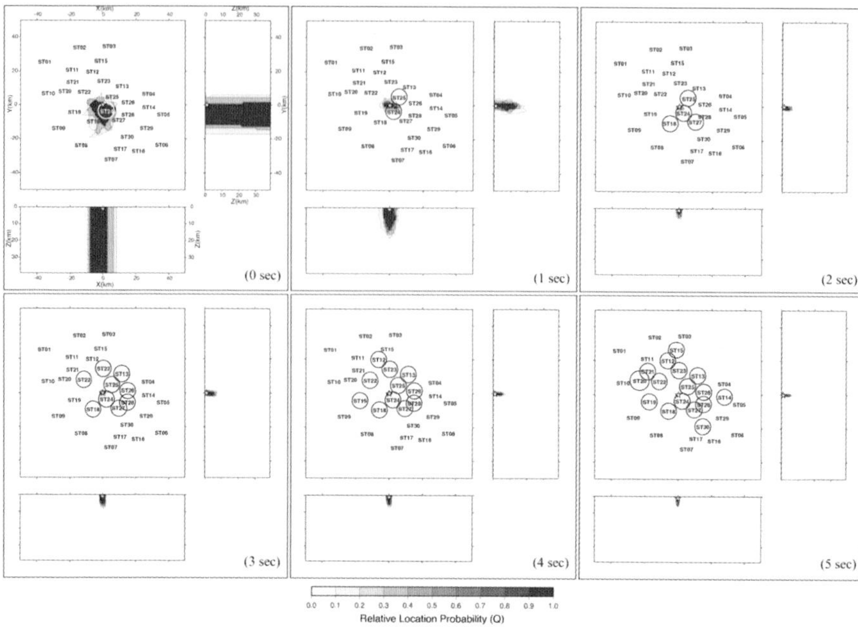

Fig. 6.2 Location test for an event occurring at the center of the ISNet network. The probability function is projected along three planes passing through the true hypocenter (identified with a star). T=0sec is the time at which the first station triggers. For each snapshot, stations which have triggered are marked with a circle. Location is performed using only P picks.

Recently in Italy there have been large earthquakes characterized by multiple event ruptures and intense seismic activity related to foreshocks and aftershocks, namely in Friuli in 1976 (Zollo et al. 1997), Irpinia in 1980 (Bernard and Zollo 1989), and Umbria-Marche in 1997 (Amato et al. 1998). The major instrumental event in the Irpinia region, the Mw=6.9, 1980 earthquake, had multiple sub-events with three main shocks occurring within about 20 seconds of each other. It is, therefore, important to check how our evolutionary location method performs when two or more events occur close in time.

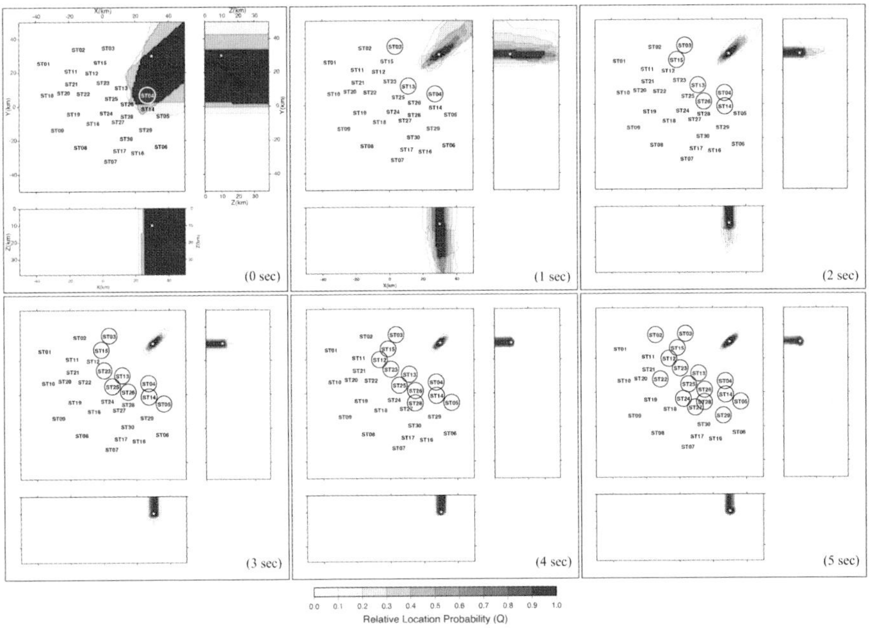

Fig. 6.3 Location test for an event occurring outside the network (see Fig. 6.2 for notes).

We made a synthetic test for two events occurring at different places within the Irpinia Seismic Network with origin times separated by 3 seconds, using both P and S picks for location (Fig. 6.4a). If an S pick from the first event (S1) comes after a P pick from the second event (P2), we assume a probability of 20% for the triggering system to erroneously interpret P2 as S1. For instance, ST13 is the first station recording the second event, but it does not trigger correctly. The first trigger comes from station ST25, biasing the hypocenter estimation in the very beginning of the location process (Figs. 6.4b, 6.4c). The bias is however strongly reduced after 1 sec circa, as soon as new stations trigger consistently. There are other misinterpreted picks at T = 4.6s (ST13, S1 as S2), 8.7s (ST14, S1 as P2), 8.8s (ST04, S1 as S2), and 12.5s (ST02, S1 as P2). These outliers, however, do not significantly influence the quality of location, since this is already constrained by a large number of consistent picks (Fig. 6.4c). The hypocentral position is correctly estimated, with no more strong oscillations, after about 2 seconds from the first trigger, while the depth is properly identified after 4-5 seconds. For both events, the uncertainty on x and y components becomes smaller than 1 km after about 4 seconds from the first trigger, while the uncertainty on z is lower than 2 km after 5 seconds.

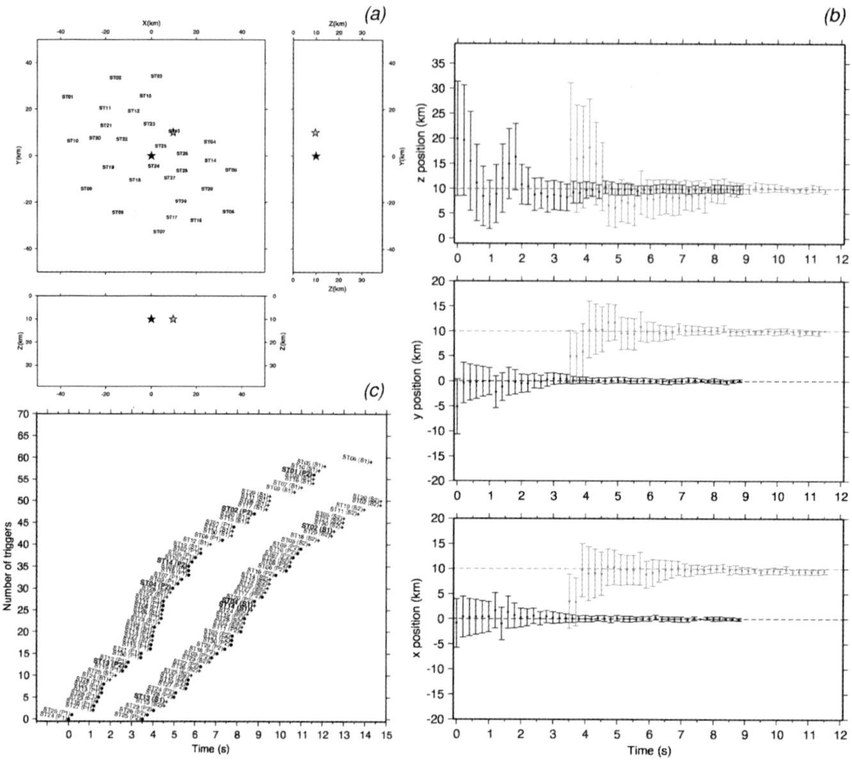

Fig. 6.4 Location test for two events occurring at different places within the Irpinia Seismic Network with origin times separated by 3 seconds. T=0sec is the time at which first station triggers. **(a)** Actual position of the first (black star) and the second (gray star) hypocenter. **(b)** Mean value and standard deviation for location along the three axes for the first (black bars) and the second (gray bars) event. The dashed lines represent the true values. **(c)** Triggering sequence for the first (upper sequence) and the second (lower sequence) event. P triggers are marked with dots, S triggers with stars. Misinterpreted arrivals are evidenced in bold.

6.4 Discussion

We have presented a real-time evolutionary location technique based on the equal differential-time (EDT) approach which makes it very robust in the presence of outliers. At each time step, this algorithm makes use of information from triggered arrivals and not yet triggered stations. Constraint on the hypocenter location is obtained as soon as the first station has triggered and is updated at fixed time intervals or when a new station triggers.

The hypocenter location is estimated as a probability density function defined in a gridded search volume. This makes it easy to incorporate the location results into a decision system for seismic early warning. Such a system can base a decisional rule on the evaluation of the probability that a certain ground motion intensity measurement (IM), like PGA or PGV, exceeds a given threshold (Iervolino et al. 2007, this issue). The probability density function for IM is calculated from the evaluation of the *hazard integral*:

$$\hat{f}_{IM}(im) = \int_M \int_R f_{IM|M,R}(im|m,r)\hat{f}_M(m)\hat{f}_R(r)\, dm\, dr$$

where $f_{IM|M,R}$ is an attenuation law, f_M is the *pdf* for the magnitude [estimated in real-time (Zollo et al. 2007, this issue)], and f_R is the *pdf* for the hypocentral (or epicentral) distance, which can be obtained directly from our location technique.

Synthetic location tests show that good accuracy, very close to standard "off-line" algorithms, is achieved after 4-5 seconds. The test on two quasi-simultaneous events demonstrates that, as long as the triggering system has a good detecting capability, the two locations can be handled as separate processes and wrong picks treated as outliers, whose bias is greatly reduced when several consistent arrivals are available.

6.5 Acknowledgements

Research carried out in the framework of AMRA Center of Competence.

References

Amato A et al. (1998) The 1997 Umbria-Marche, Italy, earthquake sequence: a first look at the main shocks and aftershocks. Geophys Res Lett 25(15):2861-2864

Bernard P, Zollo A (1989) The Irpinia (Italy) 1980 earthquake: Detailed analysis of a complex normal faulting. J Geophys Res 94(B2):1631-1647

Font Y, Kao H, Lallemand S, Liu C-S, Chiao L-Y (2004) Hypocentral determination offshore Eastern Taiwan using the Maximum Intersection method. Geophys J Int 158:655-675

Horiuchi S, Negishi H, Abe K, Kamimura A, Fujinawa Y (2005) An Automatic Processing System for Broadcasting Earthquake Alarms. Bull Seism Soc Am 95:708-718

Iervolino I, Convertito V, Giorgio M, Manfredi G, Zollo A (2007) The crywolf issue in earthquake early warning applications for the Campania region. In: Gasparini P, Manfredi G, Zschau J (eds) Earthquake Early Warning Systems. Springer

Lomax A (2005) A Reanalysis of the Hypocentral Location and Related Observations for the Great 1906 California Earthquake. Bull Seism Soc Am 95:861-877

Milne J (1886) Earthquakes and Other Earth Movements. Appelton, New York, 361 pp

Rydelek P, Pujol J (2004) Real-time seismic warning with a 2-station subarray. Bull Seism Soc Am 94:1546-1550

Weber E, Iannaccone G, Zollo A, Bobbio A, Cantore L, Corciulo M, Convertito V, Di Crosta M, Elia L, Emolo A, Martino C, Romeo A, Satriano C (2007) Development and testing of an advanced monitoring infrastructure (ISNet) for seismic early-warning applications in the Campania region of southern Italy. In: Gasparini P, Manfredi G, Zschau J (eds) Earthquake Early Warning Systems. Springer

Zhou H (1994) Rapid 3-D hypocentral determination using a master station method. J Geophys Res 99:15439-15455

Zollo A, Lancieri M (2007) Real-time estimation of earthquake magnitude for seismic early warning. In: Gasparini P, Manfredi G, Zschau J (eds) Earthquake Early Warning Systems. Springer

Zollo A, Bobbio A, Emolo A, Herrero A, De Natale G (1997) Modelling of ground acceleration in the near source range: the case of 1976, Friuli earthquake (M = 6.5), northern Italy. Journal of Seismology 1(4):305-319

7 The Virtual Seismologist (VS) Method: a Bayesian Approach to Earthquake Early Warning

Georgia Cua[1], Thomas Heaton[2]

[1] Swiss Seismological Service, Swiss Federal Institute of Technology (ETH) Zurich, Switzerland
[2] Department of Civil Engineering, California Institute of Technology, Pasadena, USA

Abstract

The goal of earthquake early warning is to provide timely information to guide damage-mitigating actions that can be taken in the few seconds between the detection of an earthquake and the onset of large ground motions at a given site. From a subscriber's perspective, effective early warning consists of both real-time information about the expected ground motions, as well as a methodology of how to use this information, and the inherent uncertainties, to guide decision-making. The Virtual Seismologist (VS) method is a Bayesian approach to early warning that provides a unified framework for the real-time earthquake source estimation, as well as the subscriber's decision-making problem. The introduction of prior information into the source estimation problem via Bayes' Theorem distinguishes the VS method from other paradigms for earthquake early warning. Station locations, previously observed seismicity, and known fault traces are among the type of information that can be used to resolve trade-offs in magnitude and location that are unresolved by the ground motion observations alone at the initial stages of earthquake rupture. The benefits of prior information are most evident in regions of low station density, where large inter-station distances result in source estimates based on a relatively sparse set of observations. The drawback of prior information is the increased complexity of information that must be communicated to the user, as the resultant earthquake source estimates can no longer be adequately described by Gaussian distributions. We illustrate the performance of the VS method in regions of high and low stations density, and discuss how subscriber requirements ultimately dictate how the real-time source estimation problem must be addressed.

7.1 Introduction

Subscribers attempting to use early warning information to reduce earthquake-related losses typically have a selected sequence of actions they would like to complete before damaging ground motions reach their site. These subscribers decide whether or not to initiate their damage-mitigating actions based on uncertain information. How to make optimal decisions with uncertain information is a fundamental question to these subscribers. Answering this question properly requires addressing a seismological and an economics problem in tandem. The seismological question is that of real-time earthquake source estimation, which can be phrased as follows: what are the best estimates of magnitude and location given the available data? The economics question is the user response problem, which can be phrased as follows: what is the optimal decision or course of action, given the current source estimates and its uncertainties?

The Virtual Seismologist (VS) method is a Bayesian approach to earthquake early warning that addresses the source estimation and user response problems in tandem. In the source estimation problem, the VS method shares with other proposed methodologies (Nakamura 1988; Allen and Kanamori 2003; Wu and Kanamori 2005a, 2005b) the use of relative frequency content or predominant period and attenuation relationships to estimate magnitude and/or location from available ground motion observations. The introduction of prior information into the earthquake source estimation problem distinguishes the VS method from other paradigms for early warning. Bayes' theorem allows the use of a "background" state of knowledge to help in resolving trade-offs in magnitude and location that cannot be resolved due to the scarcity of observations at the initial stages of the earthquake rupture. Types of information that can be included in the Bayes prior include: state of health of the seismic network, previously observed seismicity, known fault locations, and the Gutenberg-Richter magnitude-frequency relationship. The benefits of prior information are largest in regions with low station density, where large inter-station distances result in initial early warning estimates based on a sparse set of observations. We illustrate the performance of the VS method with high station density using ground motions recorded from the $M = 4.75$ Yorba Linda, California earthquake, and with low station density using data from the $M = 7.1$ Hector Mine, California earthquake. The examples shown in this paper approximate the earthquake as a single point source. While this seems adequate for earthquakes with $M < 6$, this methodology needs to be modified to be effective for long ruptures where near-source ground motions are expected even at large distance from the epicenter. Yamada and Heaton (2006) present a strategy for extending VS to handle long ruptures.

Aside from a method to estimate magnitude and location from sparse information from the initial stages of the earthquake rupture, it is equally important for earthquake early warning studies to address how subscribers might make optimal decisions using early warning information. Different subscribers will require different types of early warning estimates, depending on subscriber tolerance to missed and/or false alarms. Ultimately, the subscriber requirements dictate how the source estimation problem must be phrased. This is not consistent with the traditional separation of the source estimation and user response problems in earthquake early warning research. The VS method facilitates an integrated approach that recognizes the role of the user decision-making process in formulating the source estimation problem.

7.2 Real-time Earthquake Source Estimation

7.2.1 Review of Bayes' Theorem

Consider that we want to estimate the magnitude and location of an earthquake given an available set of observed ground motions. According to Bayes theorem, the state of belief regarding magnitude and location (*M,loc*) given a set of available observations Y_{obs} is given by

$$P(M,loc\,|\,Y_{obs}) = \frac{P(Y_{obs}\,|\,M,loc)\times P(M,loc)}{P(Y_{obs})} \tag{7.1}$$

where *M* is magnitude, *loc* is a location parameter (epicentral distance, or epicentral location), and Y_{obs} is the available set of observed ground motions. We can use the proportional form of Bayes' theorem since $P(Y_{obs})$ is not a function of the parameters being estimated (*M,loc*):

$$P(M,loc\,|\,Y_{obs}) \propto P(Y_{obs}\,|\,M,loc)\times P(M,loc) \tag{7.2}$$

$P(M,loc|Y_{obs})$, is the posterior probability density function (pdf); it is the conditional probability that an earthquake of magnitude and location *M,loc* generated the set of observations Y_{obs}. The VS estimates, $(M,loc)_{VS}$, are the most probable source estimates given the available observations; they maximize $P(M,loc|Y_{obs})$. The spread of $P(M,loc|Y_{obs})$ yields the uncertainties on the VS source estimates. On the right hand side, $P(Y_{obs}|M,loc)$ is the likelihood function; it is the conditional probability of observing a set of ground motions Y_{obs} given an earthquake with magnitude and location *M,loc*. The likelihood function requires ground motion models relating source descriptions (*M,loc*) to observed ground motion amplitudes. The VS method uses 1) relationships between ratios of peak ground motions and magnitude, and 2) ground motion attenuation relationships describing observed amplitudes as functions of magnitude and distance, to define the likelihood function $P(Y_{obs}|M,loc)$. *P(M,loc)* is the Bayes prior; it represents a background state of knowledge, independent of the observations, on relative earthquake probabilities that we want to include in the estimation process. The degree of complexity that can be incorporated into the prior is flexible. The simplest prior we can use is the assumption that all magnitudes and all locations are equally probable. This simplifies the necessary calculations, at the cost of being an inaccurate representation of the general state of knowledge regarding earthquake occurrence. Alternatively, we could incorporate into the prior some generally accepted beliefs about earthquake occurrence: 1) the magnitude-frequency relationship of earthquakes follows the Gutenberg-Richter law, 2) many earthquakes occur on known faults, 3) and earthquakes often cluster in time and space. We can also include information about the state of health of the seismic network. The choice of prior information is most influential in the initial VS early warning estimates; trade-offs between magnitude and location in the initial estimates are resolved in favor of the prior information. In real-time source estimation, the use of prior information is perhaps the most important distinction between the VS method and other proposed paradigms for early warning.

The evolution of VS early warning source estimates as a function of available information mimics how reasonable and educated humans (as opposed to unreasonable and prejudiced humans) modify their beliefs in light of new information. Early VS estimates are typically based on a scarce set of observations, for instance, on the first few seconds of peak amplitude data available at the first triggered station. In these early estimates, there will be trade-offs in magnitude and location that cannot be resolved by the available observations alone. At any given time, the VS estimates $(M,loc)_{VS}$ are the most probable source estimates; they maximize $P(M,loc|Y_{obs})$, and are consistent with the available observations, with trade-offs in magnitude and location resolved in favor of the given prior information. Initially, when observations are sparse, $(M,loc)_{VS}$ are strongly influenced by the prior; the observations become the dominant contributor as ground motions propagate to further stations. When sufficient observations are available to fully constrain the magnitude and location estimates, the choice of prior is irrelevant; $(M,loc)_{VS}$ are completely determined by the observations.

7.2.2 Defining the Likelihood Function, P(Y$_{obs}$|M,loc)

Let us initially assume a uniform prior, $P(M,loc) = c$, a constant. Eq. (7.2) yields

$$P(M,loc\,|\,Y_{obs}) \propto P(Y_{obs}\,|\,M,loc) \tag{7.3}$$

Let Y_{obs} be the set of log peak acceleration, velocity, and filtered (3-second high pass) displacement on horizontal and vertical channels available at a given time. Assuming the Y_{obs} are independent and log normally distributed, maximizing $P(M,loc\,|\,Y_{obs})$ is equivalent to maximizing the log of the likelihood function, $L = \log(P(Y_{obs}\,|\,M,loc))$, which, for the most general case of P- and/or S-wave amplitudes available at multiple stations, we define as follows

$$L(Y_{obs},M,loc) = \sum_{i=1}^{n}\sum_{j=1}^{P,S} L(M,loc)_{ij} \tag{7.4}$$

$$L(Y_{obs},M,loc)_{ij} = \frac{(Z_{obs_{ij}} - \bar{Z}_j(M))^2}{2\sigma_Z^2} + \sum_{k=1}^{4}\left(\frac{\left(Y_{obs_{ij}} - \bar{Y}_{ijk}(M,loc)\right)^2}{2\sigma_{ijk}^2}\right). \tag{7.5}$$

$L(Y_{obs},M,loc)$ incorporates 6 channels of ground motion (maximum vertical and root mean square of the maximums of 2 horizontal channels of acceleration, velocity, and filtered displacement) per station. We assume that each station has 1 vertical and 2 horizontal sensors with acceleration or broadband velocity output, and that recursive filters can be used to provide acceleration, velocity, and filtered displacement in real-time. In Eqs. (7.4) and (7.5), i is an index over the n stations with P-wave triggers, j is an index over the P- and S-wave phases, and k is an index over observed vertical velocity, and horizontal acceleration, velocity, and filtered displacement, the channels that contribute to the likelihood function via ground motion attenuation relationships. The remaining two channels, observed vertical acceleration and filtered displacement, are accounted for by the term involving $Z_{obs_{ij}}$. $Z_{obs_{ij}}$ is the ratio of peak available vertical acceleration to peak available vertical filtered displacement found by Cua and Heaton (2006b) to be optimally indicative of magnitude; it is given by

$$Z_{obs_{ij}} = \log\left(\frac{PVA_{ij}^{0.36}}{PVD_{ij}^{0.93}}\right) = 0.36\log(PVA_{ij}) - 0.93\log(PVD_{ij}) \tag{7.6}$$

In Eq. (7.6), PVA_{ij} and PVD_{ij} denote peak available vertical acceleration and peak vertical (filtered) displacement, respectively at the ith station and for the jth body wave. $Z_{obs_{ij}}$ is a measure of the relative frequency content of ground motion. Similar to methods based on predominant period (Nakamura 1988; Allen and Kanamori 2003; Wu and Kanamori 2005ab), it is based on the idea that the relative frequency content of ground motions can be indicative of magnitude since small earthquakes involve small patches of slip and will radiate more high frequency energy, while large earthquakes involve finite rupture dimensions that contribute to longer period energy.

The details on the linear discriminant analysis used to determine $Z_{obs_{ij}}$ and its relationship to magnitude are found in Cua (2005) and Cua and Heaton (2006b). There are different coefficients characterizing the magnitude dependence of Z_{obs} (dropping the subscripts for brevity) for P- and S-waves.

$$\begin{aligned} \bar{Z}(M) &= -0.615M + 5.495, \quad \sigma_Z = 0.17 \quad \text{for P-waves} \\ &= -0.685M + 5.517, \quad \sigma_Z = 0.193 \quad \text{for S-waves} \end{aligned} \tag{7.7}$$

To use Eqs. (7.5), (7.6), and (7.7), it is necessary to estimate or assume whether the peak amplitudes at a given station are from a P- or S-wave. Cua (2005) and Cua and Heaton (2006b) developed the following criteria to distinguish between P- and S-waves

$$\begin{aligned} &PS = 0.43\log(PVA) + 0.55\log(PVV) - 0.46\log(PHA) - 0.55\log(PHV) \\ &\mathit{if}\ \ PS > 0,\ \text{amplitudes are most likely from a P-wave} \\ &\mathit{if}\ \ PS < 0,\ \text{amplitudes are most likely from an S-wave} \end{aligned} \tag{7.8}$$

In Eq. (7.8), *PVA* denotes peak vertical acceleration, *PVV* peak vertical velocity, *PHA* peak horizontal acceleration, and *PHV* peak horizontal velocity. This criteria correctly identified P-waves 88% of the time in a database of ground motions from 70 Southern California earthquakes with $2 \le M \le 7.3$, and epicentral distances $R < 200$ km. Figure 7.1 shows magnitude against Z_{obs} from peak P-wave amplitudes from this dataset. Cua and Heaton (2006a) also used this database to develop the P- and S-wave amplitude attenuation relationships $\bar{Y}_{ijk}(M,R)$, which define the second term of the likelihood function.

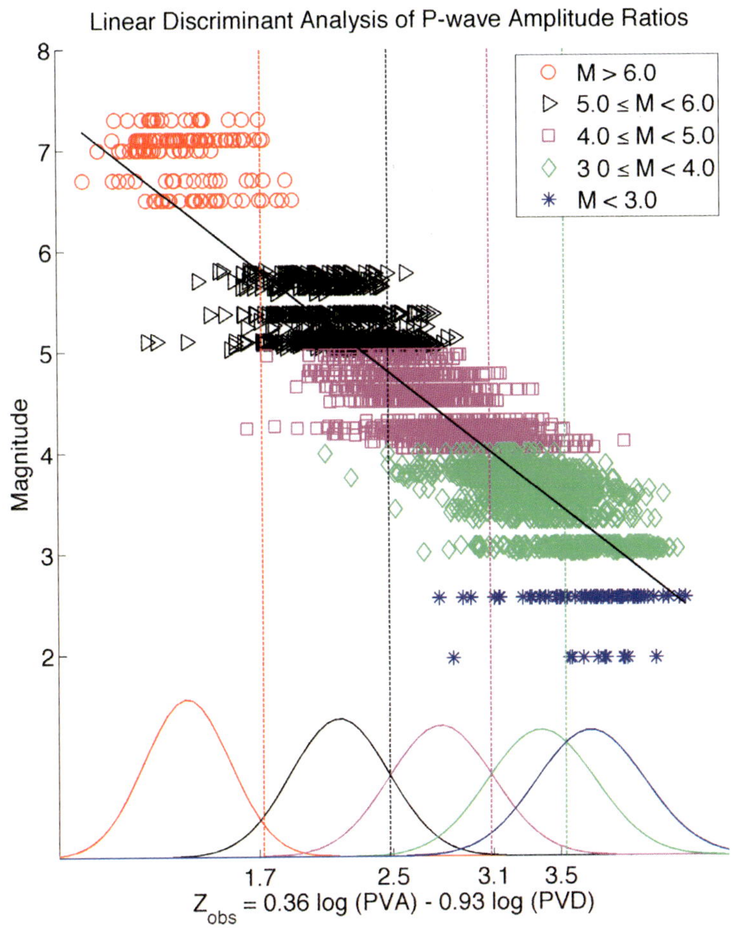

Fig. 7.1 Magnitude plotted against ground motion ratio Z_{obs} = 0.36log(*PVA*) - 0.93log(*PVD*) for P-wave amplitudes.

The second term of the likelihood function represents the contribution of observed peak vertical velocity, peak horizontal acceleration, velocity, and displacement and their respective attenuation relationships $\bar{Y}_{ijk}(M,R)$ in constraining both magnitude and location. Eq. (7.5) is expressed in terms of the source parameters M, loc; in the most general case, loc represents the epicentral latitude and longitude coordinates. Given, epicentral latitude and longitude coordinates, the epicentral distance R can be calculated. Thus, $\bar{Y}_{ijk}(M, loc)$ is equivalent to $\bar{Y}_{ijk}(M, R)$, which Cua and Heaton (2006a) modeled as

$$\bar{Y}_{ijk}(M,R) = a_{jk}M - b_{jk}(R_{1i} + C_{jk}(M)) - d_{jk}\log(R_{1i} + C_{jk}(M)) + e_{jk} + \alpha_{ijk} \tag{7.9}$$

where $R_{1i} = \sqrt{R_i^2 + 9}$, where R_i is the epicentral distance of the ith station

and $C_{jk}(M) = c_{1jk}(\arctan(M-5) + \frac{\pi}{2}) \times \exp(c_{2jk}(M-5))$.

In VS method, the regression coefficients (a, b, c_1, c_2, d, e) - dropping the subscripts - are known quantities (Table 7.1). There are different sets of regression coefficients for different channels of ground motion, for P- and S-waves, and for rock and soil sites. The station-specific site correction factors α_{ijk} take into account systematic amplification or deamplification of ground motions observed at a given site relative to the mean ground motion levels predicted by the attenuation relationships. Cua and Heaton (2006b) determined α_{ijk} for 135 SCSN stations for horizontal and vertical acceleration, velocity, and filtered displacement; these are available online at http://resolver.caltech.edu/CaltechETD:etd-02092005-125601.

The VS source estimates, $(M, loc)_{VS}$, are the source parameters that maximize the posterior pdf $P(M, loc \mid Y_{obs})$; in general, these estimates are a function of prior information and the available observations. Let $(M, loc)_L$ refer to the source parameters that maximize the log likelihood, $L = \log(P(Y_{obs} \mid M, loc))$; these are the source estimates most consistent with the available observations, whether data is available at single or multiple stations. A single-station estimate involves setting $n = 1$ and, if desired, using epicentral distance R as the loc parameter. With a uniform prior, $(M,loc)_{VS} = (M,loc)_L$. The VS method requires a minimum of 3 seconds of data following the P- arrival (or 2 seconds of data following the S-arrival) before a station contributes its amplitudes to the likelihood function. It is assumed that P-waves can be detected via short-term over long-term average methods. The S-wave arrival can be determined using Eq. (7.8).

Table 7.1 Attenuation relationships for peak P- and S-wave amplitudes

$$\bar{Y}(M,R) = aM + b(R_1 + C(M)) + d(R_1 + C(M)) + e + \alpha$$

$$R_1 = \sqrt{R+9}$$

$$C(M) = c_1\left(\arctan(M-5) + \frac{\pi}{2}\right) \times \exp(c_2(M-5))$$

where $\bar{Y}$ = acceleration (cm/s^2), velocity (cm/s), or displacement (cm)

R = epicentral distance (km) when $M < 5$

closest fault distance (km), when available for $M \geq 5$

				a	b	c_1	c_2	d	e	σ
Root mean square horizontal amplitudes	P-wave	Acceleration	rock	0.72	3.3×10^{-3}	1.6	1.05	1.2	-1.06	0.31
			soil	0.74	3.3×10^{-3}	2.41	0.95	1.26	-1.05	0.29
		velocity	rock	0.80	8.4×10^{-4}	0.76	1.03	1.24	-3.103	0.27
			soil	0.84	5.4×10^{-4}	1.21	0.97	1.28	-3.13	0.26
		displacement	rock	0.95	1.7×10^{-7}	2.16	1.08	1.27	-4.96	0.28
			soil	0.94	-5.17×10^{-7}	2.26	1.02	1.16	-5.01	0.3
	S-wave	acceleration	rock	0.78	2.6×10^{-3}	1.48	1.11	1.35	-0.64	0.31
			soil	0.84	2.3×10^{-3}	2.42	1.05	1.56	-0.34	0.31
		velocity	rock	0.89	4.3×10^{-4}	1.11	1.11	1.44	-2.60	0.28
			soil	0.96	8.3×10^{-4}	1.98	1.06	1.59	-2.35	0.30
		displacement	rock	1.03	1.01×10^{-7}	1.09	1.13	1.43	-4.34	0.27
			soil	1.08	1.2×10^{-6}	1.95	1.09	1.56	-4.1	0.32
Vertical amplitudes	P-wave	acceleration	rock	0.74	4.01×10^{-3}	1.75	1.09	1.2	-0.96	0.29
			soil	0.74	5.17×10^{-7}	2.03	0.97	1.2	-0.77	0.31
		velocity	rock	0.82	8.54×10^{-4}	1.14	1.11	1.36	-0.21	0.26
			soil	0.81	2.65×10^{-6}	1.4	1.0	1.48	-2.55	0.30
		displacement	rock	0.96	1.98×10^{-6}	1.66	1.16	1.34	-4.79	0.28
			soil	0.93	1.09×10^{-7}	1.5	1.04	1.23	-4.74	0.31
	S-wave	acceleration	rock	0.78	2.7×10^{-3}	1.76	1.11	1.38	-0.75	0.30
			soil	0.75	2.47×10^{-3}	1.59	1.01	1.47	-0.36	0.30
		velocity	rock	0.90	1.03×10^{-3}	1.39	1.09	1.51	-2.78	0.25
			soil	0.88	5.41×10^{-4}	1.53	1.04	1.48	-2.54	0.27
		displacement	rock	1.04	1.12×10^{-5}	1.38	1.18	1.37	-4.74	0.25
			soil	1.03	4.92×10^{-6}	1.55	1.08	1.36	-4.57	0.28

The $(M, loc)_L$ at any given time is the point source that best fits (in a least squares sense) the geographical distribution of observed P- and S-wave amplitudes and ground motion ratios. The location estimate from the likelihood function is an amplitude-based location; no arrival time information or velocity structure is included. It is comparable to the strong motion centroid (Kanamori 1993). Amplitude-based locations, while not very accurate, are more robust than locations based on phase arrivals, and are an efficient means to convey the spatial distribution of ground motion observations post-earthquake response (Kanamori 1993). The likelihood function develops a global maximum as the ground motions propagate to further stations. In the early stages of the estimation process, when the set of available observations is sparse (for instance, 3 seconds after the initial P detection at the first triggered station), the likelihood function may not have a global maximum; there are trade-offs between the source estimates (a small earthquake located close by, or a larger event further away) that may be unresolved by the available observations. The inclusion of prior information is most useful in these situations.

7.2.3 Defining the Prior, P(M,loc)

The Bayes prior is a statement of "background" knowledge relevant to the parameter estimation problem at hand. In the VS method, we use $P(M,loc)$ to introduce information on relative earthquake probabilities and the state-of-health of the seismic network into the real-time source estimation problem. We enumerate the different types of information that can be included in the prior.

- Long-term national hazard maps, or known fault traces, are good candidates to include in the Bayes prior since faults that have generated large earthquakes in the past, such as the San Andreas fault system in California, or the Northern Anatolian fault in Turkey, are likely to generate large events in the future.
- The Gutenberg-Richter law states that small events occur more frequently than large earthquakes. Short-term seismicity-based earthquake forecasts take into account that the magnitude-frequency distribution of earthquakes follows the Gutenberg-Richter law, and that earthquakes cluster in space and time, with Omori's law governing the decay of number of aftershocks as a function of time after the mainshock (Reasenberg and Jones 1989; Gerstenberger et al. 2003).
- The location of previously observed seismicity is important prior information since many large earthquakes have foreshocks. Abercrombie and Mori (1996) found that 44% of earthquakes in their dataset of 59 $M > 5$ California earthquakes had foreshocks. Jones (1984) found that 35% of earthquakes in a dataset of 20 San Andreas earthquakes had foreshocks within 1 day and 5 km of the mainshock.

- For a region with no known faults or previously observed seismicity, we can assume that all locations have equal probability of being an earthquake epicenter. This implies that epicentral distances are more likely to be larger than smaller.
- Nearest neighbor regions, or Voronoi cells, of operating stations provide useful constraints on earthquake locations. The Voronoi cell of a given station is the set of all location coordinates that are closer to the said station than any other station in the network. The initial P-wave detection from an event implies that the event nucleated within the Voronoi cell of the first triggered station. The denser the deployment of stations within a seismic network, the smaller the area of the average station Voronoi cell, and the stronger the constraint on earthquake location.
- The concept of not-yet-arrived data, as described by Horiuchi et al. (2004) and Rydelek and Pujol (2004), can be used along with station Voronoi cells to describe the evolution of regions of possible location following the initial P detection. In this article, we refer to the region consistent with observed arrivals as the region of possible location. It is independent of the amplitude-based location estimate obtained from maximizing the likelihood function. From Rydelek and Pujol (2004), the locations consistent with the first two P-arrivals satisfy

$$R_2 - R_1 = \bar{V}_P \times (t_2 - t_1) \quad (7.10)$$

where R_1 and R_2 are the respective epicentral distances of stations with the first 2 P-wave arrivals, t_1 and t_2 are the corresponding P-wave arrival times, and V_P is an average P-wave velocity. Given the time between the first 2 P detections, Eq. (7.10) constrains the earthquake location to a hyperbola passing between the two stations.
As originally described by Horiuchi et al. (2004) and Rydelek and Pujol (2004), not-yet-arrived data can be used once two P-wave detections are available. In contrast, Voronoi cells, in conjunction with a slightly modified use of not-yet-arrived data, can be used to describe continuously evolving constraints on earthquake locations even before the second P-wave arrival. Consider the following situation: there is an initial P-wave detection at station 1, and Δt seconds later, there are no subsequent P-wave detections at the m stations sharing a Voronoi edge with station 1. The non-arrivals at each of the $i = 1,\ldots,m$ stations provide the inequality constraints

$$R_i - R_1 > \bar{V}_P \times \Delta t \quad (7.11)$$

The region of possible location is the intersection of the Voronoi cell of station 1 (constraint on location from the first P detection) and the areas consistent with the m inequality constraints described by Eq. (7.11); the area of this region of possible location is inversely proportional to Δt. Once the P-wave arrives at the second closest station, the region of possible location collapses to the Rydelek and Pujol hyperbola. A third arrival locates the epicenter.

Voronoi cells are strictly prior information since they are derived from station locations, which are independent of the earthquake rupture process. In contrast, not-yet-arrived data is not strictly prior information, since the time elapsed since the initial P detection, Δt, is an observed quantity. However, we do not include Δt in the likelihood function since it does not involve observed amplitudes.

Prior information is particularly useful in regions with low station density, where time between the first and second P-wave detections may be relatively large.

7.3 Applications of the VS Method to Selected Southern California Earthquake Datasets

7.3.1 3 September 2002 M=4.75 Yorba Linda, California Earthquake: High Station Density

The Yorba Linda earthquake occurred on 3 September 2002 in suburban Los Angeles, a region with a high density of real-time Southern California Seismic Network (SCSN) stations. The mainshock was located by SCSN at 33.92N, -117.78W, at a depth of 12.92 km (Hauksson et al. 2002). Two foreshocks (with magnitudes $M=2.66$ and $M=1.6$) occurred within 1 km of the mainshock epicenter in the 24 hours preceding the mainshock.

The application of the VS method illustrates how the approach works in regions with high station density. Station Voronoi cells, previously observed seismicity, and the Gutenberg-Richter relationships will be included in the Bayes prior, although the prior information is for the most part unnecessary. Due to the high station density in the epicentral region, there are quickly enough observations to constrain the source estimates without resort to prior information.

Figure 7.2 shows SCSN stations in the epicentral region. The triangles are the stations operational at the time of the mainshock; the polygons are the associated Voronoi cells. Circles are locations of $M > 1$ earthquakes recorded by SCSN in the 24 hours preceding the mainshock. Due to the high station density in this region, areas of Voronoi cells (those away from edges of the network) are relatively small, in the range of 250 to 700 km^2.

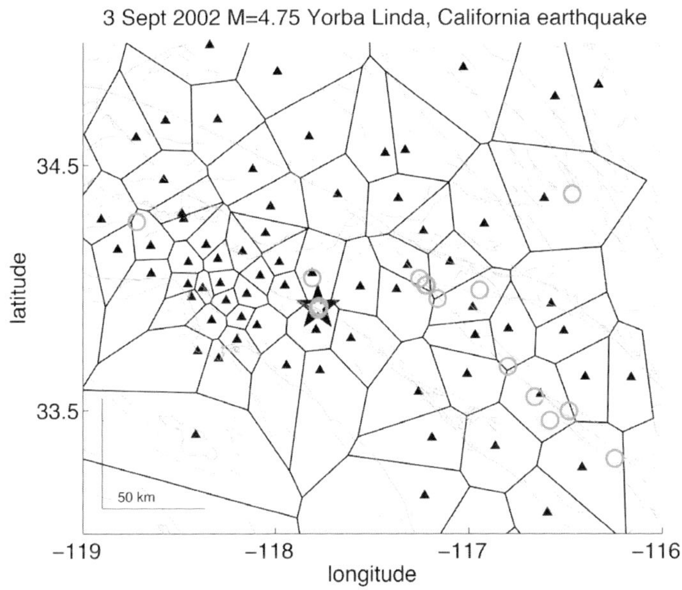

Fig. 7.2. Prior information for Yorba Linda, California earthquake: SCSN station locations (triangles), associated Voronoi cells (polygons), and seismicity in the preceding 24 hours (open circles) within 200 km of the mainshock (star). Two foreshocks occurred within the Voronoi cell of station SRN (shaded), the station closest to the epicenter.

The initial arrival at station SRN constrains the location to the area within its Voronoi cell (shaded polygon). Figure 7.3 shows the relative probabilities of epicentral distances consistent with the initial arrival at SRN. One of the simplest assumptions we can make given an initial P detection is that the event is located at the first triggered station. From Fig. 7.3, the largest possible error with this assumption is 15 km.

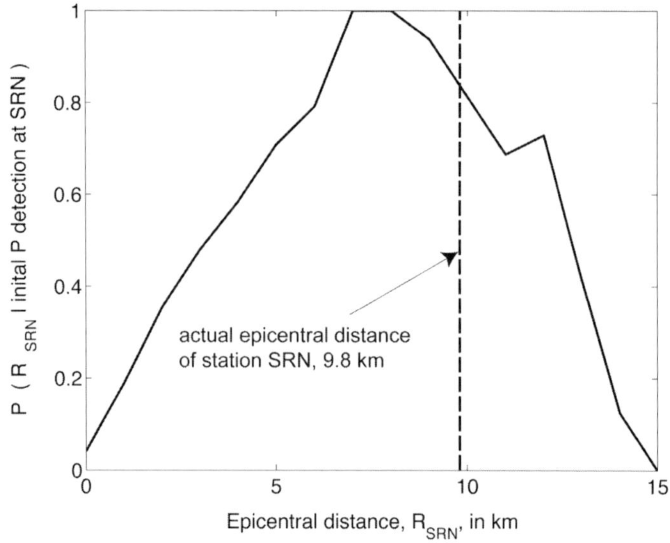

Fig. 7.3 Range of possible Yorba Linda earthquake epicentral distances from station SRN consistent with a first P detection at SRN (scaled such that maximum on y axis is 1). The weights on various distances (y axis) are obtained by giving equal weight to all locations within SRN's Voronoi cell. The most probable epicentral distance (8 km) is that which occurs most often within SRN's Voronoi cell. If we had included seismicity information in the prior, that there were two foreshocks at the mainshock epicenter means this distribution would include an impulse function situated at the distance of the mainshock.

The initial VS estimate is based on peak amplitudes available 3 seconds after the initial P detection at station SRN. With data from a single station ($n = 1$ in Eq. (7.4)), the source estimation problem can be parameterized in terms of magnitude and epicentral distance. Figure 7.4 shows contours of the likelihood function (no prior information) using peak amplitudes from 6 channels (horizontal and vertical acceleration, velocity, and filtered displacement) available at SRN 3 seconds after the initial P detection. The elongated contours of the likelihood function are indicative of trade-offs in magnitude and epicentral distance that are not fully resolved by the peak amplitudes at a single station. $M = 5.5$ and $R = 33$ maximize the likelihood function. With the likelihood expressed in terms of M, R, we can include as prior information the range of epicentral distances consistent with the first arrival at SRN (Fig. 7.3), as well as the Gutenberg-Richter relationship.

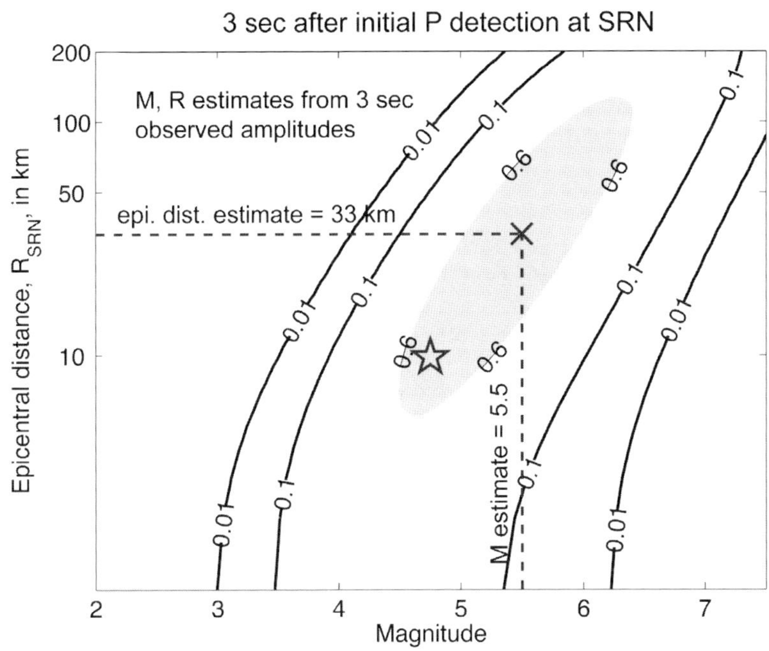

Fig. 7.4 Contours of the likelihood function for the $M = 4.75$ Yorba Linda earthquake expressed in terms of magnitude and epicentral distance using the peak amplitudes at SRN 3 seconds after the initial P detection (no prior information, only the peak amplitudes observed at SRN). The peak of the likelihood function is scaled to 1, and contours are drawn at 0.6, 0.1, and 0.01 levels. Regions where the likelihood function has a scaled value exceeding 0.6 are shaded. The likelihood function has a maximum at $M = 5.5$, $R = 33$ km. Elongated regions of equal probability indicate trade-offs in M, R unresolved by the available amplitudes and the attenuation relationships. The ground motion ratio term of the likelihood function constraints the possible magnitude to the approximate range $5 < M < 6$. The star marks the actual magnitude and epicentral distance to SRN of the Yorba Linda mainshock, $M = 4.75$, $R = 9.8$ km.

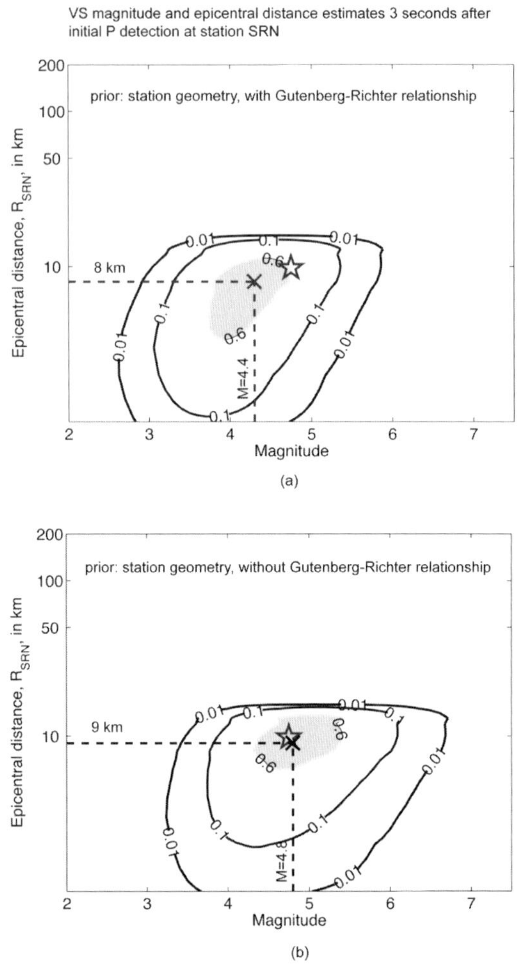

Fig. 7.5 Contours of the posterior density function showing the effect of including the Gutenberg-Richter magnitude-frequency relationship in the prior on initial *M, R* estimates 3 seconds after the initial P detection. Figures 4a and 4b include the constraint on epicentral distance from the Voronoi cell of station SRN. Figure 4a includes the Gutenberg-Richter relationship as a constraint on magnitude, while Figure 4.b does not. The star marks the actual magnitude and epicentral distance of the Yorba Linda mainshock. The cross marks the VS estimates in *M.R* space. The Gutenberg-Richter relationship favors smaller magnitude events located at smaller epicentral distances from the station. While inclusion of the Gutenberg-Richter relationship (a) produces magnitude estimates that are systematically smaller than the actual magnitudes, Bayesian statistics assure us that this is the most probable solution.

Figure 7.5 shows the effects of including the Gutenberg-Richter relationship, in addition to the constraints on epicentral distance by the Voronoi cells, on contours of the posterior density function. Recall that the posterior pdf is the product of the likelihood and the prior pdfs, and that the source estimates that maximize the posterior pdf are the most probable source estimates. $(M,R)_{VS}$, the most probable magnitude and epicentral distance estimates, are indicated in Fig. 7.5. The 3-second VS estimates without the Gutenberg-Richter relationship in the prior are closer to the actual magnitude and epicentral distance obtained by using the SCSN arrival-based location than VS estimates including the Gutenberg-Richter relationship in the prior. As the ground motions propagate to other stations, it is more convenient to parameterize the early warning location estimate in terms geographic coordinates (latitude, longitude) as opposed to epicentral distances of n stations.

Figure 7.6 shows locations consistent with the 3-second peak amplitudes at SRN for different magnitude ranges. The trade-offs here are similar to those in Fig. 7.4; we cannot unambiguously distinguish between small events in close and large events at distance using only amplitude information at a single station. The situation is much improved when previously observed seismicity and Voronoi cell constraints are taken into account. The best way to include previously observed seismicity into the estimation process would be to use short-term, seismicity-based earthquake forecasts such as STEP (Gerstenberger et al. 2003) as part of the prior. In this example, we simply increase the probability that a particular location is the event location by a factor of 2 if it is within 5 km of an event that occurred in the preceding 24 hours. Figure 7.7 shows contours of the posterior pdf (as a function of magnitude, latitude, and longitude) using peak amplitudes at SRN 3 seconds following the initial P detection to define the likelihood and including previously observed seismicity and the Voronoi cell information in the prior. The VS location estimate agrees with the SCSN location. The VS magnitude estimate without the Gutenberg-Richter relationship in the prior is $M = 4.8 \pm 0.4$, with the Gutenberg-Richter in the prior, it is $M = 4.4 \pm 0.4$. The SCSN-reported magnitude is $M = 4.75$.

Figure 7.8 shows the evolution of magnitude estimates as a function of duration of data from the seismic network. Estimates labeled "amplitudes only" maximize the likelihood function (no prior information). We distinguish between 2 VS magnitude estimates (both with station geometry and previously observed seismicity included in the prior) differing in whether or not the Gutenberg-Richter relationship was used in the prior. All magnitude estimates converge to the SCSN solution at large times, when there are sufficient available observations. The differences in the magnitude estimates in the earlier estimates are due to the differences in the prior information included. The VS estimate without the Gutenberg-Richter relationship is within 0.05 magnitude units of the SCSN-reported magnitude 3 seconds after the initial P detection.

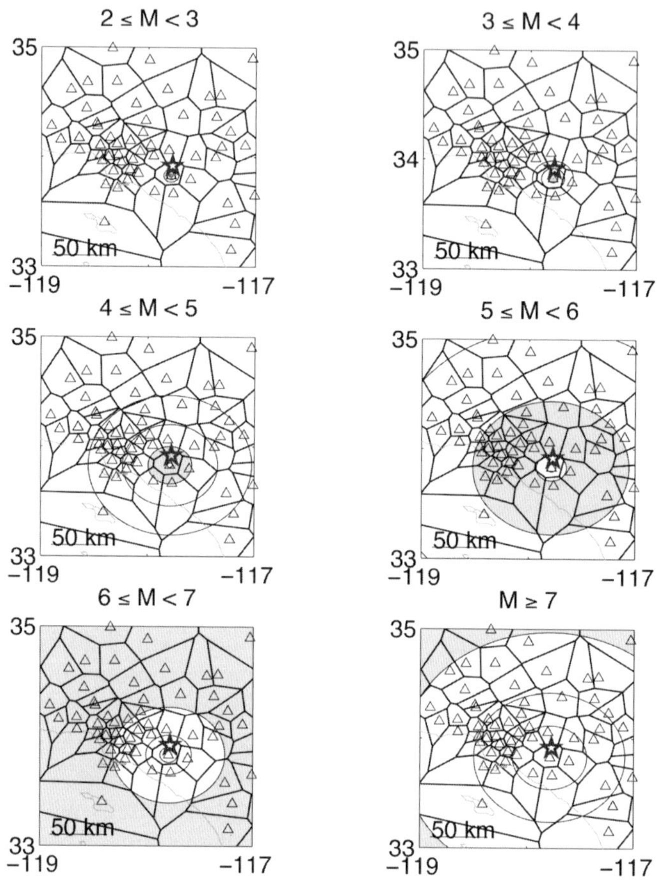

Fig. 7.6 Snapshots of the likelihood function expressed as a function of magnitude and epicentral location at various magnitude ranges. The shaded regions in each subplot are locations consistent with the peak amplitudes available at station SRN 3 seconds after the initial P detection. The trade-offs in magnitude and location are similar to those shown in Fig. 7.3. If no prior information regarding station geometries, previously observed seismicity, or the Gutenberg-Richter relationship is included, we cannot distinguish between a small event at small epicentral distance from a large event at large distance.

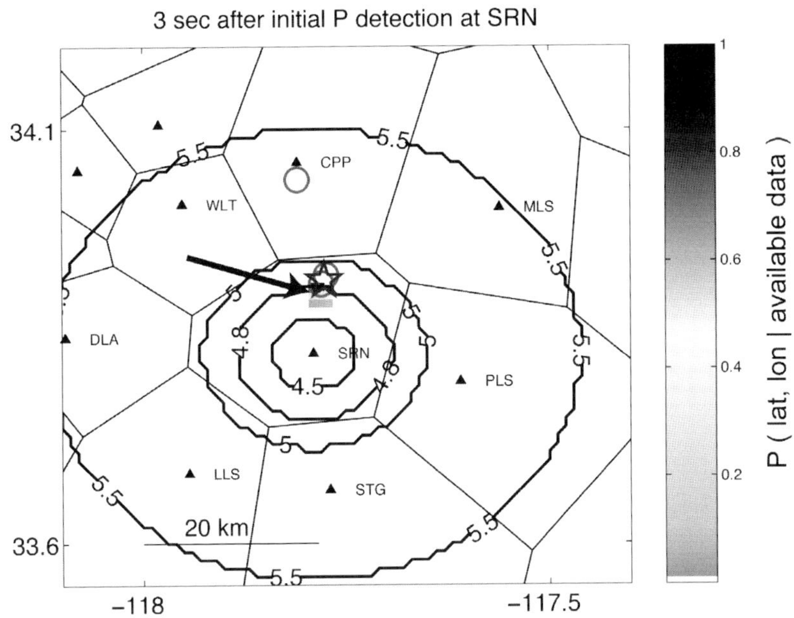

Fig. 7.7 VS estimates 3 seconds after the initial P detection at SRN for the Yorba Linda earthquake. Shading scales with the probability that the event is located at the given location. The VS location estimate (indicated by an arrow) is within 2 km of the SCSN reported location. Prior information includes previously observed seismicity (open circles), Voronoi cell information, and available arrivals at adjacent stations. Contours show the VS magnitude estimates at a given location without the Gutenberg-Richter relationship in the prior. The VS magnitude estimate without the Gutenberg-Richter relationship ($M = 4.8 \pm 0.425$) is in good agreement with the SCSN-reported magnitude of $M = 4.75$.

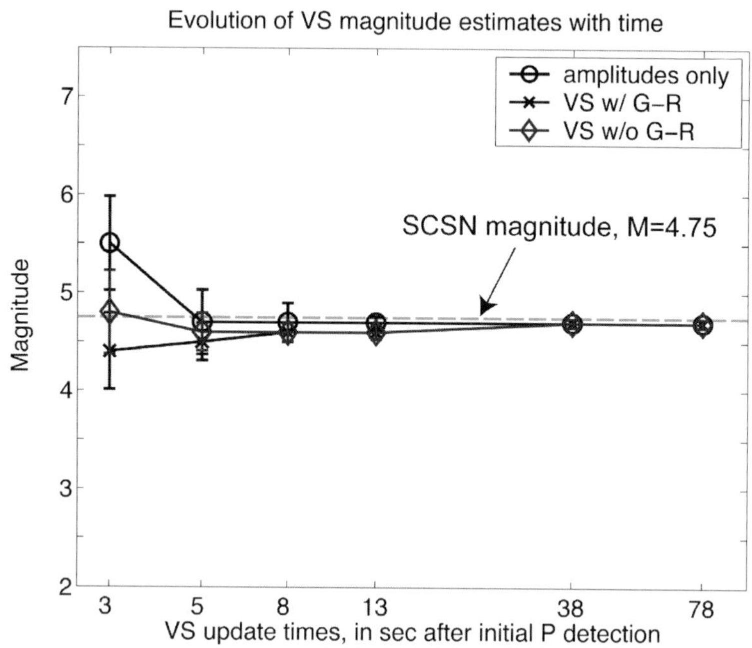

Fig. 7.8 The evolution of various magnitude estimates for the Yorba Linda earthquake as a function the duration of data from the seismic network. Estimates labeled "amplitudes only" correspond to magnitude estimates obtained by maximizing the likelihood function (no prior information). We distinguish between 2 VS magnitude estimates (both with station geometry and previously observed seismicity included in the prior) with and without the Gutenberg-Richter magnitude-frequency relationship in the Bayes' prior. All magnitude estimates converge to the SCSN solution at large times, when there are sufficient available observations. The differences in the magnitude estimates in the earlier estimates are due to the differences in prior information included. The VS estimate without the Gutenberg-Richter relationship is within 0.05 magnitude units of the SCSN-reported magnitude 3 seconds after the initial P detection.

At large time, *t,* after the event origin time, the location estimates that maximize the posterior pdf (which are identical to those that maximize the likelihood function at large *t*) are robust amplitude-based locations can be used to verify arrival-based locations. In Fig. 7.9, the amplitude-based location (green contours) is derived from the distribution of peak P- and S-wave amplitudes at 89 stations. The arrival-based location is obtained from 89 P-wave arrivals and an average P-wave velocity of 6 km/sec. The star marks the SCSN-reported location. There is general agreement between the amplitude- and arrival-based location estimates, indicating that the arrival-based location is most likely correct. These estimates are independent of each other, as they are derived from different types of data.

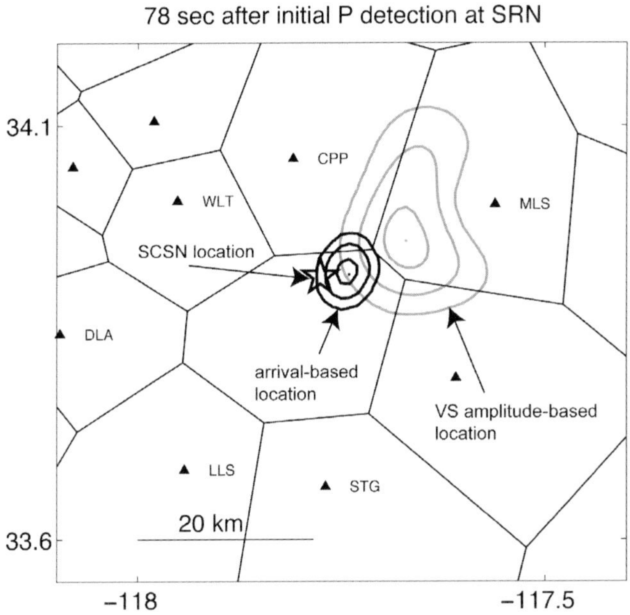

Fig. 7.9 Comparison of amplitude- and arrival-based location estimates 80 seconds after the origin time of the Yorba Linda earthquake. The amplitude-based location (green contours) is derived from the distribution of peak P- and S-wave amplitudes at 89 stations. The arrival-based location is obtained from 89 P-wave arrivals and an average P-wave velocity of 6 km/sec. The star marks the SCSN-reported location. There is general agreement between the amplitude- and arrival-based location estimates. These estimates are independent of each other, as they are derived from independent datasets.

7.3.2 16 October 1999 M7.1 Hector Mine, California Earthquake: Low Station Density

The M7.1 Hector Mine, California earthquake occurred in a region with relatively low density of SCSN stations. The closest station, HEC, was located 27 km north of the hypocenter. The mainshock, located by SCSN at 34.59N, 116.27W, at a depth of 5 km, was preceded by a cluster of 18 $1.5 \leq M \leq 3.8$ foreshocks within 1 km of the hypocenter in the previous 24 hours (Hauksson 2002). The application of the VS method on the Hector

Mine dataset illustrates the importance of prior information for regions with relatively low station density.

Figure 7.10 shows the operating SCSN stations (triangles), Voronoi cells (polygons), and seismicity in the preceding 24 hours (open circles) at the time of the mainshock. The areas of Voronoi cells of station HEC and adjacent stations range from 880 km^2 to 8020 km^2; these are an order of magnitude larger than the Voronoi cells in the epicentral region of the Yorba Linda earthquake.

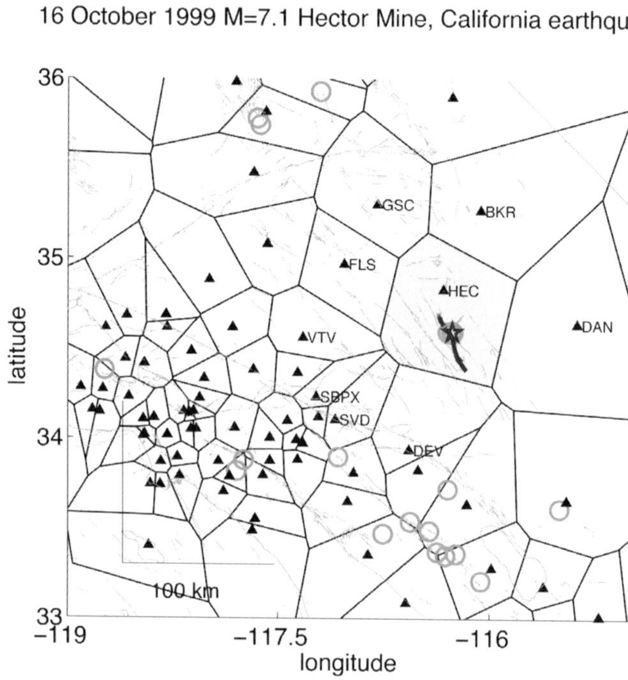

Fig. 7.10 Operational SCSN stations (triangles), associated Voronoi cells (polygons), and seismicity in the preceding 24 hours (open circles) within 200 km of the M=7.1 Hector Mine, California mainshock (star). Eighteen earthquakes occurred within 1-km epicentral distance of the mainshock hypocenter.

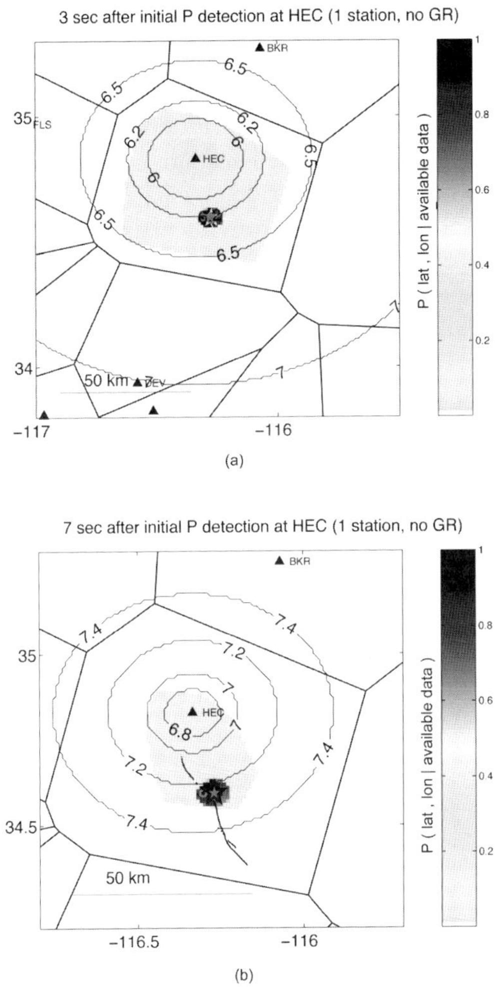

Fig. 7.11 VS estimates 3 and 7 seconds after the initial P detection at HEC for the Hector Mine earthquake. Shading scales with the probability that the earthquake is located at a given location. The shaded regions are the regions of likely location consistent with the initial P arrival at HEC, and no subsequent arrivals at adjacent stations 3 and 7 seconds later. Within the regions of likely location, the most probable locations are where there was a concentration of seismic activity in the preceding 24 hours. With Voronoi cells and not-yet-arrived data, the VS location estimates continually evolve, even before the second P arrival. The likelihood function is based on peak amplitudes available at HEC 3 seconds after the initial P detection. Contours show the VS magnitude estimates at a given location without the Gutenberg-Richter relationship in the prior.

The VS estimation process begins 3 seconds after the initial P detection at HEC, the first triggered station. Due to the large inter-station distances, the time between the initial P detection at station HEC and the P detection at the next closest station is 8 seconds. Figures 7.11(a) and 7.11(b) show the VS estimates at $\Delta t = 3$ and $\Delta t = 7$ seconds after the initial P detection at HEC. The regions of possible location consistent with HEC's Voronoi cell and the non-arrival information from the adjacent stations are shaded; these regions vary continuously as a function of time since the initial P detection. Within the regions of possible location, the most probable location estimates correspond to the locations where previously observed seismicity was concentrated. Thus, as early as 3 seconds after the initial P detection, with only one P arrival at the first triggered station, the VS location estimate agrees with the actual location (reported by SCSN). The 3-second magnitude estimate is $M = 6.2 \pm 0.45$; the 7-second magnitude estimate is $M = 7.2 \pm 0.45$.

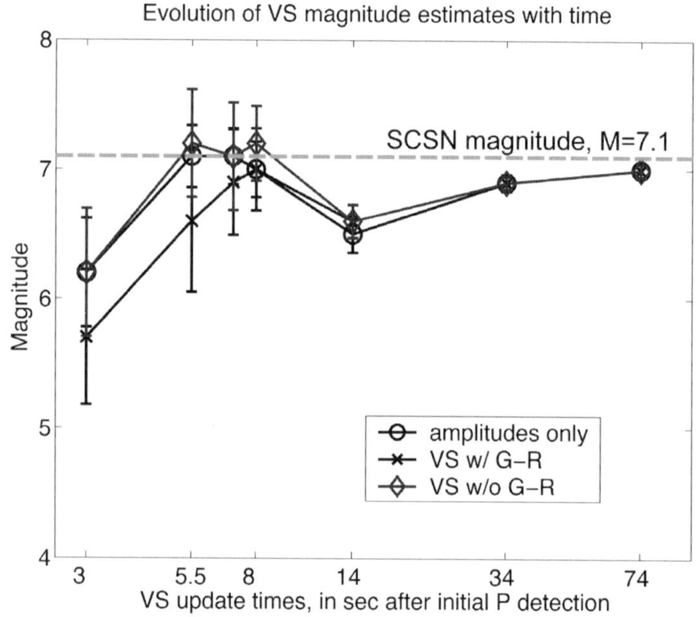

Fig. 7.12 The evolution of various magnitude estimates for the Hector Mine earthquake as a function of duration of data from the network. The estimates labeled "amplitude only" correspond to the magnitudes that maximize the likelihood function (that is, no prior information included). VS magnitude estimates with and without the Gutenberg-Richter relationship in the Bayes prior are also shown.

Figure 7.12 shows the evolution of different magnitude estimates for the Hector Mine mainshock based on: 1) observed amplitudes only with no prior, 2) observed amplitudes with Voronoi cells, seismicity, and Gutenberg-Richter in the prior, and 3) similar to 2) but without the Gutenberg-Richter relationship. There are significant differences in these types of estimates in the early part of the estimation process, when observations are scarce; these different estimates converge and approach the SCSN-reported magnitude of M7.1 as more observations become available and the prior becomes less important with increasing t. When sufficient data is available, the estimates are driven by the observations (amplitudes and arrivals) and the choice of prior is irrelevant. The prior is important only in the early stages of the estimation process, where there is not sufficient data to adequately constrain the estimation process.

7.4 How Subscribers Might Use Early Warning Information

Ultimately, the goal of seismic early warning is to provide users with information that can be used to determine the optimal course of action in the few seconds before the onset of damaging ground motion levels at their sites of interest.

Consider the case of a subscriber who wants to initiate a predetermined set of damage-mitigating actions if the peak ground motions at the site of interest exceed a threshold level, $Y_{max} > Y_{thresh}$. Given the early warning source estimates and their uncertainties, the expected ground motion levels can be predicted using attenuation relationships. The uncertainty on these predicted ground motion levels, σ_{pred}, is a combination of uncertainties from the early warning source estimates and uncertainties from the attenuation relationships. As additional observations become available, the uncertainties from the source estimates decrease; σ_{pred} approaches the uncertainty on the ground motion attenuation relationships. From Table 7.1, σ_{pred} for various amplitude types (horizontal and vertical acceleration, velocity, and displacement for P- and S-waves on rock and soil sites) is on the order of 0.3 log units, or a factor of 2. The probability of observing a maximum ground motion level Y_{max}, given the predicted ground motions Y_{pred} from the early warning estimates, is

$$P(Y_{\max} \mid M, loc) = \frac{1}{\sqrt{2\pi}\sigma_{pred}} \exp\left[-\frac{(Y_{\max} - Y_{pred}(M, loc))^2}{2\sigma_{pred}^2}\right]. \tag{7.12}$$

In Eq. (7.12), Y_{pred} is the predicted maximum amplitude expected at the site of interest. These are given by the horizontal S-wave amplitude envelope attenuation relationships in Table 7.1. The probability of the ground motion threshold Y_{thresh} being exceeded, P_{ex}, given the early warning source estimates, is

$$P_{ex} = P(Y_{\max} > Y_{thresh} \mid Y_{pred}) = \int_{Y_{thresh}}^{\infty} P(Y_{\max} \mid M, loc) \tag{7.13}$$

At the time when the subscriber must decide whether to initiate actions or not, the actual peak ground motions at the site, $Y_{\max}$, are of course unknown; the decision to initiate actions or not must be based on some function of the predicted ground motions, Y_{pred}. The uncertainty in the relationship between $Y_{\max}$ and Y_{pred} gives rise to the possibility of less than optimal decisions: 1) a false alarm corresponds to initiating action when it is ultimately unnecessary, or $Y_{\max} < Y_{thresh}$, and 2) a missed alarm corresponds to not initiating action when it is ultimately necessary, or $Y_{\max} > Y_{thresh}$.

The following is a simple cost-benefit analysis using basic decision theory for a subscriber who wants to initiate a set of actions if $Y_{\max} > Y_{thresh}$ (Grigoriou 1979).

Let $H = h_i$, $i = 1,\ldots,n$ be the (exhaustive and mutually exclusive) set of possible states of nature. In our example, $n = 2$; the only possibilities are 1) $Y_{\max} > Y_{thresh}$, or 2) $Y_{\max} < Y_{thresh}$. Let $B = b_j$, $j = 1,\ldots,m$ be the set of possible actions. In our example, $m = 2$; the possible actions we consider are 1) “initiate actions”, and 2) “do nothing”. Let $C(b_j,h_i)$ be the cost of action b_j if the state of nature is h_i. Let P_i be the probability of h_i. C_{damage} is the cost of damage incurred if no actions were taken and the peak ground motions exceeded the threshold, $Y_{\max} > Y_{thresh}$; it is the cost of a missed alarm. C_{act} is the cost of performing the damage-mitigating actions; it is also the cost of a false alarm. For simplicity, assume that C_{damage} and C_{act} are known. In practice, these are also uncertain, and probability models are required to describe these quantities. The cost table, expressed in terms of $C_{ratio} = \frac{C_{damage}}{C_{act}}$ is shown in Table 7.2. The expected cost, $E\,[C_j]$, of a particular action is given by

$$E\,[C_j] = \sum_{i=1}^{n} C(b_j, h_i) P_i \tag{7.14}$$

Table 7.2 Cost table for early warning subscriber

h_j	$P_i = P(h_j \mid Y_{pred})$	"Do nothing"	"Initiate actions"
$Y > Y_{thresh}$	P_{ex}	C_{ratio}	1
$Y < Y_{thresh}$	$1 - P_{ex}$	0	1

The optimal action given a particular early warning source estimate is the action with the minimum cost. If we set $E[\text{"initiate action"}] \quad E[\text{"do nothing"}]$, we find the critical probability of exceedance, above which it is optimal to initiate action

$$P_{ex,crit} = \frac{1}{C_{ratio}}. \tag{7.15}$$

Since $P_{ex,crit}$ is a probability, it takes on values between 0 and 1. This implies that $C_{ratio} \geq 1$; the cost of damage as a consequence of not acting must be equal to or greater than the cost of performing the actions, else early warning information provides no benefit to the subscriber.

We can relate $P_{ex,crit}$ to the predicted level of ground motion, $Y_{pred,crit}$, above which it is optimal to act

$$Y_{pred,crit} = Y_{thresh} - \sigma_{pred}\sqrt{2}\left[erf^{-1}\left(1 - \frac{\sqrt{2\pi}\sigma_{pred}}{C_{ratio}}\right)\right]. \tag{7.16}$$

Thus, taking into account that there are uncertainties in the ground motion predictions from an early warning source estimate, the appropriate criteria for a subscriber to initiate action is $Y_{pred} > Y_{pred,crit}$ where $Y_{pred,crit}$ depends both on subscriber-specific values Y_{thresh} and C_{ratio}, as well as the uncertainty on the predicted ground motions σ_{pred}.

Figure 7.19 shows $Y_{pred,crit}$ as a function of σ_{pred} for various values of C_{ratio}. Y_{thresh} is simply a constant offset; in this plot, we set $Y_{thresh} = 0$. For subscribers with $C_{ratio} \sim 1$, false alarms are relatively expensive; when σ_{pred} is large, it is sometimes optimal to "do nothing" even when the predicted ground motions exceed the threshold, $Y_{pred} > Y_{thresh}$. For subscribers with $C_{ratio} >> 1$, false alarms are relatively inexpensive; it is optimal to initiate action even when $Y_{pred} < Y_{thresh}$. This highlights the importance of C_{ratio} in optimizing use of early warning information. Simple applications, such as opening fire station doors, or stopping elevators at the closest floor, would have relatively high values of C_{ratio}. Considerable application-specific efforts are required to determine what values of C_{ratio} would be appropriate for more complex applications such as diverting airport traffic, putting nuclear plants into safe mode, or stopping sensitive manufacturing equipment. Each subscriber needs to invest efforts into determining its appropriate C_{ratio}, and whether it is in fact a suitable candidate for early warning.

7.5 Station Density and the Evolution of Estimate Uncertainties

The uncertainties on early warning source estimates translate to uncertainties in predicted ground motions, which play an important role for users in decision-making using early warning information. In the VS method, the posterior density function is a three dimensional function in magnitude, latitude, and longitude. How should the network transmit the source estimates and their uncertainties to the subscribers? If the posterior density function can be described as a Gaussian, it may be sufficient to transmit 6 parameters (3 means, and 3 standard deviations) to the subscribers. The marginal probability density (integrating over latitude and longitude) of the magnitude estimate can always be described as a Gaussian, whether or not the Gutenberg-Richter law is included (Figs. 7.13a and 7.14a).

This is not always the case for the latitude and longitude estimates. The simplest location estimate is to assume that the earthquake is located at the first triggered station. In regions with relatively high station density, this is a relatively good assumption to make. The maximum possible errors on such location estimates depend on inter-station distances. In regions where instrumentation density is high (for instance, the epicentral region of the Yorba Linda mainshock), this is a reasonable assumption, with maximum possible errors on the order of 10 km. In contrast, this assumption is inappropriate in regions with low station density or regions on the outer boundaries of the network. In the epicentral region of the Hector Mine earthquake, the maximum possible error in assuming the event is located at the first triggered station is 60 km (the maximum epicentral distance consistent with the Voronoi cell geometry).

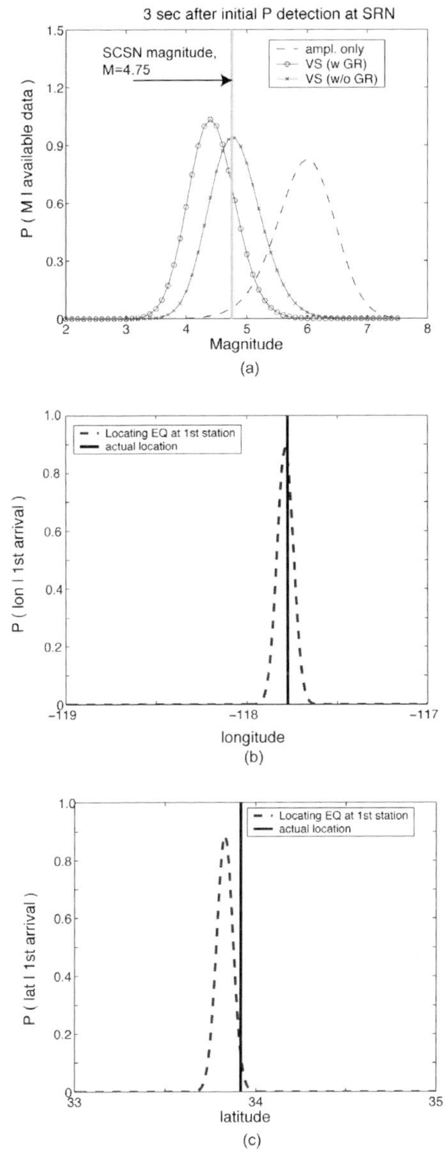

Fig. 7.13 In regions with high station density, such as the epicentral region of the Yorba Linda earthquake, assuming that the earthquake is located at the first triggered station is a valid assumption. The marginal pdfs for the VS magnitude and location estimates can be approximated as Gaussian distributions. The relevant information about the early warning estimates can be summarized by 6 parameters (3 means, and 3 standard deviations), which can be easily transmitted to subscribers.

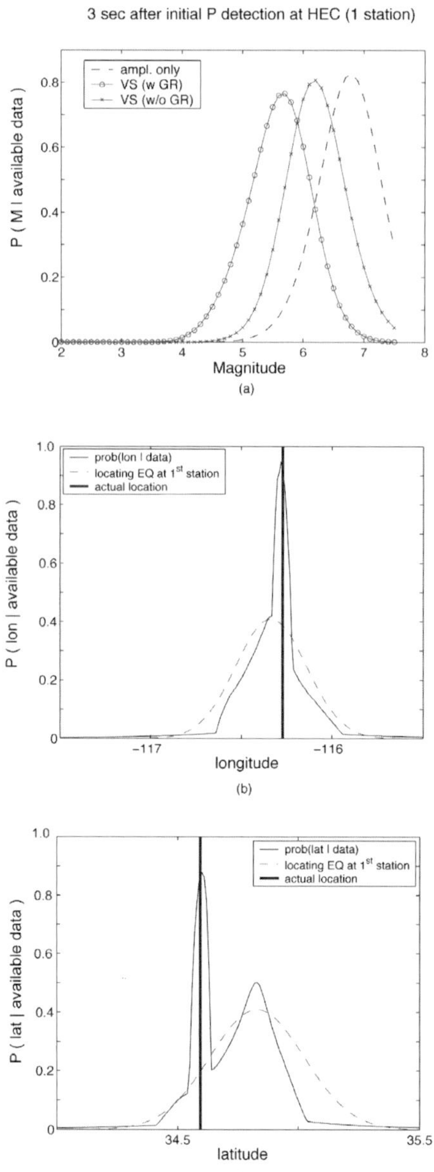

Fig. 7.14 In regions with low station density, such as the epicentral region of the Hector Mine earthquake, the initial VS estimates are heavily influenced by prior information. While the marginal pdf for the magnitude estimate can still be expressed as a Gaussian distribution, the marginal pdfs for the location estimates are non-Gaussian, due to the influence of previously observed seismicity.

In less instrumented regions within SCSN, the maximum possible location error associated with such an assumption can be as large as 120 km. When station density is high, the posterior density function can be described by 3 Gaussian functions since 1) assuming a location at the first triggered station is appropriate, and 2) prior information is not necessary, since there are quickly enough arrivals to properly constrain the location.

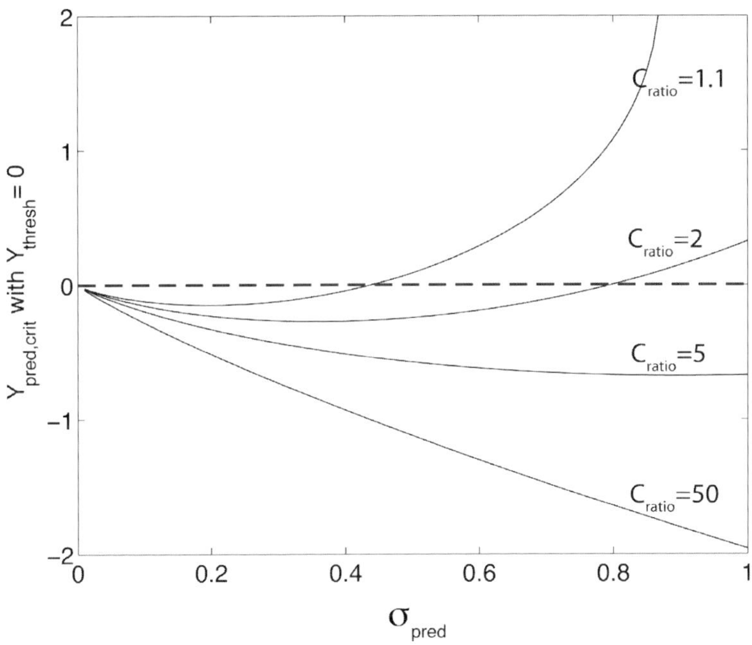

Fig. 7.15 The critical predicted level of ground motion above which early warning subscribers should initiate action, $Y_{pred,crit}$, as a function of uncertainty on predicted ground motions σ_{pred} and C_{ratio}. Depending on the value of C_{ratio}, it may be cost-effective for an early warning subscriber to initiate action even when the predicted ground motion levels are below the threshold at which damage occurs, Y_{thresh}.

When station density is low, assuming a location at the first triggered station can have very large errors. Prior information is important for the initial estimates in such regions, since large inter-station distances mean that it takes a while before there are enough observations to properly constrain the magnitude and location estimates. Prior information such as previously observed seismicity and known fault locations make the latitude and longitude marginal density functions highly irregular and multimodal.

It is prohibitive in terms of time and communications bandwidth for the network to transmit the full 3d posterior density function to subscribers. Nevertheless, it is necessary to provide this information to subscribers. An attractive alternative is for the network to transmit the likelihood function to the subscribers, and for the subscribers combine the prior and likelihood on site. This is alternative provides the subscriber added flexibility in: 1) defining its own prior (for instance, whether or not to include the Gutenberg-Richter relationship), 2) determining computational resources (for instance, maximizing the posterior density function could be implemented on parallel processors).

It is particularly important for subscribers to have a control over whether or not the Gutenberg-Richter relationship should be included in the prior. Figure 7.16 shows the evolution of different magnitude estimates (amplitudes only, VS with Gutenberg-Richter in prior, VS without Gutenberg-Richter in prior) for 4 Southern California earthquakes. In all cases, the VS magnitude estimates converge to the magnitudes reported by SCSN once there are enough observations to properly constrain the estimates, regardless of the choice of prior. When there are trade-offs in the source parameters unresolved by the observed amplitudes, the VS magnitude estimates without the Gutenberg-Richter relationship have a smaller error compared to when the Gutenberg-Richter relationship is included. This seems to indicate that the information provided by the Gutenberg-Richter relationship is not useful. However, the Gutenberg-Richter relationship has been observed to hold in general worldwide. This apparent inconsistency is resolved by taking into account the user considerations. In these 4 cases, VS magnitude estimates with the Gutenberg-Richter relationship in the prior are smaller than the actual magnitude. Subscribers basing decisions on VS estimates with the Gutenberg-Richter relationship will lower their incidence of false alarms, at the cost of increasing their vulnerability to missed warnings.

Thus, subscribers with $C_{ratio} \sim 1$ should consider using the Gutenberg-Richter relationship in their prior. In contrast, subscribers with $C_{ratio} \gg 1$ can exclude the Gutenberg-Richter relationship from their prior, and benefit from the smaller errors in the magnitude estimates. Such subscribers have relatively high costs for missed warnings. Since they place a premium on making the appropriate decisions during the large, infrequent events, they need to accept a certain level of false alarms dictated by C_{ratio}. More studies regarding how VS estimates evolve with time are necessary.

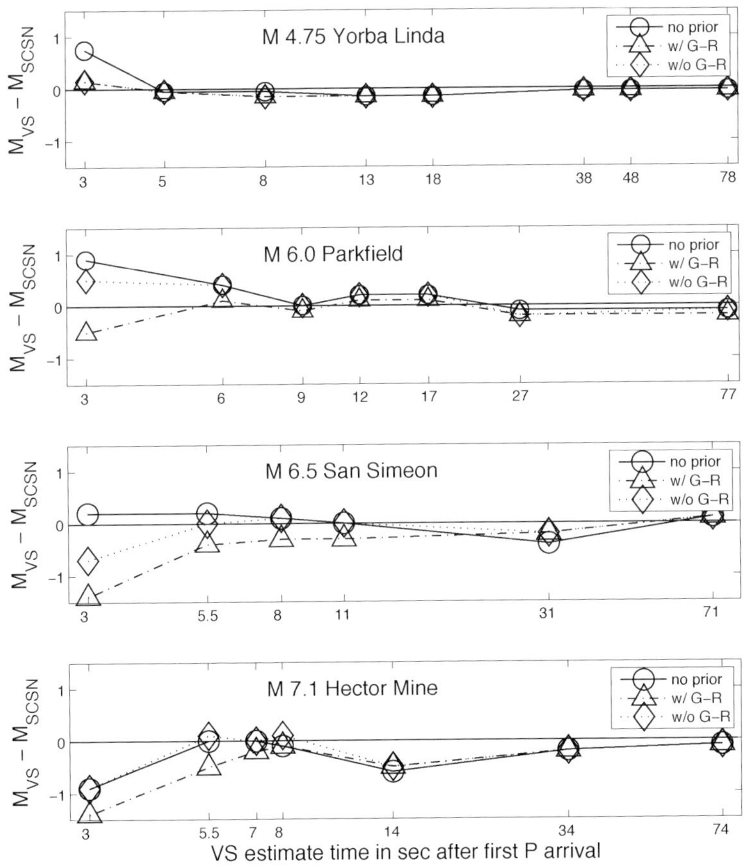

Fig. 7.16 The evolution of different magnitude estimates (amplitudes only, amplitudes with prior including Gutenberg-Richter relationship, and amplitudes with prior excluding the Gutenberg-Richter relationship) as a function of duration of data from the network for selected Southern California earthquake datasets. Estimates with the Gutenberg-Richter relationship in the prior reduce the probability of false alarms. Subscribers should have the flexibility to decide which type of magnitude estimates are best suited to their applications, based on their relative cost of missed to false alarms.

7.6 Conclusions

The Virtual Seismologist (VS) method is a Bayesian approach that provides a unified framework for addressing both the real-time source estimation and the user-response problems in earthquake early warning. Disparate types of information, such as previously observed seismicity, station locations, and the Gutenberg-Richter relationship, are included in the source estimation problem via the Bayes prior. The trade-offs in the initial source estimates that cannot be resolved by the available data are resolved in favor of the prior information. The evolution of source estimates as a function of available data is similar to how humans modify their opinions or judgments in light of new information; prior information is important when data is sparse, but decreases in influence as additional observations become available.

What type of information to include in the Bayes prior, and hence, what type of source estimates to solve for, is dependent on user-specific considerations, in particular, the relative cost of missed to false alarms. Users with relatively expensive false alarms should include the Gutenberg-Richter relationship in the prior, at the cost of increasing their vulnerability to missed warnings. Users with relatively expensive missed alarms should use source estimates excluding the Gutenberg-Richter relationship. A certain level of false alarms must be accepted if the aim is to act appropriately during the large, infrequent earthquakes.

7.7 Acknowledgements

We wish to thank Hiroo Kanamori, Egill Hauksson, James Beck, and John Clinton for their stimulating discussions and for their comments on the thesis work on which this manuscript is based. This work was supported by the Department of Civil Engineering at the California Institute of Technology. The completion of the manuscript was made possible with support of the Puerto Rico Seismic Network, the Puerto Rico Strong Motion Program, and the Swiss Seismological Service.

References

Allen RM, Kanamori H (2003) The potential for earthquake early warning in Southern California. Science 300:786-789

Cua G (2005) Creating the Virtual Seismologist: developments in ground motion characterization and seismic early warning. PhD thesis, California Institute of Technology. http://resolver.caltech.edu/CaltechETD:etd-02092005-125601

Cua G, Heaton T (2006) Characterizing average properties of Southern California ground motion envelopes. (in preparation)

Cua G, Heaton T (2006) Linear discriminant analysis in earthquake early warning. (in preparation)

Gerstenberger M, Wiemer S, Jones L (2003) Real-time forecasts of tomorrow's earthquakes in California: a new mapping tool. United States Geological Survey Open File Report, 2004-1390

Goltz JD (2002) Introducing earthquake early warning in California: a summary of social science and public policy issues. Technical Report, Governor's Office of Emergency Services

Grazier V, Shakal A, Scrivner C, Hauksson E, Polet J, Jones L (2002) TriNet strong-motion data from the M7.1 Hecor Mine, California earthquake of 16 October 1999. Bull Seism Soc Am 92:1525-1542

Grigoriu M, Veneziano D, Cornell CA (1979) Probabilistic modeling as decision making. Journal of the Engineering Mechanics Division, ASCE EM4, 585-596

Hauksson E, Hutton K, Jones L, Given D (2002) The September 03, 2002 earthquake M4.6 near Yorba Linda. http://www.trinet.org/eqreports

Hauksson E, Jones L, Hutton K (2002) The 1999 M_w7.1 Hector Mine, Caifornia earthquake sequence: complex conjugate strike-slip faulting. Bull Seism Soc Am 92:1154-1170

Heaton T (1985) A model for a seismic computerized alert network. Science 228:987-990

Horiuchi S, Negishi H, Abe K, Kimimura A, Fujinawa Y (2004) An automatic processing system for broadcasting earthquake alarms. Bull Seism Soc Am 95:708-718

Jones L (1984) Foreshocks (1966-1980) in the San Andreas system, California. Bull Seism Soc Am 74:1361-1380

Kanamori H (1993) Locating earthquakes with amplitude: application to real-time seismology. Bull Seism Soc Am 83:264-268

Nakamura Y (1988) On the urgent earthquake detection and alarm system (UrEDAS). Proceedings of 9th World Conference in Earthquake Engineering, Tokyo-Kyoto, Japan

Rydelek P, Pujol J (2004) Real-time seismic warning with a 2-station subarray. Bull Seism Soc Am 94:1546-1550

Sivia DS (1996) Data Analysis: a Bayesian tutorial. Oxford University Press

Wu YM, Kanamori H (2005a). Experiment on an onsite early warning method for the Taiwan early warning system. Bull Seism Soc Am 95:347-353

Wu YM, Kanamori H (2005b) Rapid assessment of damaging potential of earthquakes in Taiwan from the beginning of P waves. Bull Seism Soc Am 95:1181-1185

Yamada M, Heaton T (2006) Early warning systems for large earthquakes: estimation from fault location using ground motion envelopes. Bull Seism Soc Am (submitted)

8 A Strong Motion Attenuation Relation for Early-warning Application in the Campania Region (Southern Apennines)

Vincenzo Convertito[1], Raffaella De Matteis[2], Annalisa Romeo[3], Aldo Zollo[3], Giovanni Iannaccone[1]

[1] Istituto Nazionale di Geofisica e Vulcanologia, Osservatorio Vesuviano, Napoli, Italy
[2] Dipartimento di Studi Geologici ed Ambientali, Università degli Studi del Sannio, Benevento, Italy
[3] Dipartimento di Scienze Fisiche, Università degli Studi di Napoli Federico II, Napoli, Italy

Abstract

For early-warning applications in particular, the reliability and efficiency of rapid scenario generation strongly depend on the availability of reliable strong ground-motion prediction tools. If shake maps are used to represent patterns of potential damage as a consequence of large earthquakes, attenuation relations are used as a tool for predicting peak ground-motion parameters and intensities. One of the limitations in the use of attenuation relations is that these have only rarely been retrieved from data collected in the same tectonic environment in which the prediction has to be performed. As a consequence, strong ground motion can result in underestimations or overestimations with respect to the recorded data. This also holds for Italy, and in particular for the Southern Apennines, due to limitations in the available databases, both in terms of distances and magnitude. Moreover, for “real-time” early-warning applications, it is important to have attenuation models for which the parameters can be easily upgraded when new data are collected, whether this has to be done during the earthquake rupture occurrence or in the post-event, when all the strong motion waveforms are available.

Here we present a strong-motion attenuation relation for early-warning applications in the Campania region (Southern Apennines), Italy. The model has a classical analytical formulation, and its coefficients were retrieved from a synthetic strong-motion database created by using a stochastic approach. The input parameters for the simulation technique were obtained through the spectral analysis of waveforms of earthquakes recorded by the Istituto Nazionale di Geofisica e Vulcanologia (INGV) network for a magnitude range M_d (1.5,5.0) in the last fifteen years, and they have been extrapolated to cover a larger range.

To validate the inferred relation, comparisons with two existing attenuation relations are presented. The results show that the calibration of the attenuation parameters, i.e., geometric spreading, quality factor Q, static stress drop values along with their uncertainties, are the main concern.

8.1 Introduction

The prediction of strong ground-motion parameters in both the time and frequency domains is of fundamental importance for seismic hazard analyses and for seismic early-warning applications. The reliability of the predictions depends mainly on the ability to model all of the aspects that can affect the radiated energy at the source during its propagation to the sites of interest.

Although many prediction methods exist (e.g., empirical Green's functions, semi-empirical, semi-theoretical, stochastic and theoretical), the most widely used in almost all earthquake-prone regions are those based on an empirical approach. Empirical models, which are generally referred to as attenuation relations, are mathematical functions that relate the selected strong ground-motion parameter (peak ground acceleration, velocity and displacement or spectral ordinates) to parameters characterizing the source, the medium (in terms of geometric spreading and anelastic absorption and scattering) and the local site geology (Campbell 1985). Once the functional form has been selected, the coefficients are retrieved using regression analysis on an existing strong-motion database.

When attenuation relations are used for predicting ground motion, one of the major prerequisites is that the estimates should be used only for regions in which the data are collected or for regions that based on geophysical and seismological data, have similar source and propagation characteristics (Reiter 1990). However, the absence of large and complete databases, in terms of magnitude, distances and fault mechanisms, represents a serious problem in such cases. Moreover, the intrinsic limitations of a point-like description of the seismic source must be considered, in particular when ground motion due to a large earthquake in the near-source range has to be estimated. In fact, these motions are strongly affected by source duration and geometry, which cannot be taken into account in attenuation relations.

The Campania region is situated in Italy's Southern Apennine chain, where there have been several large destructive earthquakes in recent times. The last of these was the 1980 M 6.9 Irpinia earthquake, which caused several thousand deaths and resulted in severe economic losses. This region is now covered by an advanced network that is equipped with highly dynamic and densely spaced instruments that will allow the registration of non-saturated time histories from a broad-band spectrum of magnitude (Weber et al. 2007, this issue). The network is mainly devoted to early-warning seismic purposes, and due to its geometric characteristics it is designed to provide rapid damage scenarios through the calculation of shake maps. These maps are simple or sophisticated representations of the ground shaking in terms of a selected strong ground-motion parameter that results from large earthquakes. As a consequence, the formulation of a regional attenuation relation is of great concern.

The present study is aimed at retrieving a strong motion relation for Campania as a possible alternative to those already existing that were developed recently for Italy (Sabetta and Pugliese 1987, 1996; Malagnini et al. 2000). The results presented here refer to the peak ground acceleration (Pga) and peak ground velocity (Pgv), but they can be extended to other time or spectral parameters. The major difficulty was the absence of a strong ground-motion database recorded in the area of interest. Thus, to partially overcome this difficulty, we analyzed a database of waveforms recorded from 1988-2003 by the INGV network for a magnitude range M_d (1.5,5.0). Although these data do not provide the required strong-motion database, they do allow the retrieval of local scaling laws (i.e., seismic moment vs corner frequency; static stress drop vs corner frequency) and anelastic characteristics that can be extrapolated for larger earthquakes and the production of a synthetic database using a stochastic simulation procedure. This database can thus be used both to complete existing limited databases and to directly retrieve local attenuation relations.

8.2 Database and Scaling Laws

One of the main problems to be faced when a local attenuation relation is to be retrieved is the absence of a large and complete strong-motion database. This can be partially overcome using a simulation technique that is able to generate waveforms with the spectral and temporal characteristics of interest. The technique adopted in the present study is the stochastic simulation method that was proposed by Boore (1983), which requires as input parameters the static stress drop and corner frequency for a selected range of earthquake magnitudes. To infer these parameters for Campania, local scaling laws were obtained by analysing the earthquake waveforms recorded by the INGV network from 1988 to 2003 with magnitude between 1.5 and 5.0 (Fig. 8.1). We selected only the earthquakes recorded at least at six seismic stations with a clear P-wave phase. Then, starting from an original catalogue of 788 earthquakes, we extracted for the following analyses 2,774 waveforms of earthquakes in the magnitude range M_d (1.8,4.5). We picked the first P- and S-wave arrival times and re-located the events in a flat-layered velocity model proposed by Bernard and Zollo (1989).

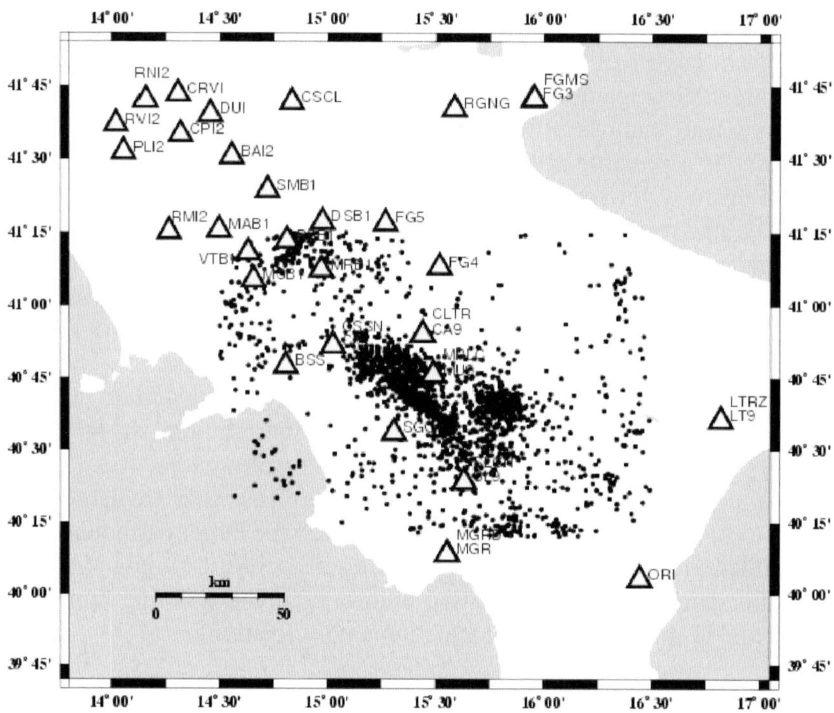

Fig. 8.1 Map of the INGV seismic stations (black triangles) and the epicentral distribution of the earthquakes analysed (black dots).

The classic ω^{-2} Brune (1970) spectral model is used to estimate the source parameters Ω_0 (low-frequency spectral amplitude), f_c (corner frequency) and Q_P (quality factor) by the inversion of displacement spectra. Since the seismic stations of the INGV network are, for the most part, equipped with single vertical component sensors, only the P-wave spectra were inverted.

The displacement spectrum was calculated on a 1-s window, around the first P-wave arrival time (panels a and b of Fig. 8.2). The window width was chosen taking into account the distance and magnitude ranges covered in the database. The non-linear inversion technique used is based on the Simplex optimization method (Nelder and Mead 1965), which requires a reference starting model. It is a local search method based on the minimization of a cost function defined as the absolute value of the observed minus the predicted spectrum, multiplied by the square of the frequency. To overcome the problem of falling into local minima, several inversions were performed with different initial model parameters that were chosen randomly in reasonable ranges. This avoided an arbitrary selection criterion for the starting model. As an example, the initial and final values for Ω_0, f_c and Q_P obtained after the P-wave spectral inversion are shown in Fig. 8.3.

The retrieved parameters were finally used to estimate the relations between seismic moment, stress drop and corner frequency (Fig. 8.4), which can be used as input parameters for the stochastic simulation technique when large earthquakes need to be simulated.

For the quality factor Q_P, the results of the inversion procedure show a linear source-to-station distance dependence (Fig. 8.5). This could be ascribed to incorrect modeling of the geometric attenuation effect in our spectral model. To minimize this effect in the simulation of the strong ground motion, the dependence was introduced into the stochastic approach, as detailed further in the next section.

8.3 Peak Ground-motion Simulation

To set up a database for the distance and magnitude ranges of interest regarding the seismotectonics of the Campania region, the stochastic simulation technique introduced by Boore (1983) was used to generate synthetic time histories of motion, from which the Pga and Pgv values were calculated. This technique works in both the time and frequency domains, and it is based on the generation of a time sequence of filtered and windowed, band-limited, Gaussian noise with zero expected mean and variance chosen to give unity as spectral amplitude. This spectrum is then multiplied by a specified spectrum (e.g., Brune 1970) and the transformation back to the time domain yields the final time series.

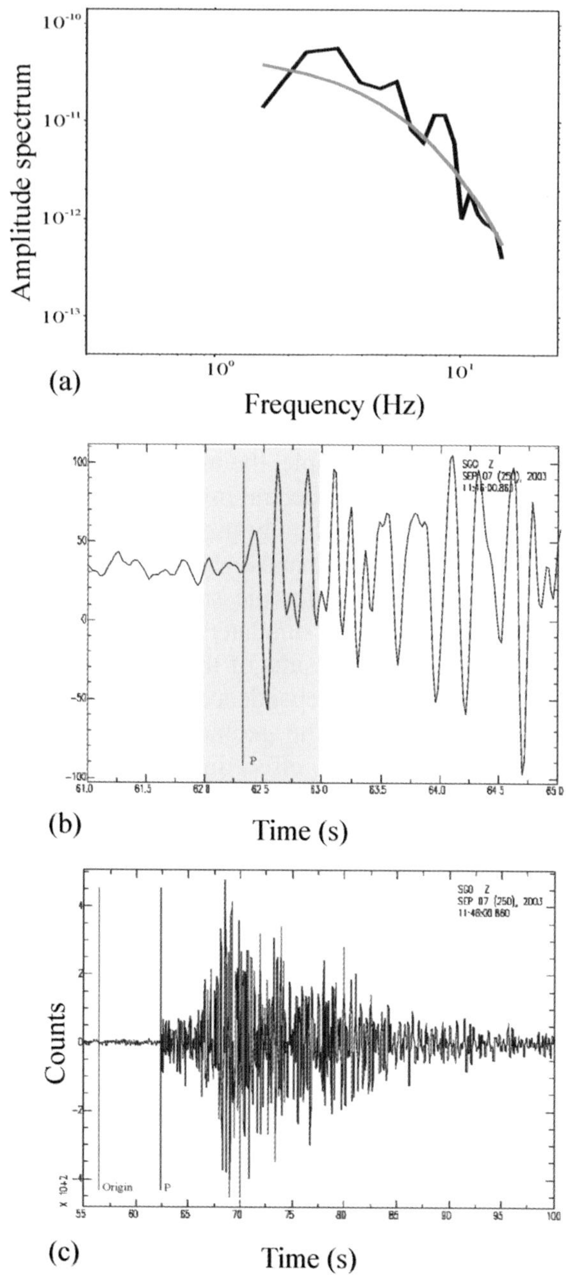

Fig. 8.2 (**a**) Observed (black line) and inverted (grey line) amplitude displacement spectra (**b**) Selected time window; (**c**) Vertical component of velocity.

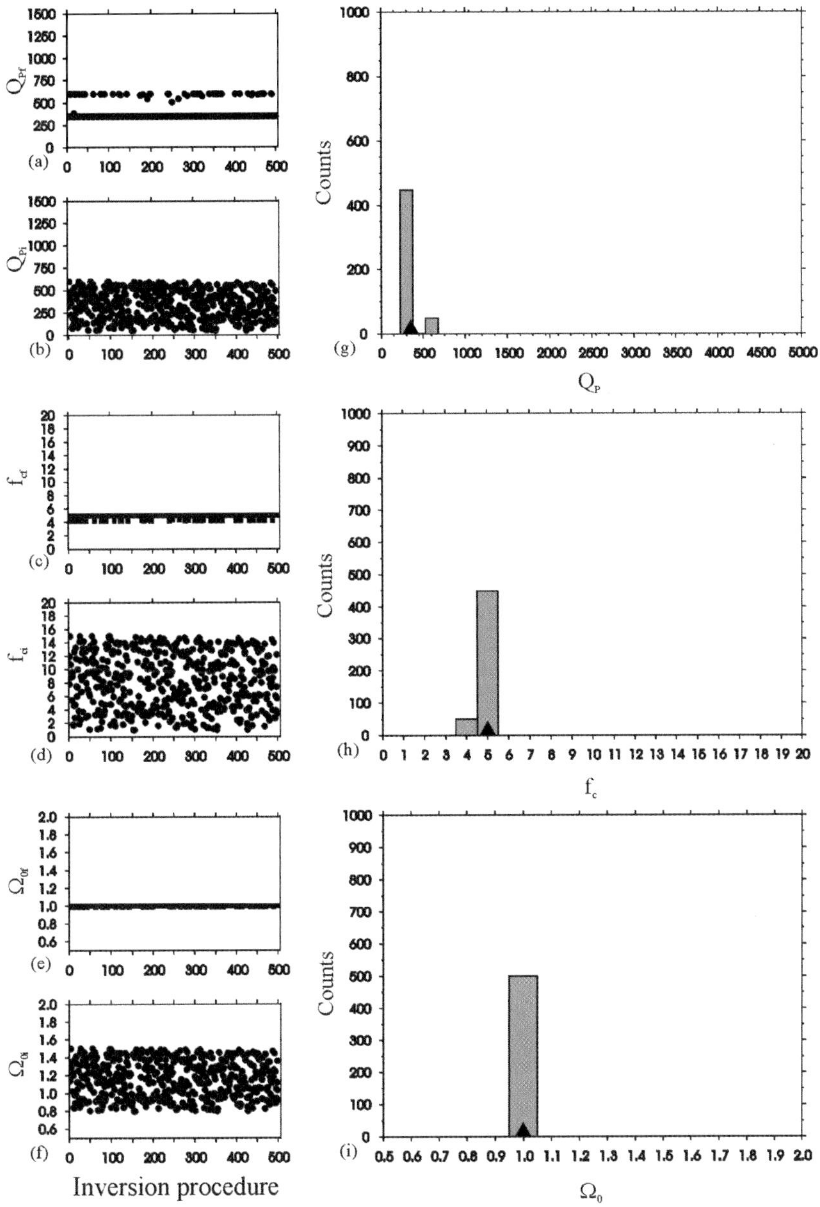

Fig. 8.3 (**a**) (**c**) (**e**) Final values for Q_P, f_c and Ω_0, respectively, after inversion; (**b**) (**d**) (**f**) Starting values for the Q_P, f_c and Ω_0 parameters, respectively, used in the inversion procedures; (**g**) (**h**) (**i**) Histograms of the final values for Q_P, f_c and Ω_0, respectively (red bars) and their corresponding modal values (black triangles).

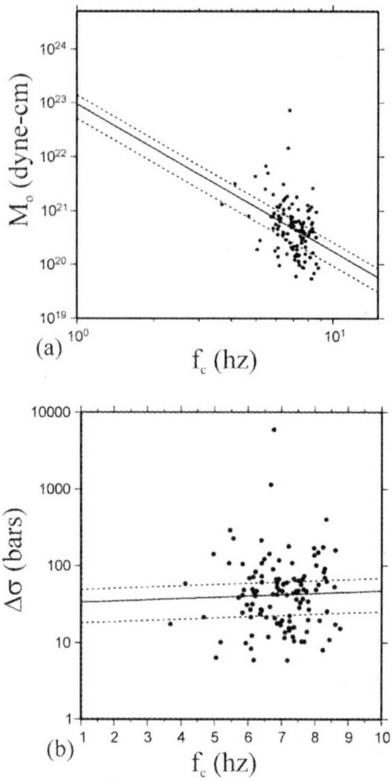

Fig. 8.4 Retrieved scaling laws: (**a**) Seismic moment vs corner frequency. The continuous line represents the median and the dotted lines represent ±1 standard deviations. (**b**) Static stress ($\Delta\sigma$) drop vs corner frequency. The continuous line represents the median and the dotted lines represent ±1 standard deviations.

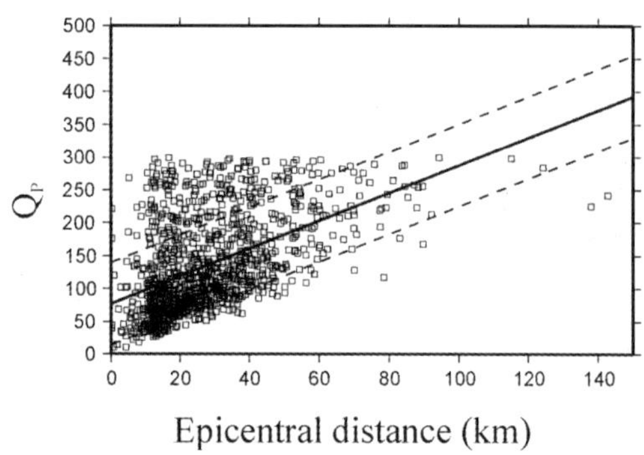

Fig. 8.5 Q_P values retrieved from the spectral P-waveform inversion as a function of epicentral distance. Continuous and dotted lines as for Fig. 8.4.

The technique proposed by Boore (1983) allows for the selection of several acceleration spectrum shapes, taking into account the source, attenuation (geometric, anelastic and superficial) and site effects. In the present application, the following formula for the acceleration spectrum *A(f)* was used:

$$A(f) = CM_0 S(f, fc) \frac{e^{-\frac{\pi R}{Q\beta}}}{R} \tag{8.1}$$

where R is the epicentral distance and C is a constant given by:

$$C = \frac{R_{\theta\phi} \cdot FS \cdot P}{4\pi\rho\beta^3} \tag{8.2}$$

$R_{\vartheta\phi}$ is the mean radiation pattern, which is assumed to be 0.63 for shear waves; *FS* is the amplification due to the free surface (taken as 2.0 here); *P* is the reduction factor that accounts for the partitioning of the energy into two horizontal components (taken as $1/\sqrt{2}$ here); and ρ and β are the density and shear-wave velocity.

The shape of the source spectrum is the classical ω^{-2} Brune (1970) model with a single corner frequency f_c:

$$S(f, fc) = \frac{f^2}{1 + \left(\frac{f}{f_c}\right)^2}. \tag{8.3}$$

For the geometric spreading, due to the aim of the present study, it was assumed to have a functional form that accounts for the *S*-wave geometric spreading up to 300 km, and then for the surface-wave geometric spreading. This is specified through a distance function *g(r)* that follows the following expression:

$$g(r) = \begin{cases} r^{-1.0} & 1 \le r \le 300\, km \\ r^{-0.5} & r > 300\, km \end{cases}. \tag{8.4}$$

The effect of anelastic attenuation was considered by allowing for a frequency dependence of the type:

$$Q(f) = Q_0 \left(f / f_0 \right)^n \tag{8.5}$$

where f_0 is a reference frequency, and n is a parameter controlling the difference in the attenuation of low and high frequency components. Following previous studies for the same region (i.e., Rovelli et al. 1988; Malagnini et al. 2000), f_0 was assumed to be 1.0 Hz and n was fixed at 1.0 [both of these previous studies assumed f_0 to be 1.0 Hz, while Malagnini et al. (2000) assumed n to be 0.1].

The selection of the Q_0 reference value in Eq. (8.5) was based on the analysis of the results obtained from the P-wave spectra inversion, as shown in Fig. 8.5. From Fig. 8.5, a distance dependence of the Q factor can be seen, which must be taken into account in Eq. (8.5). As mentioned in the previous section, this result can be ascribed to an intrinsic trade-off between the implicitly assumed $1/r$ geometric spreading and a frequency independence of the Q factor in the spectrum inversion. As a consequence, Eq. (8.5) was reformulated to the following expression:

$$Q(r, f) = (Q_0 + Kr) \left(f / f_0 \right)^n . \tag{8.6}$$

This can simply be introduced into the Boore (1983) simulation technique. In particular, Q_o and K were estimated using linear regression analysis, and were 77 ±13 and 2.10 ±0.01 km^{-1}, respectively. No kappa filter (Anderson and Hough 1984) was used to account for attenuation of ground motion due to surface layers because of the aim of the present study: to obtain ground motion estimates relative to rock-site conditions.

The last parameters to be specified for the simulation are the seismic moment M_0 and static stress drop $\Delta\sigma$, which can be considered either magnitude dependent or independent. Based on the results shown in Fig. 8.4a, a magnitude independence can be assumed for $\Delta\sigma$. On the other hand, for each fixed magnitude, three values of stress drop were used. The median value and the values corresponding to ±1 standard deviation (see Fig. 8.4a) were selected to take into account uncertainties in the estimated stress drop values and to introduce further variability into the peak ground-motion values.

The synthetic database for Pga and Pgv for M 5, 6 and 7 and for distances ranging between 5 and 150 km is shown in Fig. 8.6a, b, respectively. The values were calculated with reference to rock-site conditions. To partially verify the correctness of the parameters selected for the stochastic simulation procedure, the same figure includes the Pga and Pgv values (black crosses) corresponding to two aftershocks of the 23 November 1980/18:34 M 6.9 Irpinia earthquake: the 01 December 1980/19:04 M 4.6 and 16 January 1981/00:37 M 4.7 earthquakes. As seen from Fig. 8.6, the simulated peak values are in good agreement with the observed data for both Pga and Pgv. The generated database allowed us to formulate the attenuation model and compute the coefficients for predicting strong ground motion.

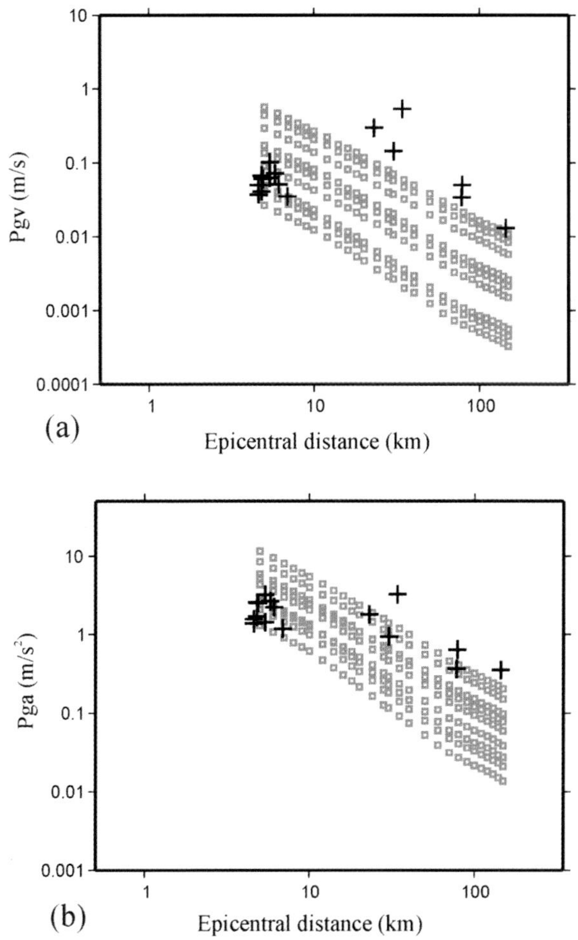

Fig. 8.6 (**a**, **b**) Synthetic databases for Pgv and Pga, respectively, as functions of the epicentral distances for M 5, 6 and 7. The crosses refer to the data of the November 1980/18:34 M 6.9 Irpinia earthquake and the 01 December 1980/19:04 M 4.6 and 16 January 1981/00:37; M 4.7 aftershocks.

8.4 Regression Analysis

Once the simulation had been calibrated for the region of interest, in terms of attenuation (geometric and anelastic) and static stress drop conditions, a synthetic database for a fixed range of magnitude and distances of interest was generated. Such synthetic data can be used both to integrate existing databases and to retrieve *ad-hoc* regional attenuation relations for strong motion parameters.

This approach was preferred to those based on attenuation models for the Fourier spectrum (Toro et al. 1997; Rovelli et al. 1988; Malagnini et al. 2000) due to the aim of the present study. In fact, for early-warning applications, scenarios that represent parameters which can be correlated to structural damage need to be provided rapidly to the civil protection authorities. Although the response spectrum provides the most interesting engineering parameters, Pga and Pgv still remain the most commonly used.

In the present study, a synthetic database was used to retrieve an *ad-hoc* attenuation relation that can be easily upgraded when new data are available, whether during the earthquake occurrence or in a post-event stage, for the formulation of shake maps.

A classical attenuation model for the peak ground parameter was assumed (e.g., Joyner and Boore 1981; Campbell 1997; Sabetta and Pugliese 1996; Abrhamson and Silva 1997; Boore et al. 1997), the formulation for which is as follows:

$$\log_{10} Pgx = a + bM + c\log_{10}\sqrt{R^2 + h^2} \pm \sigma \tag{8.7}$$

where Pgx corresponds to both Pga and Pgv; M is magnitude; R is the epicentral distance in kilometres; h is a fictitious depth value in kilometres; and σ is the standard deviation of the logarithm of Pgx. The selected model assumes an exponential function of the magnitude that comes from the basic definition of magnitude as the logarithmic measure of the ground-motion amplitude (Campbell 1985), and a geometric spreading *1/R*, and h accounts for the property normally referred to as "saturation with distance" (Joyner and Boore 1981; Campbell 1985). The model assumed does not account for site effect, in so far as this effect is introduced *a posteriori* in the shake maps using particular soil classifications (e.g., QTM) and coefficients of amplification for the selected strong ground-motion parameter (Wald et al. 1999).

To fit the database for each magnitude, the c parameter was fixed during the inversion, while the a and b parameters were allowed to vary. Using a trial-and-error procedure, c was found to be -1.4 and h was found to be 5.5

km for Pga and 5.0 km for Pgv. The difference in the h values accounts for the different attenuation when the different ranges of frequencies to which Pga and Pgv belong are considered.

The best estimates for parameters a and b, with the relative uncertainties obtained using the ordinary least-squares analysis, are given in Table 8.1, along with the standard deviation of the logarithm of Pgx.

Table 8.1 Regression coefficients and standard errors of Eq. (8.7) for Pga and Pgv.

Pgx	a	b	c	h	σ
Pga (m/s^2)	-0.514 ± 0.007	0.347 ± 0.001	-1.4	5.5	0.145
Pgv (m/s)	-3.04 ± 0.01	0.552 ± 0.002	-1.4	5.0	0.154

Figures 8.7 and 8.8 show the attenuation relations for M 5, 6 and 7 (continuous lines) and the databases for Pga and Pgv (grey squares), respectively. For comparison, the other two attenuation laws for the same magnitude are given on the same figures. The dotted lines represent the attenuation curves of Sabetta and Pugliese (1996) (hereinafter: SP96), while the bold dashed lines represent the attenuation curves of Campbell (1997) (hereinafter: CA97). The selection of these two attenuation relations was based on the fact that the SP96 attenuation relation was retrieved from an Italian strong motion database that contained chiefly normal and thrust fault mechanisms, while the CA97 attenuation relation was retrieved from a worldwide database that included different seismotectonic environments and a larger number of faulting styles. The comparisons between the different curves show that the attenuation relation retrieved in the present study and in CA97 have similar behaviours for distances greater than 20 km in terms of attenuation, but they provide different ground-motion estimates. Thus, CA97 provides larger estimates for Pga and Pgv for each magnitude, except for Pgv and M 7. By contrast, comparison with SP96 shows different behaviours in terms of both the peak ground-motion estimates and the trend of the attenuation. This further stresses the need to retrieve attenuation relations from data collected in the same region in which the estimates are performed.

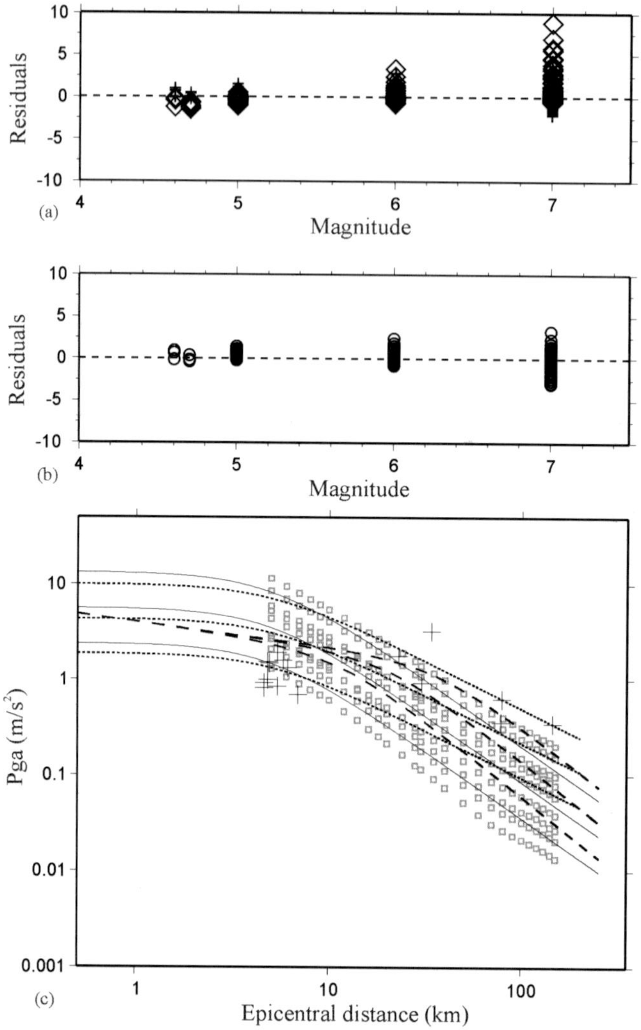

Fig. 8.7 (**a**) Residual analysis for the SP96 (crosses) and CA97 (rhombuses) attenuation relations. (**b**) Residual analysis for the attenuation relation retrieved in the present study. (**c**) Attenuation curves for Pga for M 5, 6 and 7: The continuous lines correspond to the attenuation relation retrieved in the present study, the dotted to that of SP96, and the bold dashed to CA97.

Panels b and c of Figs. 8.7 and 8.8 show the residuals (observed minus predicted Pga values) versus magnitude for each of the attenuation relations, where the circles in panels b represent the residuals relative to the attenuation relation calculated in the present study, with panels c showing the residuals relative to the SP96 (crosses) and CA97 (diamonds) attenuation relations. The results show that the attenuation relation inferred in the present study provides estimates that are consistent with those of SP96, but smaller than those of CA97, in particular, for larger magnitude.

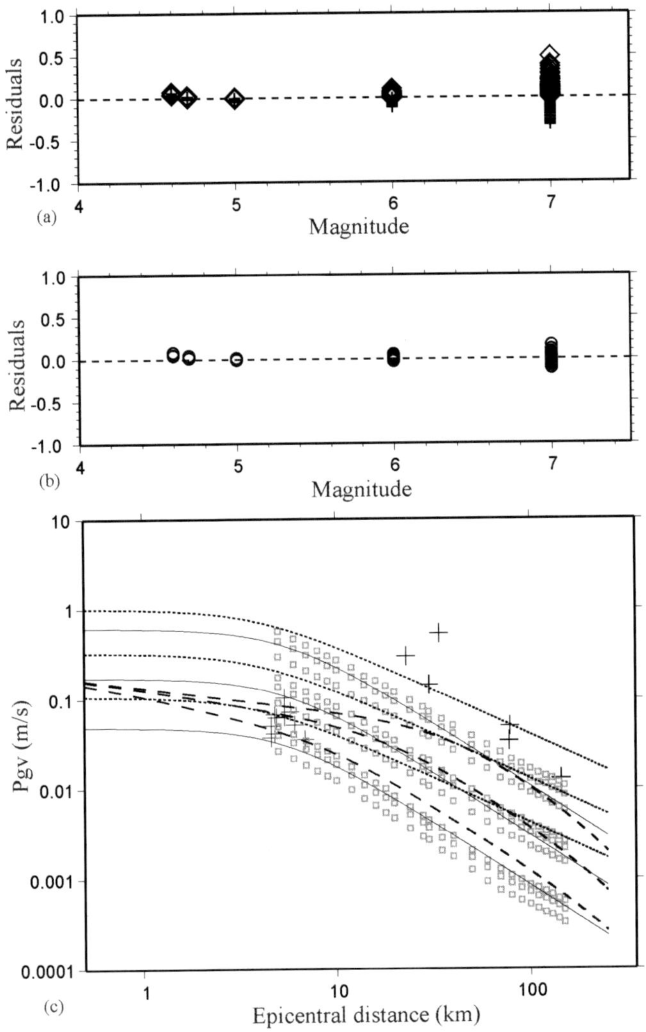

Fig. 8.8 As for Fig. 8.7, but for Pgv.

To validate the assumed regression model of Eq. (8.7) and the computed coefficients given in Table 8.1, a further test was performed using the data relative to the 23 November 1980/18:34 M 6.9 Irpinia earthquake, where residuals analysis was re-performed in a two-stage approach. In the first stage, the estimates obtained using the retrieved regression model in the present study for magnitude M 6.9 and epicentral distances for which data are available were compared. The results are shown in Figs. 8.9 and 8.10 for Pga and Pgv, respectively. In panels a of both these figures, the triangles represent the available data and the circles represent the estimates, for the same distances, obtained with the attenuation relation retrieved in the present study. On the same panels, the estimates obtained using the attenuation relations of SP97 (crosses) and CA97 (diamonds) are also shown, while the lines refer to a continuous range of distances and a magnitude M 6.9. The style used for these lines corresponds to that shown in Fig. 8.7. The corresponding residuals are shown in Figs. 8.9b and 8.10b.

In the second stage of the analysis, the data of the 23 November 1980/18:34 M 6.9 Irpinia earthquake were added to the synthetic database and the regression analysis re-performed, to have the same starting information for the three different attenuation relations. The new retrieved coefficients and relative uncertainties are given in Table 8.2, both for Pga and Pgv. It can be seen that there are only slight variations in the coefficients, while the uncertainties remain the same. The corresponding attenuation curves for Pga and Pgv are shown in Figs. 8.9 and 8.10, respectively (dashed bold lines). The estimates for each single data point and the residuals are also given (black inverted triangles on panels b of the same figures). As expected, from the analysis of the residuals, the introduction of the data relative to the 23 November 1980/18:34 M 6.9 Irpinia earthquake into the synthetic database partially improved the estimates.

Table 8.2 Regression coefficients and standard errors of Eq. (8.7) for Pga and Pgv following the addition to the synthetic database of the data from the 23 November 1980/18:34 M 6.9 Irpinia earthquake.

Pgx	a	b	c	h	σ
Pga (m/s^2)	-0.559 ± 0.007	0.383 ± 0.001	-1.4	5.5	0.155
Pgv (m/s)	-3.13 ± 0.01	0.570 ± 0.002	-1.4	5.0	0.185

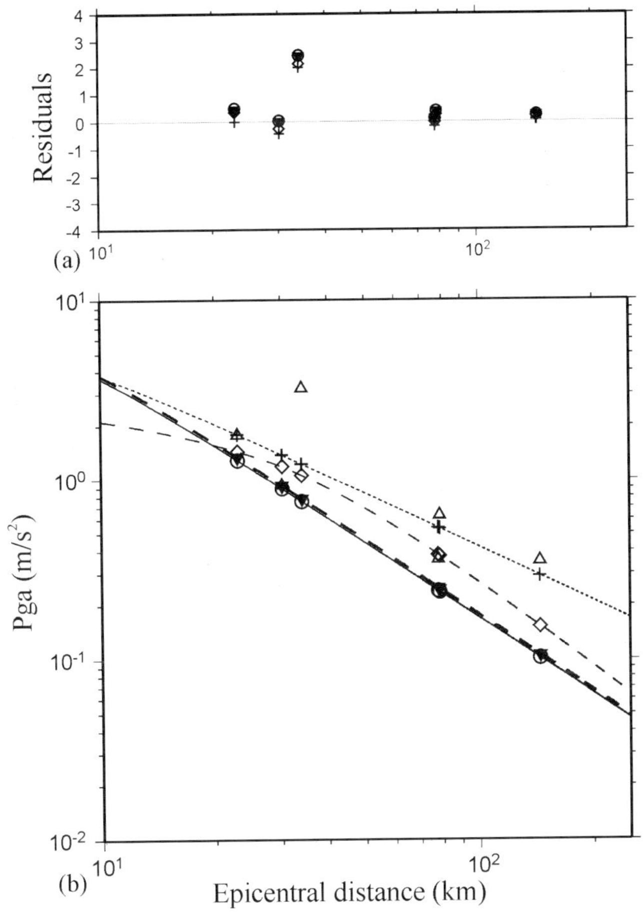

Fig. 8.9 Validation test for Pga: (**a**) Residual analysis for the validation test. Symbols as for panel (**b**). (**b**) The triangles represent the observed data of the November 1980/18:34 M 6.9 Irpinia earthquake, the circles represent the estimates of the attenuation relation retrieved in the present study, the crosses correspond to the estimates of the SP96 attenuation relation, and the rhombuses correspond to the estimates of the CA97 attenuation relation. The inverted black triangles correspond to the estimates of the attenuation relation retrieved in the present study when the observed data of the November 1980/18:34 M 6.9 Irpinia earthquake were added to the synthetic database. The lines show the continuous ranges of distances; see legend to Fig. 8.7 for details.

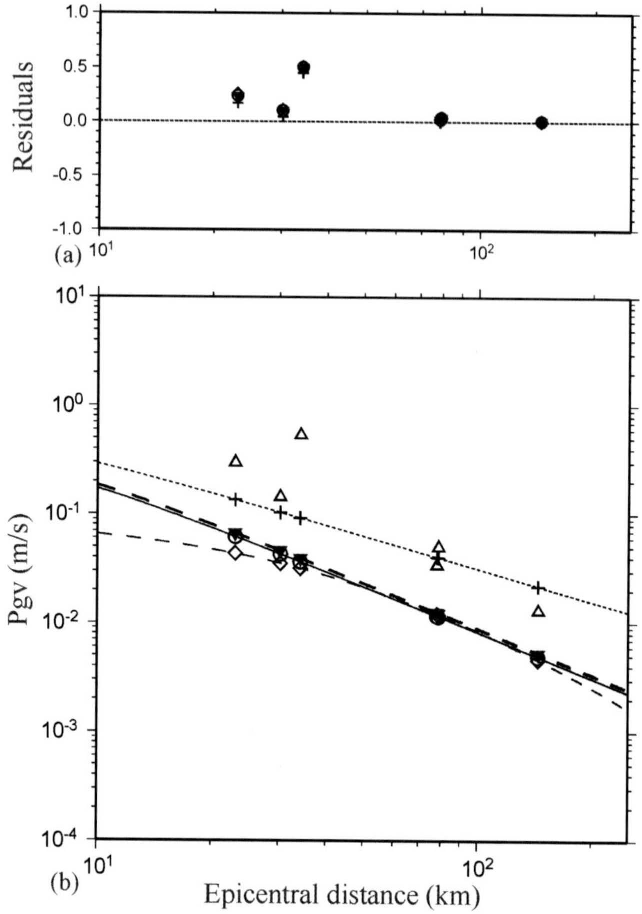

Fig. 8.10 As for Fig. 8.9, but for Pgv.

8.5 Conclusions

In the present study, an *ad-hoc* attenuation relation for the Campania region of the Southern Apennines (Italy) was retrieved. The regression model is reported in Eq. (8.7) and the coefficients to use for the prediction are reported in Table 8.2. The aim of the present study was driven by ongoing realization of a dense seismic array in Campania for early warning and rapid shake map evaluation purposes.

Due to the absence of a large and complete strong motion database, spectral analysis was performed on earthquake waveforms recorded from 1988-2003 by the INGV network for a magnitude range M_d (1.5,5.0). This allowed scaling laws to be obtained, i.e., seismic moment vs corner frequency, and static stress drop vs corner frequency, which can be extrapolated for larger magnitude values. The inferred parameters were then used as the input for the stochastic simulation technique proposed by Boore (1983), to calculate the synthetic waveforms and the associated Pga and Pgv values for the magnitude range of interest (M 5, 6 and 7).

Once the database had been built, regression analysis and comparison tests of the residuals were performed. The estimates obtained by the attenuation relation retrieved in the present study were compared with those obtained by two other existing relations. The results show that Pga and Pgv are substantially in agreement, while there are different trends in the attenuation (geometric and anelastic), thus stressing the need to retrieve attenuation relations with data that are collected in the same region in which the estimates are to be performed.

8.6 Acknowledgments

The figures were prepared with Generic Mapping Tools (Wessel and Smith 1991).

References

Abrahamson NA, Silva WJ (1997) Empirical response spectral attenuation relations for shallow crustal earthquakes. Seism Res Lett 68:94-127

Anderson JG, Hough SE (1984) A model for the shape of the Fourier amplitude spectrum acceleration at high frequencies. Bull Seism Soc Am 74:1969-1993

Bernard P, Zollo A (1989) The Irpinia (Italy) 1980 earthquake: detailed analysis of a complex normal faulting. J Geophys Res 94:1631-1648

Boore DM (1983) Stochastic simulation of high-frequency ground motion based on seismological models of the radiated spectra. Bull Seism Soc Am 73:1865-1893

Boore DM, Joyner WB, Fumal TE (1997) Equations for estimating horizontal response spectra and peak acceleration from Western North American earthquakes: a summary of recent work. Seism Res Lett 68:128-153

Brune J (1970) Tectonic stress and spectra of seismic shear waves from earthquakes. J Geophys Res 75:4997-5009

Campbell KW, Eeri M (1985) Strong Motion Attenuation Relations: A Ten-Years Perspective. Earthquake Spectra 1(4)

Campbell KW (1997) Empirical Near-Source Attenuation Relationships for Horizontal and Vertical Components of Peak Ground Acceleration, Peak Ground

Velocity, and Pseudo-Absolute Acceleration Spectra. Seism Res Lett 68;154-179

Joyner WB, Boore DM (1981) Peak horizontal acceleration and velocity from strong-motion records including records from the 1979 Imperial Valley, California earthquake. Bull Seism Soc Am 71:2011-2038

Malagnigni L, Herrmann RB, Di Bona M (2000) Ground-Motion Scaling in the Apennines (Italy). Bull Seism Soc Am 90:1062-1081

Nelder JA, Mead R (1965) A simplex method for function minimization. Computer Journal 7:308

Reiter L (1990) Earthquake hazard analysis-Issues and Insights. Columbia University Press, New York, 254 pp

Rovelli A, Bonamassa O, Cocco M, Di Bona M, Mazza S (1988) Scaling laws and spectral parameters of the round motion in active extensional areas in Italy. Bull Seism Soc Am 78:530-560

Sabetta F, Pugliese A (1987) Attenuation of peak horizontal acceleration and velocity from Italian strong-motion records. Bull Seism Soc Am 77:1491-1513

Sabetta F, Pugliese A (1996) Estimation of Response Spectra and Simulation of Nonstationary Earthquake Ground Motions. Bull Seism Soc Am 86:337-352

Toro R, Abrahamson NA, Schneider JF (1997) Model of Strong Ground Motions from Earthquakes in Central and Eastern North America: Best Estimates and Uncertainties. Seism Res Lett 68:41-57

Wald DJ, Quitoriano V, Heaton TH, Kanamori H, Scrivner CW, Worden BC (1999) TriNet "ShakeMaps": Rapid Generation of Peak Ground Motion and Intensity Maps for Earthquake in Southern California. Earthquake Spectra 15:537-555

Weber E, Iannaccone G, Zollo A, Bobbio A, Cantore L, Corciulo M, Convertito V, Di Crosta M, Elia L, Emolo A, Martino C, Romeo A, Satriano C (2007) Development and testing of an advanced monitoring infrastructure (ISNet) for seismic early-warning applications in the Campania region of southern Italy. In: Gasparini P, Manfredi G, Zschau J (eds) Earthquake Early Warning Systems. Springer

Wessel P, Smith WHF (1991) Free software helps map and display data, EOS Trans. AGU 72(441):445-446

9 Quantitative Seismic Hazard Assessment

Jean Virieux[1], Pierre-Yves Bard[2], Hormoz Modaressi[3]

[1] Géosciences Azur, Valbonne, France
[2] LGIT – Maison des Géosciences, Saint-Martin-d'Hères, France
[3] BRGM, ARN, Orléans, France

Abstract

We analyze the ingredients required for deterministic wave propagation simulation in order to estimate more accurately ground motion in terms of amplitude, frequency content and duration. Building maps of expected ground motion before a catastrophic event for various scenarios may help design ways to mitigate impacts of ground vibrations as well as fast calibration of these maps once the event occurs. Reconstruction of 3D structures requires collection of information at two different scales: the regional one (a few tens of kilometres) and the local (a few tens of metres to hundreds of metres). Different techniques from permanent to temporary deployments of seismic stations and from active to passive source excitations together with other geophysical and geotechnical investigations will provide the necessary information. Characterisation of possible seismic sources is another challenge and requires careful seismotectonic analysis in the region of interest. Uncertainties may be large provided that one can reproduce wave propagation from these hypothetical spatially extended sources. Frequency content of these simulations is more limited by our knowledge of the medium than by computer resources. In fact, in order to perform such simulations, the geological structure has to be known within a resolution scale of one-tenth of the wavelength. Duration and amplitude are affected by the source mechanism and the mechanical properties of the underground structure. Variability in these ground motion estimations should be appreciated and key parameters identified. Designing approaches to calibrate these simulations with recorded data as well as necessary links with the probabilistic approach should be addressed in the future.

9.1 Introduction

Because building collapse induced by ground vibratory motion during destructive earthquakes is the main cause of casualties, correct *a priori* estimation of this ground motion is therefore a necessary, but in fact not sufficient, element of an efficient risk mitigation policy. Strategies for quantitative seismic ground estimation go from purely deterministic ones to more statistical ones.

We tentatively address issues of required observations in order to constrain the crustal structure description where waves are propagating and in order to build appropriate finite-extended source mechanisms. We shall describe different complementary approaches for better ground-motion estimation. Geophysical and geotechnical measurements may be used for model reconstruction. Both recorded seismograms of significant earthquakes and possible extrapolation of recorded seismograms from small events will provide data that a wave propagation tool might describe in a first step and might assimilate in a second more ambitious step.

For given specific frequency contents – presently still limited to low frequencies, i.e., generally below 1 Hz – numerical methods may provide estimation of ground motion using purely deterministic approaches. We shall examine the potentials of different methods which seem to lead to quite realistic seismograms once source and medium are specified at a given resolution, thanks to the impressive increase in computer efficiency.

Three major questions are still open and will be tentatively addressed in this paper, and illustrated with a few examples:

- What is the upper frequency limit of this approach, knowing that it depends on our capacity of accurate description of the propagation medium and the seismic source?
- How should such ground motion estimations be validated and what confidence should we associate to them?
- How should these deterministic approaches be combined with more stochastic approaches which seem necessary at higher frequencies where reconstruction of accurate models is not possible?

9.2 Crustal and Surface Velocity Reconstruction Using Passive and Active Data Acquisition Systems

Basically, two scales are essential for characterising the medium for ground shaking estimation: the local scale and the regional scale, respectively from a few tens of metres to a few kilometres.

Amplitudes of waves from the source up to the ground surface depend on how waves are travelling in the geological structure at a regional scale. Variable velocity contrasts with depth may lead to significant variation in wave path and amplitude. For example, channelling waves at the continental/margin boundary may change incoming waves impacting a city near a coastline. Figure 9.1 shows the density of seismic acquisition near the coast in the Ligurian Sea. Due to limited information at this scale, the reconstruction of 3D crustal structures will still be limited to long wavelength variations only and will rely on geological as well as geophysical information like gravity measurements (Truffet et al. 1993) or active and passive seismic investigations when available (Le Meur et al. 1997, Latorre et al. 2004, Paul et al. 2001). Consequently, permanent seismic networks as well as temporary networks will bring valuable information from both local seismic activity and teleseismic sources. Receiver function techniques (Bertrand 1999, Bertrand et al. 2002) will increase our knowledge of regional structures. New methods based on cross-correlation of noise signals as proposed by Shapiro et al. (2005) will allow reconstruction at that regional scale without requiring local seismic sources. Of course, seismic wave excitation will come from generic sources as oceanic storms.

Superficial local propagation and site effects contribute to the complexity of seismograms and require accurate description of tens to a few hundred metres below the surface. The largest mechanical contrasts observed in the weathered layer and sedimentary cover zone strongly alter the propagation of elastic waves (focussing, diffraction and so on), modifying relative amplitude contents of different frequency ranges with respect to what one expects from simple geometrical propagation with simple dissipative mechanisms. These modulations affect and often increase ground-motion amplitude for a specific frequency range before the shaking of buildings. Moreover, as surface soils may behave non-linearly under large strains (a few tenths of percents to a few percents), precise characterisation of this layer requires numerous parameters for adequate modelling of its complex rheological behaviour, especially the very superficial part (upper 20-30 metres). Collecting information from boreholes is usually insufficient because of the limited sampling and uncertainties associated to spatial data interpolation. Moreover, the standard geotechnical/geophysical investigations such as cross-hole or down-hole may introduce bias due to privileged wave paths.

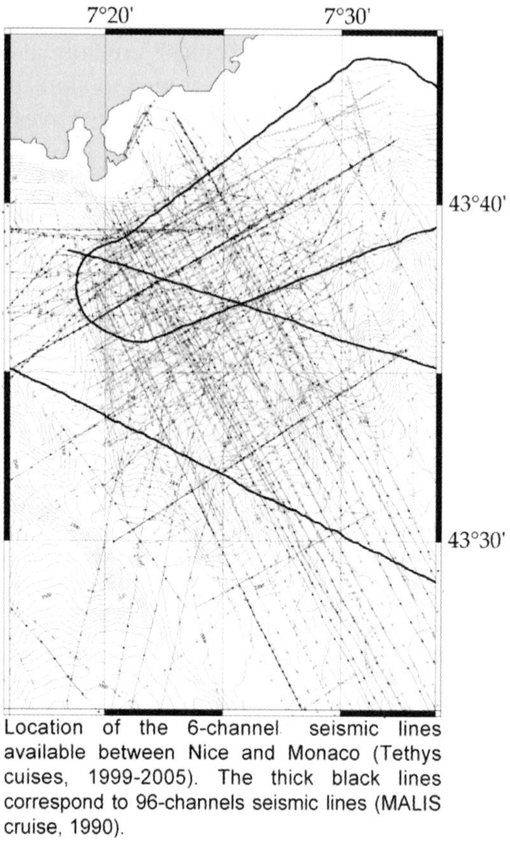

Fig. 9.1 High spatial sampling of the Ligurian Sea by high-resolution seismic profiles (Courtesy of Françoise Sage).

As an alternative, one may consider deriving quantitative properties at shallow depths from active seismic experiments when available (Zollo et al. 2005) as well as passive seismic experiments (Tiziana et al. 2005). Seismic analysis of surface waves (Stokoe 1989, Park 1999, Miller 1999, Liu 2000) allows reconstruction of vertical profiles. Whatever the reconstruction technique, this definitely requires very dense instrumentation. In particular, promising developments are under way to retrieve the shear wave velocity profile from careful analysis of the microtremor wavefield, through various techniques, either very simple - but sometimes tricky and misleading - such as the H/V technique (Nogoshi and Igarashi 1971, Nakamura 1989, Kudo 1995, Bard 1998) or more advanced such as array techniques (Aki 1957, Horike 1985, Matsushima and Okada 1990, Tokimatsu et al. 1992, Tokimatsu 1997, Cornou 2002, Sèbe 2004, Le Bihan

2001, Schissele 2002, Schissele et al. 2004, Arai and Tokimatsu 2004, Ohrnberger et al. 2004, Wathelet et al. 2005).

High-resolution seismic profiles allow similarly at sea a precise description of different geological units (Fig. 9.2) which allows a 3D reconstruction of the shallow subsurface of about a few hundred metres. As the density of sensors increases, one may extract more information from seismograms based on lateral coherence and redundancy of converted phases. Sharp local variations in impedance for velocity may be useful for constraining model reconstruction. Both migration techniques for active sources in complex structures (Pratt et al 1998, Operto et al. 2004) and converted phase analysis (Latorre et al. 2004) will improve the resolution of our reconstruction as is done for reservoir characterisation and monitoring.

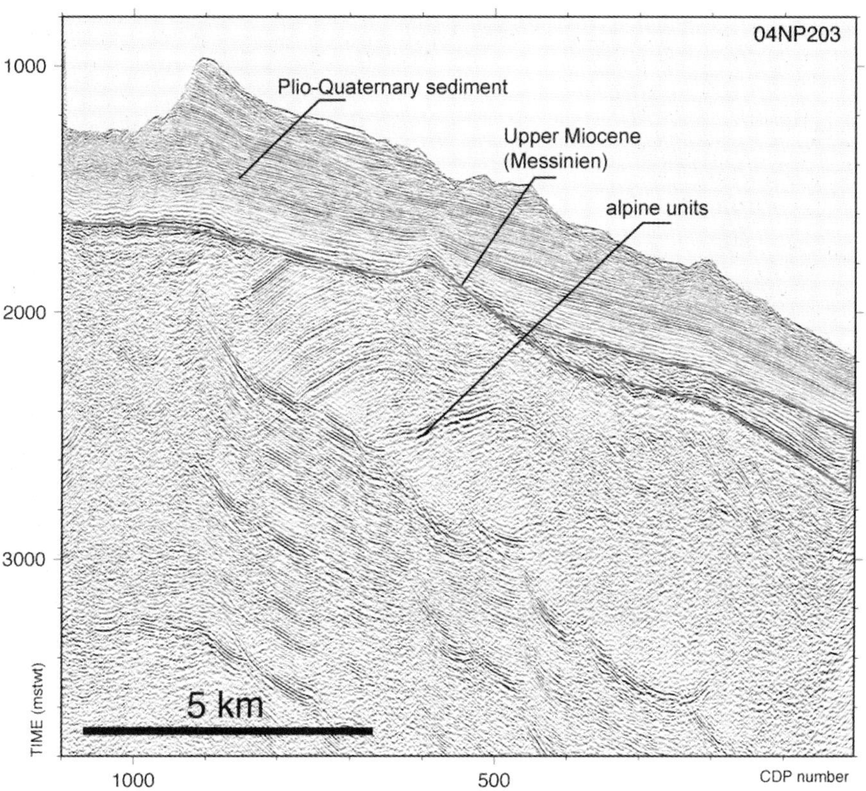

6-channel seismic line showing the seaward continuation of the alpine structures over the North Ligurian Margin.

Fig. 9.2 A description of the superficial structure in a time migration section useful for geological unit identification. This information is used for 3D model reconstruction.

In whatever way we extract the information for the description of a deterministic 3D model, this information must be inserted using a dynamic updating strategy based on combining different pieces of information from various origins with different qualities, resolutions and uncertainties (Nivlet et al. 2002, Caimon et al. 2004, Mallet 2004, Castanié et al. 2005). These model generators (http://www.geomodeller.com) should be able to integrate various pieces of information from geological to geophysical ones and must output what is needed for adequate wave propagation modelling: as an example, heterogeneous medium description is provided for the Nice city zone (Fig. 9.3). As knowledge increases in the studied area, one could upgrade the information inside the generator and provide a new model reconstruction for better modelling.

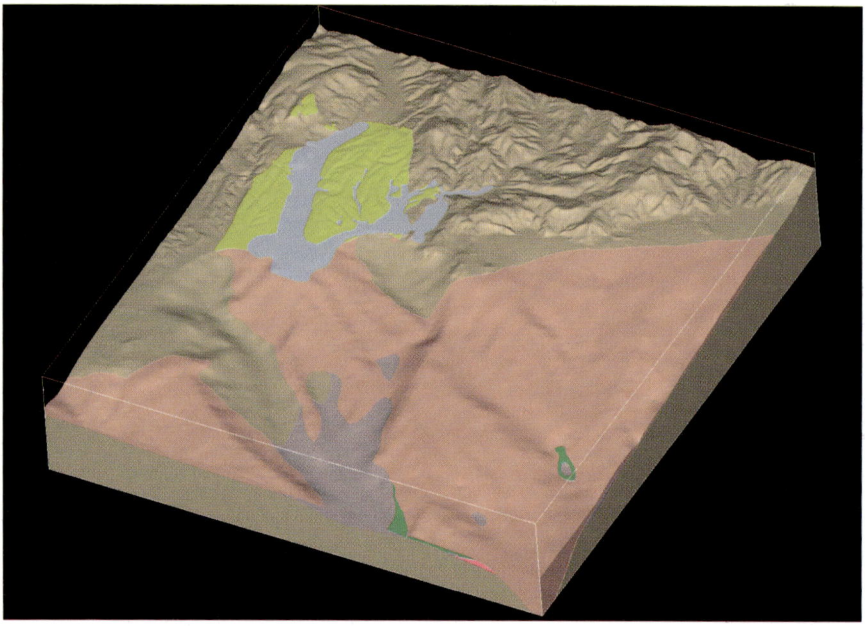

Fig. 9.3 Model description using a geomodeller of the area of Nice in France (Courtesy of BRGM).

9.3 Source Description for a Given Seismo-tectonic Zone

Whatever the medium description, we need a quantified estimation of expected sources for accurate medium excitation and wave propagation simulation. Both the geometry of the sliding zone as well as the distribution of the slip should be provided. Model numerical constraints may induce other specifications such as meshing density or the limited size of the box for simulation although they could be reduced as improved modelling is constructed.

Seismo-tectonic description must be set up in such a way as to perform this estimation of possible sources. Active fault identification is an essential ingredient in this source definition, and bridges from palaeoseismicity (Michetti et al. 2005) towards historical seismicity (Lambert and Levret-Albaret 1995) will certainly be of crucial importance as well as the tectonic framework of the target area. Uncertainties in the geometry of the faulting source as well as on the slip distribution should be provided as simulations may partially overcome this knowledge problem by performing various scenarios. One may estimate ground-motion variability at the expense of an intensive computer effort. Of course, narrowing this variability by better constraining possible seismic source scenario in a given area is of crucial importance because of the economic impact of such estimation for land management.

The current state of the art for direct source modelling is that, while low frequency seismograms are computed using kinematic or dynamic source models, high frequencies are modelled using random oscillators. A major difficulty of these simulations is to predict a physically correct directivity of the radiation at all frequencies. The threshold frequency limit between these two modelling issues should be identified but will certainly range around a few hertz. Therefore, an integrated strategy should find a way to combine these results in a common framework. This analysis will certainly put constraints on the seismic source description: small features may unnecessary for wave propagation simulations and, therefore, could be avoided.

This frequency limit is related to the insufficient knowledge of the crustal structure and source kinematics at short wavelength and consequently at high frequencies, leading us to a value around a few hertz. Alternative strategies might be the extrapolation of small events in order to estimate large hypothetical earthquakes. When recorded, associated seismograms could be used as empirical Green functions where model reconstruction is impossible as long as seismic sources are correctly described. For this purpose, tentative accurate description of small earthquakes with better localisation and an accurate focal mechanism must be performed. Magnitude estimation is essential and should be accurately estimated for seismic moment conversion. Studies of small earthquakes when recorded will help for more precise seismogenic characteristics of the investigated seismic active zone. Their source characteristics could be reconstructed from a few unsaturated seismograms (Delouis et al. 2002), providing key insights into the definition of the expected main event.

Moreover, these seismograms may help as well for adequate verification of the ground motion estimations as modelling tools are improved. Of course, differences in mechanical characteristics and behaviours between small and significant earthquakes will still be an open question which could be tackled where great earthquakes are recorded.

What will the feasible strategy be in the future: shall we rely on accurate kinematic source model where slip distributions are adequately reconstructed in time on the fault plane (Hartzell and Heaton 1983, Archuleta 1984, Beroza and Spudish 1988, Boatwright et al. 1991, Herrero and Bernard 1994, Hartzell et al. 1995, Cotton and Campillo 1995, Couboulex et al. 1997, Cocco et al. 1997, Delouis et al. 2002, Vallée and Bouchon 2004, Emolo and Zollo 2005) or shall we consider dynamic models where friction laws will be considered as well as cohesion distribution (Andrews 1972, Andrews 1976, Madariaga 1976, Virieux and Madariaga 1982, Day 1982, Harris et al. 1991, Cochard and Madariaga 1994, Olsen et al. 1997, Fukuyama and Maradiaga 1998, Madariaga et al. 1998, Oglesby et al. 2000, Peyrat et al. 2001, Aochi and Fukuyama 2002, Aochi et al. 2003, Aochi and Madariaga 2003, Peyrat et al. 2004). Both approaches are used in the literature for seismic source modelling and imaging although the dynamic approach is still computer-demanding for recovering friction and failure functions on the fault surface.

9.4 Challenging Issues of Seismic Wave Propagation in 3D Heterogeneous Media

Knowledge of crustal structures will increase in the future because new data will provide us accumulative information, mainly collected at the surface of the Earth. For different specific earthquake scenarios, one may proceed for accurate and deterministic ground-motion simulation using various techniques. Up to now, no specific technique well suited to seismic ground-motion modelling has yet emerged while climate simulation has focused on pseudo-spectral methods. These techniques have been proved to be a fairly reliable, efficient and widespread technique for partial differential integration over a spherical volume (Eliasen et al. 1970, Foster et al. 1992). Therefore, for seismic wave propagation at a regional scale, we are still investigating methods and their performances, especially for accurate implementation of extended sources: for example, Day et al. (2005) investigate respective performances of Finite-Different techniques and Boundary-element methods.

As the wave impacts the site zone, one must often consider different mechanisms of attenuation as well as possible non-linear behaviour of soils which might substantially modify ground-motion response. Topography and heterogeneous subsurface structures do increase the complexity of seismograms. Modelling should certainly take these points into consideration.

If non-linearity plays an important role, we must cautiously use seismograms of small earthquakes for empirical estimation of the ground motion for bigger earthquakes. Precise extrapolation methodologies should be constructed and tested on already available data, although there is little space to comment further in this direction.

Three standard techniques have been used by different groups for seismic wave propagation simulation. Boundary integral methods are quite appealing because they reduce by one the dimension of the discretization. Finite difference and the more versatile finite volume techniques rely on very simple principles which make these techniques quite efficient. Finally, finite element approaches handle complex geometries and the specific spectral element method allows precise simulations. Emerging techniques based on micro-scale description of simple interactions between discrete elements may mimic wave propagation features and may allow extension to more complex porous media and non-linear behaviours.

Finally, a possible strategy based on coupling between different numerical techniques may be recommended in order to benefit from the advantages of each of these techniques and to avoid their disadvantages. For instance, Aochi and co-workers (2003, 2004) model the seismic non-planar source by boundary elements imbedded in a 3D structure discretised by the finite difference method. The non-linear shallow sediments are modelled by a finite element technique where impedance on its boundary is provided by a large scale finite difference computation (Aochi et al. 2005).

9.4.1 Boundary Integral Equations

Boundary Integral methods require discretization of surfaces delimiting zones where properties of the medium are such that the Green's functions can be computed analytically. How these surfaces are discretized introduces differences between different formulations (Aki and Larner 1970, Sanchez-Sesma 1983, Bard and Bouchon 1985, Aubry and Clouteau 1991, Sanchez-Sesma and Luzon 1995). Interactions between points of these surfaces lead to singularities when points collapse at the same place. These singularities must be carefully estimated for accurate modelling (Dangla et al. 2005).

9.4.2 Finite Difference-Finite Volume Methods

Finite-difference techniques are very popular because of design simplicity related to numerical efficiency which is critical for considering 3D geometries [see Moczo et al. (2006) for a review on FD methods]. How to discretize the medium is a key issue, especially when considering density and elastic parameters: Graves (1996) has suggested a local harmonic average for better diffraction modelling while Pitarka (1999) among others has introduced spatially irregular grid meshing. Moczo et al. (2004) has carefully analyzed numerical precision for free surface in a 3D configuration. Intensive simulations have been performed by Olsen (2001). Implementation of absorbing boundary conditions has been improved dramatically in the last few years: for example, Marcinkovich and Olsen (2003) have proposed a practical implementation of the Perfectly Matched Layer (PML) method for seismic 3D configuration.

Finite-volume methods have been found more difficult to implement (Dormy and Tarantola 1995) but recent investigations by Käser and Igel (2001) have once again raised interest due to the flexibility of such an approach which has been quite successful in electromagnetism (Piperno et al. 2002).

9.4.3 Finite Element Methods (Spectral Elements Approach)

Finite-element methods have long been applied in seismic wave propagation modelling (Marfurt 1984, Aubry et al. 1985). Parallelised finite-element methods are now available for seismic wave propagation at regional scale (Foerster et al. 2005). New interpolation techniques of high-order based on Gauss-Lobato-Legendre points have shown spectral convergence of solutions while retaining local interpolation [see Chajlub et al. (2006) for references]. Various extensions at the global scale show how powerful this variational formulation is (Komatitsch and Vilotte 1998, Komatitsch and Tromp 2002a, 2002b, Komatitsch et al. 2004). Complex effects such as gravitational ones have been considered by Chajlub and Valette (2004), for example. PML boundary conditions have been implemented as well for defining an infinite medium (Komatitsch and Tromp 2003, Festa and Vilotte 2005).

9.4.4 Discrete Element Methods (Distinct Element/Lattice Approach)

Coming from fluid flow simulation, the lattice approach is suitable when considering complex media (Rothman 1988). Few attempts have been performed when considering seismic wave propagation since one must consider two speeds of propagation. Huang and Mora (1994a) have proposed acoustic formulation with potential complex non-linear behaviour (Huang and Mora 1994b). More recently, Toomey and Bean (2000) were able to reproduce elastic propagation features by considering other simple interactions between elements. Huang and Mora (1996) showed that very heterogeneous media can be considered. Moreover, Toomey et al. (2002) found that fractured medium might be considered as well. At the expense of computer resources, this method seems promising for modelling seismic wave propagation for complex mechanical behaviours of media.

In summary, as computer resources increase dramatically, performing accurate simulations of seismic wave propagation in complex heterogeneous media including extended seismic sources is a goal which can be achieved in the near future. Figure 9.4 displays the geometrical configuration for modelling wave propagation inside a layered medium for the Landers earthquake. The fault surface is a 2D complex vertically strike-slip surface which fits the rupture traces at the free surface. Figure 9.5 displays North particle velocity computed on the free surface of the Earth for the Landers Earthquake. The simulation is performed by a 3D finite-difference code using boundary conditions as defined by Cruz-Atienza and Virieux (2004). Dynamic parameters are those defined by Peyrat et al. (2001). One may see possible shock waves generated as the crack tip moves at a rupture speed higher than the shear wave speed. Synthetic seismograms can be computed anywhere in the model. Dynamic interaction between fault rupture zone and medium heterogeneities increases dramatically the complexity of seismograms.

Unfortunately, ground-motion estimation must combine these simulations and uncertainties in our knowledge of the medium and in our definition of potential seismic sources. Of crucial importance is the strategy for variability estimation coming from our limited knowledge and understanding.

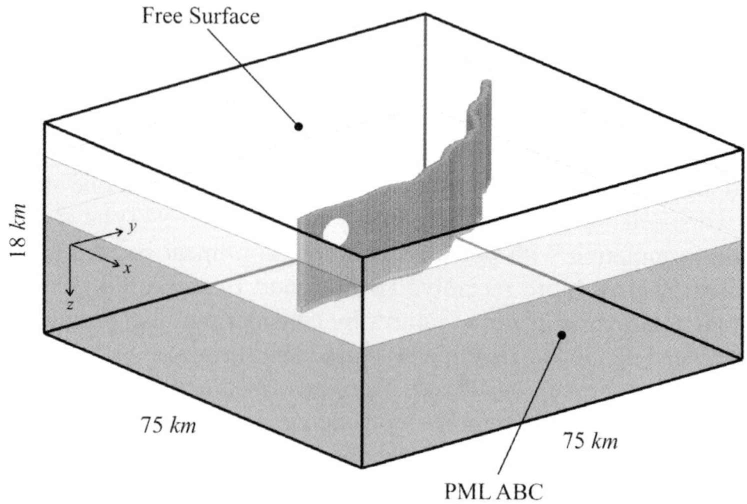

Fig. 9.4 Bloc-diagram of the fault surface of the Landers earthquake embedded in a layered structure. The red zone is the nucleation zone (Courtesy of Victor Cruz-Atienza).

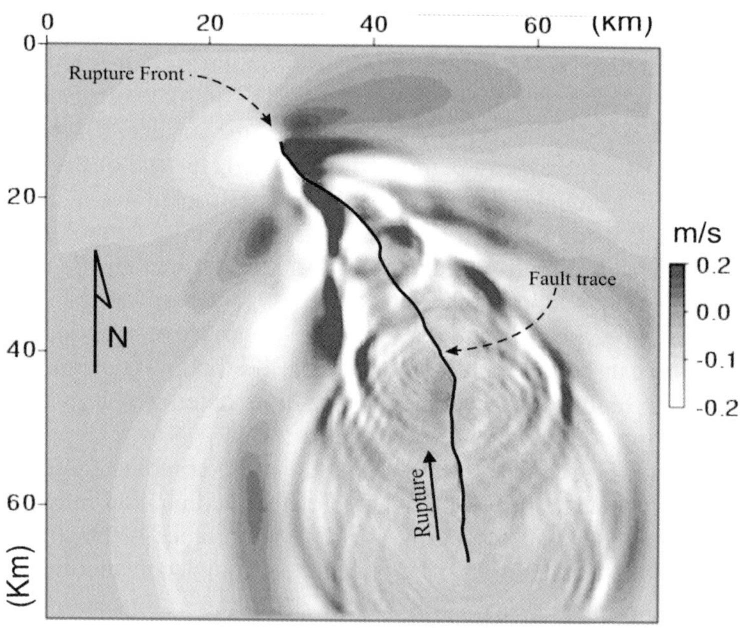

Fig. 9.5 Vertical particle velocity computed at the free surface of the Earth at a given time after the nucleation starts for the Landers earthquake. Dynamic parameters are such that shock waves are emitted by the the fault rupture zone (Courtesy of Victor Cruz-Atienza).

9.5 Quantification of the Dispersion of Ground Motion Estimation

These uncertainties are usually classified in two categories: the epistemic uncertainty is related to our incomplete knowledge (propagation medium for instance), and should be reduced in the future, while aleatory variability is considered to be associated with some quasi-random or stochastic processes (related, for instance, to unpredictable source heterogeneities, short wavelength); this distinction may change in the future since it is hoped that, 100 years from now, knowledge will have increased so much that some variability now considered as "quasi-random" will be explained deterministically. Both types result in significant variability in ground motion estimates, which need to be estimated – though it is a far from easy task.

9.5.1 Deterministic Modelling Approach and Sensitivity Studies

A significant amount of epistemic variability thus arises from our poor knowledge of the medium and of the source in the deterministic modelling approach. "Exact" modelling of the propagation effects would indeed require the knowledge of the geological structure with a resolution scale of one tenth of the wavelength. This is unfortunately unachievable in practice and will probably remain so for decades. Therefore, one has to allow for a random geological model with known mean properties and average geometry and large spatial fluctuations. The effect of these fluctuations is essential in the propagation of the wave field since they control the multiple scattering at high frequency (and thus the attenuation with space and time of the signal), the location of constructive and destructive interferences, and also possibly some focusing or defocusing effects, phenomena all with noticeable consequences for frequencies of engineering interest. Possible approaches based on direct numerical modelling are usually addressed in two ways; classical theories of wave propagation in random media usually consider a reference (often homogeneous) media with a random distribution of scatterers which often act as single diffractors (Antwerpen et al. 2002), whereas others consider a heterogeneous mean medium with strong spatial fluctuations (Krüger et al. 2005). These approaches represent fundamentally different physical assumptions and use different simulation techniques.

A similar comment can also be made for some source parameters where uncertainties in the location of the nucleation point as well as subsequent slip time evolution lead to quite different directivity effects for kinematic sources. Similarly, location of asperities or barriers as well as various friction laws may allow different radiation pattern for dynamic extended sources. Estimating the consequences on ground motion parameters presently requires running many different models to capture the whole variability of source characteristics, which may rapidly become prohibitive in terms of computer time cost. Applying this sensitivity approach to the Empirical Green's Function technique, Pavic et al. (2000) proposed a way to limit the number of runs through the use of the "Latin Hypercube Sampling" technique for the estimation of "standard" uncertainty in source parameters, the resulting final variability in ground motion characteristics is slightly larger than that directly observed in data (for instance in the standard deviation of "empirical attenuation relationships").

Whatever trends are simulated in synthetic ground motion, they should mimic the attenuation law deduced from observations, and their variability as well (standard deviation). As the number of observation points increases and better constraints are given on fitted attenuation law from recorded seismograms, one may hope that source and propagation features in attenuation laws may be reproduced using modelling inside the variability zone.

9.5.2 Variability Estimates in Empirical Approaches

As the knowledge of the medium and the source might be poor, a promising strategy comes from the use of recorded signals, either on a site-specific basis (empirical Green's function technique), or in a statistical way. The latter is indeed the standard engineering practice, and is associated with now rather well-identified and quantified uncertainties ("attenuation relationships" or "Ground-Motion Prediction Equations" – namely GMPE – and their standard deviation).

These GMPEs are derived through a least square fit between a few simple ground motion parameters obtained from a selection of (strong motion) recordings, and a generic equation accounting for magnitude and distance dependence, and possibly site conditions. Hundreds of such GMPEs are presently available in the scientific literature, depending on the original data set, the generic functional form chosen to capture magnitude and distance dependence, and the fitting procedure.

In addition to the inherent (aleatory) uncertainty embedded in the standard deviation, two different GMPEs may lead to largely different ground estimations for the same values of magnitude, distance and site conditions.

Estimating ground motions with such equations thus requires some caution and warnings. To begin with, the independent parameters used, i.e. the magnitude, style-of-faulting, distance metrics and site classes, should be consistent with the original data set of each GMPE. This often implies some parameter conversions (inducing also the appropriate uncertainties involved in each conversion). Several procedures have been proposed to undertake these conversions [Douglas et al. (2004) with the computer code CHEEP or Scherbaum et al. (2005) among others].

Then, in order to truly capture epistemic uncertainties within a GMPE approach, one should never use one single GMPE, but always use several, best suited to one's particular case: Cotton et al. (2005) have proposed a procedure for selecting such "best" GMPEs. The first generation of such "GMPEs" (Ambraseys et al. 1996) used a magnitude-independent spatial decay rate, and is therefore appropriate only for a limited magnitude range [see the review by Douglas et al. (2004), or Ambraseys et al. (2005)]. Taking into account the theoretical investigations by Anderson (2000), a new generation of equations has recently emerged (Abrahamson and Silva 1997, Bragato and Slejko 2005, Cotton et al. 2005, Pousse 2005): they definitively show that a magnitude-dependent rate is required to have reliable ground motion estimations over a broad magnitude range: weak earthquakes have a more rapid spatial decay (faster than $1/R^2$ law)? than large magnitude events (close to 1/R law). An adequate description of site conditions should in principle reduce the data dispersion, but there are two additional sources of uncertainty. The first lies in the classification of sites due to the lack of available information. Secondly, even if the sites are correctly classified within each class, different site amplifications are possible due to the intrinsic limitations of the necessarily simple categories used, and also apparently non-predictable variation in site response (e.g. Boore 2004).

Combining existing deterministic tools used for ground-motion estimation with an optimisation procedure to calibrate the key numerical parameters (e.g. for the constitutive model) and a parametric study, in order to perform simulations involving many variables and (linear and non-linear) constraints that reproduce the observed overall dependence and variability (GMPE), is certainly a direction which has not yet been sufficiently explored.

9.5.3 Probabilistic Approach

In view of engineering applications, there is undoubtedly a strong tendency to prefer probabilistic hazard estimates to purely deterministic ones (even combined with sensitivity studies and assessment of uncertainties): such an approach indeed allows, when convolved with fragility curves, better assessment of the final risk level. Basically a probabilistic seismic hazard analysis leads to "hazard curves" representing the annual probability of a ground-motion parameter (e.g., PGA) to exceed a given level at a given site.

Accounting for uncertainties in PSHA studies implies both automatic accounting for GMPE dispersion, and epistemic uncertainties through logic trees with different branches corresponding to different "plausible" assumptions, which are assigned different weights, which may in addition vary from one expert to another.

Such approaches, which seem appealing and convincing, actually face a number of difficulties that will have to be solved as soon as possible:

- The separation between epistemic uncertainties and aleatory variability is not so clear in practical applications, and there are risks of "double counting" some uncertainties – which in the end result in too pessimistic estimates.
- The application to very low probability levels (down to 10^{-5}, or even 10^{-7}) may result in apparently completely unrealistic values (for instance, up to 5g for pga, up to 10g for spectral acceleration). This is intimately related with the exact shape of the tail distribution of residuals in GMPE, which is still unknown with the presently available data set (see Bommer et al. 2004). Another direction of future effort is to seek physical upper bounds of seismic ground motion.
- Another challenge is to develop a way to "calibrate" PSHA studies on available data, i.e., instrumental recordings over a very short duration (no more than a few decades), (fuzzy) intensities over a few centuries: is it possible to have a model valid both at short and long return periods that can be calibrated only on one side and used at the other side ?

Finally, one of the most difficult issues is to match and to reconcile if possible viewpoints of earth scientists and engineers: the former are very pleased and proud if they can predict the observations within a factor of 2, while economic constraints force the latter to "optimize" their design, which leads them to ask (to urge ?) seismologists to predict ground motion within 10-20% ...

9.6 Discussion and Conclusions

As observations of small and moderate earthquakes become available in many different zones, estimations of ground motion, defined as the seismic hazard prediction, may be performed with increasing accuracy for better economic appraisal. One must be aware, however, that there will still exist for decades a very large level of uncertainty, and that it is also the duty of seismologists to provide the engineering community with a way to quantify this uncertainty and allow for it in engineering applications.

Different strategies have to be investigated and we should identify on what frequency range we may rely for a deterministic approach for probabilistic estimation of ground motion. Above this frequency range, we need to consider more statistical characteristics and information for probabilistic estimation and we should design ways to bridge these two different approaches.

One area to be considered with attention is how to narrow ground-motion prediction as soon as an event has occurred. As information becomes increasingly precise for a specific event, could we use estimations performed by deterministic and probabilistic procedures for better ground estimations of the specific event? If the answer is affirmative, there may be major implications for seismic early warning.

9.7 Acknowledgments

Many thanks are due to those involved in the QSHA project and who made this article possible through fruitful scientific discussions. We are grateful to the partial support of the Agence Nationale de la Recherche through project ANR-05-CATT-011-01.

References

Aki K (1957) Space and time spectra of stationary stochastic waves, with special reference to microtremors. Bull Earthq Res Inst 35:415-456

Aki K, Larner KL (1970) Surface motion of a layered medium having an irregular interface due to incident plane SH waves. J Geophys Res 75:1921-1941

Aki K, Richards PG (1980) Quantitative Seismology: theory and methods. W.H. Freeman

Ambraseys NN, Simpson KA, Bommer JJ (1996) Prediction of horizontal response spectra in Europe. Earthquake Eng. & Structural Dyn 25:371-400

Ambraseys NN, Douglas J, Sarma SK, Smit PM (2005) Equations for the estimation of strong ground motions from shallow crustal earthquakes using data from Europe and the Middle East: Horizontal peak ground acceleration and spectral acceleration. Bull Earthquake Eng 3

Abrahamson NA, Silva WJ (1997) Empirical response spectral attenuation relations for shallow crustal earthquakes. Seism Res Lett 68:94-127

Anderson JG (2000) Expected shape of regressions for ground-motion parameter on rock. Bull Seim Soc Am 90(6B):S42-S52

Andrews DJ (1973) A numerical study of tectonic stress release by underground explosions. Bull Seism Soc Am 63:1375-1391

Andrews DJ (1976) Rupture velocity of plane-strain shear cracks. J Geophys Res 81;5679-5687

Andrews DJ (1985) Dynamic plane-strain shear rupture with a slip-weakening friction law calculated by a boundary integral method. Bull Seism Soc Am 75:1-21

Antwerpen VV van, Mulder WA, Herman GC (2002) Finite-difference modeling of twodimensional elastic wave propagation in cracked media. Geophys J Int 149:169-178

Aochi H, Fukuyama E (2002) Three-dimensional nonplanar simulation of the 1992 Landers earthquake. J Geophys Res 107(B2), doi:10.1029/2000JB000061

Aochi H, Olsen KB (2004) On the effects of non-planar geometry for blind thrust faults on strong ground motion. Pure Appl Geophys, in press

Aochi H, Madariaga R (2003) The 1999 Izmit, Turkey, earthquake: Non-planar fault structure, dynamic rupture process and strong ground motion. Bull Seism Soc Am 93:1249-1266

Aochi H, Fukuyama E, Madariaga R (2003) Constraints of Fault Constitutive Parameters Inferred from Non-planar Fault Modeling. Geochem, Geophys, Geosyst 4(2), doi:10.1029/2001GC000207

Aochi H, Seyedi M, Douglas J, Foerster E, Modaressi H (2005) A complete BIEM-FDM-FEM simulation of an earthquake scenario – dynamic rupture process, seismic wave propagation and site effects, presentation at the Meeting/Conference: General Assembly of the European Geosciences Union

Arai H, Tokimatsu K (2004) S-Wave Velocity Profiling by Inversion of Microtremor H/V Spectrum. Bull Seism Soc Am 94(1):53-63

Archuleta RJ (1984) A Faulting Model for the 1979 Imperial Valley Earthquake. J Geophys Res 89:4559-4585

Aubry D, Clouteau D (1991) A regularized boundary element method for stratified media. In: Cohen G, Halpern L, Joly P (eds) Proceedings of the First International Conference on Mathematical and Numerical Aspects of Wave Propagation Phenomena, Strasbourg, France. SIAM, Philadephia, pp 660-668

Aubry D, Chouvet D, Modaressi H, Mouroux P (1985) Local Amplification of a Seismic Incident Field through an Elastoplastic Sedimentary Valley. In: Kawamoto, Ichikawa (eds) Numerical Methods in Geomechanics. Balkema, pp 421-428

Bard P-Y (1998) Microtremor measurements: a tool for site effect estimation?, State-of-the-art paper. In: Irikura K, Kudo K, Okada H, Satasini T (eds) Effects of Surface Geology on Seismic Motion. Yokohama, Rotterdam, Balkema 3:1251-1279

Bard P-Y, Bouchon M (1985) The two-dimensional resonance of sediment-filled valleys. Bull Seism Soc Am 75:519-541

Beroza GC, Spudich P (1988) Linearized inversion for fault rupture behaviour: application to the 1984 Morgan Hill, California, earthquake. J Geophys Res 93:6275-6296

Bertrand E, Deschamps A (2000) Lithospheric structure of the southern French Alps inferred from broadband analysis. Phys Earth Planet Interiors 122(1-2):79-102 (special issue)

Bertrand E, Deschamps A, Virieux J (2002) Crustal structure deduced from receiver functions via single-scattering migration. Geophys J Int 150(2):524-541

Boatwright J, Fletcher JB, Fumal TE (1991) A general inversion scheme for source, site, and propagation characteristics using multiply recorded sets of moderate-sized earthquakes. Bull Seism Soc Am 81:1754-1782

Bommer J, Abrahamson NA, Strasser FO, Pecker A, Bard P-Y, Bungum H, Cotton F, Fäh D, Sabetta F, Scherbaum F, Studer J (2004) The challenge of defining upper bounds on earthquake ground motions. Seismological Research Letters 75(1):82-95

Boore DM (2004) Can site response be predicted? J Earthquake Eng 8:1-41 (special issue 1)

Bragato L, Slejko D (2005) Empirical ground-motion attenuation relations for the eastern Alps in the magnitude range 2.5-6.3. Bull Seism Soc Am 95:252-276

Chaljub E, Valette B (2004) Spectral element modelling of three-dimensional wave propagation in a self-gravitating Earth with an arbitrarily stratified outer core. Geophys J Int 158:131-141

Chaljub E, Komatitsch D, Capdeville Y, Vilotte J-P, Valette B, Festa G (2006). In: Wu R-S, Maupin V (eds) Advances in Wave Propagation in Heterogeneous Media. "Advances in Geophysics" series, Elsevier, in press

Castanié L, Lévy B, Bosquet F (2005) Advances in seismic interpretation using new volume vizualisation techniques. First Break 23:69-72

Caimon G, Lepage F, Sword C, Mallet J-L (2004) Building and Editing a Sealed Geological Model. Mathematical Geology 36:405-424

Cochard A, Madariaga R (1994) Dynamic Faulting under Rate-dependent Friction. Pure Appl Geophys 142:419-445

Cocco M, Pacheco J, Singh SK, Courboulex F (1997) The Zihuatanejo, Mexico, earthquake of 1994 December 10 (M = 6.6): Source characteristics and tectonic implications. Geophys J Int 131:135-145

Cornou C (2002) Contribution du traitement d'antenne et de l'imagerie sismique à la compréhension des effets de site dans l'agglomération grenobloise. Thèse de doctorat, Université Joseph Fourier, Grenoble I

Cotton F, Campillo M (1995) Frequency domain inversion of strong motions: Application to the 1992 Landers earthquake. J Geophys Res 100:3961-3975

Cotton F, Scherbaum F, Bommer J, Bungum H, Sabetta F (2006) Criteria for selecting and adapting ground-motion models for specific target regions Application to Central Europe and rock sites. In press

Courboulex F, Singh SK, Pacheco JF, Ammon CJ (1997) The 1995 Colima-Jalisco, Mexico, earthquake (Mw 8): A study of the rupture process. Geophys Res Lett 24:1019-1022

Cruz-Atienza VM, Virieux J (2004) Dynamic rupture simulation of non-planar faults with a finite-difference approach. Geophys J Int 158:939-954, doi:10.1111/j.1365-246X.2004.02291.x

Dangla P, Semblat J-F, Xiao H, Delépine N (2005) A simple and efficient regularization method for 3D BEM: application to frequency-domain elastodynamics. Bull Seism Soc Am 95:1916-1927

Day SM (1982) Three-dimensional finite difference simulation of fault dynamics: rectangular faults with fixed rupture velocity. Bull Seism Soc Am 72:705-727

Day SM, Dalguer LA, Lapusta N, Liu Y (2005) Comparison of finite difference and boundary integral solutions to three-dimensional spontaneous rupture. J Geophys Res 110, B12307, doi:10.1029/2005JB003813

Delouis B, Giardini D, Lundgren P, Salichon J (2002) Joint inversion of InSAR, GPS, teleseismic and strong motion data for the spatial and temporal distribution of earthquake slip: Application to the 1999 Izmit Mainshock. Bull Seism Soc Am 92:278-299

Delouis B, Lundgren P, Salichon J, Giardini D (2000) Joint inversion of InSAR and teleseismic data for the slip history of the 1999 Izmit (Turkey) earthquake. Geophys Res Lett 27:3389-3392

Douglas J, Bungum H, Scherbaum F (2005) Composite hybrid ground-motion prediction relations based on host-to-target conversions: Case studies for Europe. J Earthquake Eng, under revision

Eliasen E, Machenhauer B, Rasmussen E (1970) On a Numerical Method for Integration of the Hydrodynamical Equations with a Spectral Representation of the Horizontal Fields. Report No. 2, Institute of Theoretical Meteorology, University of Copenhagen, Denmark

Emolo A, Zollo A (2001) Kinematic Source parameters for the 1989 Loma Prieta Earthquake from the nonlieanr inversion of accelerograms. Bull Seismo Soc Am 95:981-994

Festa G, Vilotte J-P (2005) The Newmark scheme as a Velocity-Stress Time staggering: An efficient PML for Spectral Element simulations of elastodynamics. Geophys J Int 161:789-812, doi:10.1111/j.1365-246X.2005.02601.x

Foerster E, Dupros F, Bernardie S (2005) Parallel three-dimensional finite element computations for site response analysis. In SIAM Conference on Mathematical and Computational Issues in the Geosciences, France, June 2005

Foster I, Gopp W, Stevens R (1992) Parallel Scalability of the Spectral Transform Method, in Computer Hardware, Advanced Mathematics and Model Physics, Department of Energy Report DOE/ER-0541T, 7-10

Fukuyama E, Madariaga R (1998) Rupture dynamics of a planar fault in a 3D elastic medium: rate- and slip-weakening friction. Bull Seism Soc Am 88:1-17

Graves RW (1993) Three-dimensional finite-difference modelling of the San Andreas Fault: Source parametrization and ground motion levels. Bull Seism Soc Am 83:881-897

Graves RW (1996) Simulating seismic wave propagation in 3D elastic media using staggered-grid finite differences. Bull Seism Soc Am 96:1091-1106

Harris RA, Archuleta RJ, Day SM (1991) Fault steps and the dynamic rupture process: 2-D numerical simulations of a spontaneously propagating shear fracture. Geophys Res Lett 18:893-896

Hartzell SH, Heaton TH (1983) Inversion of strong ground motion and teleseismic waveform data for the fault rupture history of the 1979 Imperial Valley, California, earthquake. Bull Seism Soc Am 73:1153-1184

Hartzell SH, Stewart GS, Mendoza C (1996) Comparison of L1 and L2 norms in teleseismic waveform inversion for the rupture history of the earthquake. In: Spudich P (ed) The Loma Prieta, California, earthquake of October 17, 1989 – main shock characteristics. U.S. Geol. Surv. Profess. Pap. 1550-A:39-57

Hernandez B, Cocco M, Cotton F, Stramondo S, Scotti O, Courboulex F, Campillo M (2004) Rupture history of the 1997 Umbria-Marche (Central Italy) main shocks from the inversion of GPS, DInSAR and near-field strong motion data. Ann Geophys 47:1355-1376

Herrero A, Bernard P (1994) A kinematic self-similar rupture process for earthquakes. Bull Seism Soc Am 84:1216-1228

Horike M (1985) Inversion of phase velocity of long-period microtremors to the S-wavevelocity structure down to the basement in urbanized area. J Phys Earth 33:59-96

Huang L-J, Mora P (1994a) The phononic lattice solid by interpolation for modelling P waves in heterogeneous media. Geophys J Int 119:766-778

Huang L-J, Mora P (1994b) The phononic lattice solid with fluids for modelling non-linear solid-fluid interactions. Geophys J Int 117:529-538

Huang L-J, Mora P (1996) Numerical simulation of wave propagation in strongly heterogeneous media using a lattice solid approach. In: Hassanzadeh S (ed) Mathematical Methods in Geophysical Imaging IV, Proc. SPIE 2282:170-179

Komatitsch D, Tromp J (2003) A Perfectly Matched Layer absorbing condition for the second-order elastic wave equation. Geophys J Int 154:146-153

Komatitsch D, Tromp J (2002b) Spectral-element simulations of global seismic wave propagation -I. Validation. Geophys J Int 149:390-412

Komatitsch D, Tromp J (2002b) Spectral-element simulations of global seismic wave propagation -II. 3-D models, oceans, rotation, and gravity. Geophys J Int 150:303-318

Komatitsch D, Vilotte J-P (1998) The spectral element method: an efficient tool to simulate the seismic response of 2-D and 3-D geological structures. Bull Seism Soc Am 88:368-392

Kudo K (1995) Practical estimates of site response, State-of-the-Art report. In: Proceedings of the Fifth International Conference on Seismic Zonation, October 17-19, Nice, France, Ouest Editions Nantes, 3, pp 1878-1907

Käser M, Igel H (2001) Numerical simulation of 2D wave propagation on unstructured grids using explicit differential operators. Geophysical Prospecting 49(5):607-619

Krüger OS, Saenger EH, Shapiro SA (2005) Scattering and diffraction by a single crack: An accuracy analysis of the rotated staggered grid. Geophys J Int 162:25-31, doi:10.111/j.1365-246X.2005.02647.x

Lambert J, Levret-Albaret A (1995) Milles ans de séismes en France. Catalogue d'épicentres. Paramètres et références; Ouest-Editions, Presses Académiques, Nantes, pp 85

Latorre D, Virieux J, Monfret T, Lyon-Caen H (2004) Converted seismic wave investigation in the Gulf of Corinth from local earthquakes. CR Geoscience 336:259-267

Le Bihan N (2001) Traitement algébrique des signaux vectoriels: application en séparation d'ondes sismiques. Thèse de Doctorat, Institut National Polytechnique de Grenoble, Grenoble, pp 135

Le Meur H, Virieux J, Podvin P (1997) Seismic tomography of the Gulf of Corinth: a comparison of methods. Ann Geofis 40:1-24

Liu H-P, Boore DM, Joyner WB, Oppenheimer DH, Warrick RE, Zhang W, Hamilton JC, Brown LT (2000) Comparison of phase velocities from array measurements of rayleigh waves associated with microtremor and results calculated from borehole shear-wave profiles. Bull Seism Soc Am 90:666-678

Madariaga R (1976) Dynamics of an expanding circular fault. Bull Seism Soc Am 66:639-666

Madariaga R, Olsen K, Archuleta R (1998) Modeling dynamic rupture in a 3D earthquake fault model. Bull Seism Soc Am 88:1182-1197

Mallet J-L (2005) Space-time mathematical framework for sedimentary geology. Mathematical Geology 36:32

Marcinkovich C, Olsen K (2003) On the implementation of perfectly matched layers in a three-dimensional fourth-order velocity-stress finite-difference scheme. J Geophys Res 108:2276, doi:10.1029/2002JB002235

Marfurt KJ (1984) Accuracy of finite-difference and finite-element modeling of the scalar and elastic wave equations. Geophysics 49:533-549

Matsushima T, Okada H (1990) Determination of deep geological structures under urban areas using long-period microtremors. Butsuri-Tansa 43(1):21-33

Michetti AM, Audemard FA, Marco S (2005) Future trends in paleoseismology: integrated study of the seismic landscape as a vital tool in seismic hazard analyses. Tectonophysics 408:3-21

Miller RD, Xia J, Park CB, Ivanov J (2000) Shear-wave velocity field from surface waves to detect anomalies in the subsurface. Geophysics 2000, FHWA and MoDOT Special Publication, 4:8-1-4:8-10

Moczo P, Kristek J, Gális M (2004) Simulation of planar free surface with near-surface lateral discontinuities in the finite-difference modeling of seismic motion. Bull Seism Soc Am 94:760-768

Moczo P, Robertsson JOA, Eisner L (2006) The finite-difference time-domain method for modelling of seismic wave propagation. In: Wu RS, Maupin V (eds) Advances in Wave Propagation in Heterogeneous Earth. Advances in Geophysics series, Elsevier Academic Press, in press

Nogoshi M, Igarashi T (1971) On the amplitude characteristics of microtremor (Part 2). Jour Seism Soc Japan 24:26-40 (in Japanese with English abstract)

Nakamura Y (1989) A method for dynamic characteristics estimation of subsurface using microtremor on the ground surface. Quarterly Report of the Railway Research Institute, Tokyo, 30:25-33

Nivlet P, Fournier F, Royer JJ (2002) A new nonparametric discriminant analysis algorithm accounting for bounded data errors. Mathematical Geology 34:223-246

Oglesby DD, Archuleta RJ, Nielsen SB (2000) The three-dimensional dynamics of dipping faults. Bull Seism Soc Am 90:616-628

Ohrnberger M, Schissele E, Cornou C, Bonnefoy-Claudet S, Wathelet M, Savvaidis A, Scherbaum F, Jongmans D (2004) Frequency wavenumber and spatial autocorrelation methods for dispersion curve determination from ambient vibration recordings. Proceedings of the 13th World Conference on Earthquake Engineering, Vancouver, August 2004, Paper # 946

Olsen KB (2001) Three-dimensional ground motion simulations for large earthquakes on the San Andreas fault with dynamic and observational constraints. Jour Comp Acoust 9(3):1203-1215

Olsen KB (2000) Site Amplification in the Los Angeles Basin from 3D Modeling of Ground Motion. Bull Seis Soc Am 90:S77-S94

Olsen KB, Madariaga R, Archuleta RJ (1997) Three-dimensional dynamic simulation of the 1992 Landers earthquake. Science 278:834-838

Operto S, Ravaut C, Improta L, Virieux J, Herrero A, Dell'Aversana P (2004) Quantitative imaging of complex structures from dense wide-aperture seismic data by multiscale traveltime and waveform inversions: a case study. Geophysical Prospecting 52:625-651 (special issue)

Park CB, Miller RD, Xia J (1999) Multichannel analysis of surface waves. Geophysics 64:800-808

Paul A, Cattaneo M, Thouvenot F, Spallarossa D, Bethoux N, Fréchet J (2001) A three-dimensional crustal velocity model of the south-western Alps from local earthquakes. J Geophys Res 106:19367-19389

Pavic R, Koller M, Bard P-Y, Lacave-Lachet C (2000) Ground motion prediction with the empirical Green's function technique: an assessment of uncertainties and confidence level. Journal of Seismology 4(1):59-77

Peyrat S, Olsen KB, Madariaga R (2004) On the estimation of dynamic rupture parameters. Pure Appl Geophys, in press

Peyrat S, Madariaga R, Olsen KB (2001) Dynamic modelling of the 1992 Landers earthquake. J Geophys Res 106:25467-25482

Piperno S, Remaki M, Fezoui L (2002) A non-diffusive finite volume scheme for the 3d maxwell equations on unstructured meshes. SIAM J Numer Anal 39:2089-2108

Pitarka A (1999) 3D elastic finite-difference modeling of seismic wave propagation using staggered grid with non-uniform spacing. Bull Seism Soc Am 89:54-68

Pousse G (2005) Analyse de données accélérométriques des réseaux accéléromériques K-net et Kik-net: implications pour la prédiction du mouvement sismique et la prise en compte des effets de site non-linéaires. Thèse de Doctorat de l'Université Joseph Fourier

Pratt RG, Shin, Changsoo, Hicks GJ (1998) Gauss-Newton and full Newton methods in frequency domain seismic waveform inversion. Geophys J Int 133:341-362

Rothman DH (1988) Cellular-automaton fluids: a model for flow in porous media. Geophysics 53:509-519

Sanchez-Sesma FJ (1983) Diffraction of elastic waves by three-dimensional surface irregularities. Bull Seism Soc Am 73(6):1621-1636

Sanchez-Sesma FJ, Luzon F (1995) Seismic response of three-dimensional alluvial valleys for incident P, S and Rayleigh waves. Bull Seism Soc Am 85:269-284

Sèbe O (2004) Déconvolution aveugle en sismologie: Applications à l'étude de la source sismique et au risque sismique. Thèse de Doctorat de l'Université Joseph Fourier

Scherbaum F, Bommer JJ, Bungum H, Cotton F, Abrahamson NA (2005) Composite ground-motion models and logic trees: methodology, sensitivities and uncertainties. Bull Seism Soc Am 95, in press

Schisselé E (2002) Analyse et caractérisation des phases sismiques régionales enregistrées par des antennes de capteurs. Thèse de doctorat de l'Université de Nice Sophia-Antipolis

Schisselé E, Gaffet S, Cansi Y (2005) Characterization of regional and local diffraction effects from low-aperture seismic arrays recording. Journal of Seismology 9:137-149

Shapiro NM, Campillo M, Stehly L, Ritzwoller MH (2005) High resolution surface wave tomography from ambient seismic noise. Science 307:1615-1618

Spudich P, Olsen KB (2001) Fault zone amplified waves as a possible seismic hazard along the Calaveras fault in central California. Seism Res Lett 28:2533-2536

Stokoe KH II, Wright SG, Bay JA, Roesset JM (1989) Characterization of geotechnical sites by SASW method. In: Woods RD, Balkema AA (eds) Geophysical Characterization of Sites. Rotterdam, pp 15-25

Vanorio T, Virieux J, Capuano P, Russo G (2005) Three-dimensional seismic tomography from P wave and S wave microearthquake travel times and rock physics characterization of the Campi Flegrei Caldera. J Geophys Res 110:B03201, doi:10.1029/2004JB003102

Tokimatsu K (1997) Geotechnical site characterization using surface waves. In: Ishiliara (ed) Earthquake Geotech. Eng. Balkerna, Rotterdam, pp 1333-1368

Tokimatsu K, Shinzawa K, Kuwayama S (1992) Use of shortperiod microtremors for VS profiling. J Geotech Eng 118(10):1544-1588

Toomey A, Bean CJ, Scotti O (2002) Fracture properties from seismic data – a numerical investigation. Geophys Res Lett 29:1050, doi:10.1029/2001GL013867

Toomey A, Bean CJ (2000) Numerical simulation of seismic waves using a discrete particle scheme. Geophys J Int 141:595-604

Truffert C, Chamot-Rooke N, Lallemant S, de Voogd B, Huchon P, Le Pichon X (1993) A crustal-scale cross section of the Western Mediterranean Ridge from deep seismic data and gravity modelling. Geophys J Int 114:360-372

Vallée M, Bouchon M (2004) Imaging coseismic rupture in far field by slip patches. Geophys J Int 156:615-630

Virieux J, Madariaga R (1982) Dynamic faulting studied by a finite difference method. Bull Seism Soc Am 72:345-369

Wathelet M, Jongmans D, Ohrnberger M (2005) Direct Inversion of Spatial Autocorrelation Curves with the Neighborhood Algorithm. Bull Seism Soc Am 95:1787-1800

Zollo A, Judenherc S, Auger E, D'Auria L, Virieux J, Capuano P, Chiarabba C, de Franco R, Makris J, Michelini A, Musacchio G (2003) Evidence for the buried rim of Campi Flegrei caldera from 3-D active seismic imaging. Geophys Res Lett 30, doi:10.1029/2003GL018173

10 Seismic Early Warning Systems: Procedure for Automated Decision Making

Veronica F. Grasso[1], James L. Beck[2], Gaetano Manfredi[3]

[1] Department of Structural Analysis and Design, University of Naples Federico II; Visiting Special Student, Caltech
[2] Department of Applied Mechanics and Civil Engineering, Caltech
[3] Department of Structural Analysis and Design, University of Naples Federico II

Abstract

Earthquake Early Warning Systems (EWS) potentially allow mitigation measures to be carried out from the moment in which a seismic event is detected, depending on the amount of warning time available. Examples of such measures are evacuation of buildings, shut-down of critical systems (nuclear reactors, industrial chemical processes, etc.) and stopping of high-speed trains.

In areas close to faults, where seismic EWS are especially relevant, only tens of seconds of warning are available. Such short warning times mean that to be effective a seismic EWS must depend on automated procedures, including those for decision making about whether to activate mitigation measures; the time is too short to require human intervention when the event is first detected. As a result of the automation, careful attention must be paid to the design of the local seismic EWS for each critical facility; in particular, a means of controlling the trade-off between false alarms and missed alarms is desirable.

To investigate this trade-off, the consequences of the two alternatives of taking mitigation actions or not acting must be analyzed, accounting for significant uncertainty in the predictions.

A feasibility assessment of any proposed regional EWS is of critical importance, and it should involve an examination of whether the requirements, in terms of warning time available and the probability of making wrong decisions, are met. A methodology is presented for estimating the probabilities of making wrong decisions that can be incorporated in a feasibility assessment of proposed EWS. To illustrate the methodology we

consider the case of feasibility assessment of a seismic EWS for Southern California.
We, also, present a decision-making procedure based on the real-time evaluation of the consequences of taking no action and of activating mitigation measures which is based on the probabilities of false and missed alerts. The threshold at which mitigating actions should be taken is quantified based on a cost-benefit analysis. The method is applied to the M 4.75 Yorba Linda seismic event occurred on 3 September 2002, in Orange County, California.

10.1 Introduction

The high social and economic vulnerability of urbanized areas to seismic risk has become evident in recent years due to severe losses as a consequence of catastrophic earthquakes. The extent of structural damage and economic loss due to these catastrophic events underlines the strong necessity of social, political and scientific cooperation for disaster prevention.

It is clear that timely warnings can mitigate the effects of natural disasters. Such warnings are commonly given for floods, hurricanes, tornados and tsunamis, but still under development for earthquakes. Effective early warning technologies for earthquakes are much more challenging to develop because warning times range from only a few seconds to a minute or so (Allen and Kanamori 2003). In areas close to faults, where seismic early warning systems (EWS) are especially relevant, only tens of seconds of warning are available. Such short warning times mean that to be effective a seismic EWS must depend on automated procedures, including those for decision making about whether to activate mitigation measures; the time is too short to require human intervention when the event is first detected. As a result of the automation, careful attention must be paid to the design of the local seismic EWS for each critical facility; in particular, a means of controlling the trade-off between false alarms and missed alarms is desirable.

Of course, as an essential part of a seismic EWS, an infrastructure must be in place that consists of a seismic sensor network with high-speed communication to some data-processing center, along with a broadcasting system to disseminate the early warning information to the local automated system that can activate the mitigation measures designed for a specific facility.

Historical lessons come of some help to appreciate the EWS potentialities. In the Indian Ocean tsunami disaster of 26 December 2004, a tsunami

EWS was in place for the Pacific Ocean that detected the large Sumatran subduction zone event but there was an inadequate broadcasting system to disseminate the early warning information to the countries at risk surrounding the Indian Ocean. Many thousands of lives could have been saved if a preventive alarm had been broadcasted to warn people in the coastal regions about the tsunami. In this case, for those locations sufficiently far from the ruptured segment of the Sumatran subduction zone, warning times of several hours were possible, so that before the tsunami arrived, many people might have been able to escape from the low-lying coastal areas to higher elevations. This mitigation action would have been enhanced by the availability of predicted inundation and damage maps to direct people to safer locations.

Although early warning technologies have been developed to provide natural hazard mitigation for many types of hazards, attention here is focused on seismic risk mitigation because the technologies for this application are not yet fully developed.

10.1.1 Potential Benefits of Seismic Early Warning Systems

The main goal of an EWS for earthquakes is the reduction or prevention of loss of life and mitigation of structural damage and economic loss. The benefits of EWS are due to the measures that can be carried out from the moment in which a seismic event is detected at a certain place until the moment in which the seismic waves arrive at a location of interest. These measures, for prevention or emergency response, can be categorized by considering the phases of the seismic event (Wieland 2001).

After event detection but before the earthquake arrives at a site, the warning provided by EWS with pre-arrival times of up to perhaps 90 seconds, could be used to evacuate buildings, shut-down critical systems (such as nuclear and chemical reactors), put vulnerable machines and industrial robots into a safe position, stop high-speed trains, activate structural control systems (Kanda et al. 1994; Occhiuzzi et al. 2004), and so on.

During an earthquake, the alarm generated by EWS could still enable such mitigation processes to be activated if there was insufficient time to do so prior to the arrival of the earthquake at the site. Within seconds after an earthquake, the information provided by EWS could be used to produce damage and loss maps based on the ground shaking intensity and could be the basis for more effective emergency response and rescue operations.

Evacuation of at-risk buildings and facilities is only feasible if the warning time is around 1 minute before the arrival of the strong shaking, which is possible only in the case where the seismic source zone is sufficiently far away. This is the situation, for example, for the threat to Mexico City from earthquakes occurring in the subduction zone along the Pacific Coast (e.g. Lee and Espinosa-Aranda 1998), where the time available is sufficient to alert large segments of the population by commercial radio and television, and for evacuation of strategic buildings, such as schools, crowded facilities, and so on.

In the case of a few seconds warning time before the shaking, it is still possible to slow down trains (e.g. Saita and Nakamura 1998), to switch traffic lights to red (as for the Lions Gate bridge EWS, Vancouver), to close valves in gas and oil pipelines, to release control rods in nuclear power plants (e.g. Wieland et al. 2000), activate structural control systems, and so on. In addition, secondary hazards can be mitigated that are triggered by earthquakes but which take more time to develop, such as landslides, tsunamis, fires, etc., by predicting the ground motion parameters for the incoming seismic waves. This could be used, for example, to initiate the evacuation of endangered areas.

Given that an appropriate EWS is in place for a local area or critical facility, its impact or effectiveness is dependent on the warning time available and the quality and reliability of the information that is provided, since these influence and constrain the utilization of the information. In most EWS applications, the available warning time is likely to be no more than tens of seconds, enabling the possibility of activating mitigation measures but meaning that automated activation is essential to fully utilize the available warning time.

10.1.2 Limitations of Effectiveness of Seismic Early Warning Systems

The benefits of an EWS for earthquakes are often not fulfilled due to limitations that depend on:

- the amount of warning time
- the probability of making wrong decisions (false alarms and missed alarms)

These parameters strongly influence EWS impact and effectiveness on seismic risk reduction. Each of these is discussed in turn.

To examine the available warning time for activation of seismic mitigation measures, the main principle on which an EWS is based is first described. This principle is that seismic waves travel through the Earth with a velocity that is much less than the velocity of the electromagnetic signals transmitted by telephone or radio to provide seismic information about the incoming event. In addition, seismic body waves can be identified as compression waves (primary waves or P-waves) and shear waves (secondary waves or S-waves) (Occhiuzzi et al. 2004), where the P-waves are characterized by a propagation velocity that is almost twice that of S-waves; the latter are stronger and give almost horizontal ground motion at the base of a structure, in contrast to P-waves, so S-waves tend to be more damaging. The time interval from the detection of P-waves in the epicentral area, and the arrival of S-waves in the area where the structure or facility is located, may be utilized to activate mitigation measures. The feasible warning time is given by:

$$T_w = T_s - T_r \tag{10.1}$$

$$T_r = T_d + T_{pr} \tag{10.2}$$

where the origin of time is the P-wave detection time; T_r is the reporting time comprised of the time T_d needed by the system to trigger and record a sufficient length of waveforms and the time T_{pr} to process the data; T_s is the S-wave travel time and T_w is the early warning time. For the warning time to be considered adequate for the activation of a mitigation measure, it has to be greater than the time necessary for activation of the measure.

Suppose that the EWS works by setting an alarm if a critical shaking intensity threshold is predicted to be exceeded at a site, where the choice of critical threshold depends on the vulnerability of the system to be protected at the site. Assuming that the warning time provided by the EWS is sufficient for activation of the mitigation measure, then based on the predictions from the first few seconds of P-wave observation, a decision has to be made of whether to activate the alarm or not. In making this decision, two kinds of errors may be committed (Wald 1947):

- Type I error: the alarm is not activated when it should have been.
- Type II error: the alarm is activated when it should not have been.

We call Type I errors *missed alarms* and type II errors *false alarms*. The probability of each of these wrong decisions is denoted by:

- P_{ma}= probability of missed alarm, that is, the probability of having critical threshold exceedance but no alarm activation.
- P_{fa} = probability of false alarm, that is, the probability of having no threshold exceedance but alarm activation.

The tolerance of a type I or II error is related to a trade-off between the benefits of a correct decision and the costs of a wrong decision and it could vary substantially, depending on the relative consequences of possible missed and false alarms. For example, the automated opening of a fire station door has minimal impact if the door is opened for a false alarm. On the contrary, an automated shutdown of a power plant because of a false alarm could cause problems over an entire city and involve expensive procedures to restore to full-operational status. In this latter case, the EWS must be designed to keep the probability of false alarms very low. In general, the automated decision process has to be designed with attention focused on the probability of false and missed alarms along with a cost-benefit analysis. Some mitigation measures could be unacceptable to operate as a result of the false or missed alarm rate being too high.

The probability of a wrong decision is due to having only partial knowledge of the phenomenon and so any prediction, as a consequence, is affected by uncertainty. A key element of an EWS is a better understanding of the parameters that play a fundamental role in this uncertainty, and hence a better understanding of the quality of the predictions on which decision making is based.

10.2 Ground Motion Prediction Process in Seismic EWS

10.2.1 Basic Idea of EWS Operation

The EWS is composed of a network of seismic stations, a dedicated real-time data communication system, central processing system, broadcasting system and information receivers at the user's end. In some cases, this network may be a dedicated local one placed at some distance around a structure to provide data about incoming seismic waves. When a seismic event occurs, the stations close to the epicentral area are triggered by P-waves, then the ground motion data is recorded and sent by the communication system to the central processor, where, based on predictive models, an evaluation is made in real-time to predict the earthquake source parameters of the event.

This information is provided to the user by a distribution network, and based on this early information, the user predicts a ground motion or structural performance parameter of interest for a facility. This parameter represents the predictor on which the decision whether to raise the alarm or not is based. In this report, the predictor is taken to be an intensity measure *IM*, which is a ground motion parameter that represents the shaking intensity at the site of the facility. Two common intensity measures are peak ground acceleration (PGA) and response spectral acceleration (S_a). The *IM* is to be predicted based on the first few seconds of data registered by some of the seismic stations in a regional network that are near the earthquake source. The parameter of interest for a facility could also be taken as some critical engineering demand parameter, such as inter-story drift in a building or floor acceleration at the location of vulnerable equipment, or even economic loss.

The prediction model for the ground motion parameters can be represented as a sequential multi-compartment model (Bates et al. 2003), composed of two sub-models, M_1 and M_2 (Grasso et al. 2005b), as illustrated in Fig. 10.1.

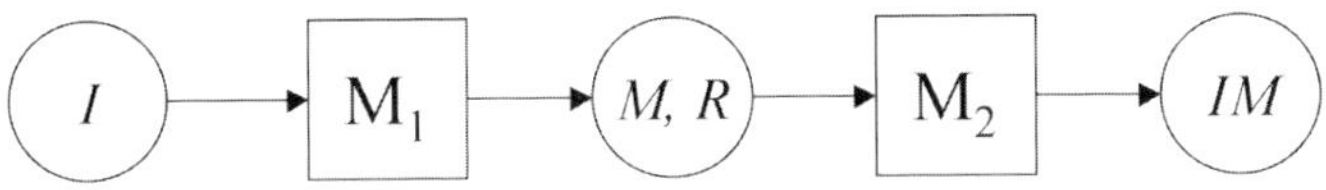

Fig. 10.1 The multi-component model representing EWS. M_1 is the earthquake prediction model and M_2 is the ground motion attenuation model

The earthquake prediction model, M_1, estimates earthquake source parameters (magnitude, *M*; epicentral distance, *R*), based on parameters, *I*, extracted from real-time measurements of the first few seconds of P-waves, e.g., *I* is the predominant period in Allen and Kanamori (2003); while *I* is the observed ground motion ratio for the Virtual Seismologist method in Cua and Heaton (2004).

The ground motion attenuation model, M_2, predicts ground motion parameters (intensity measures *IM*), based on the magnitude and epicentral distance predicted by M_1. The parameter *IM*, which could be the final outcome of the EWS prediction process, represents the predicted ground motion intensity (e.g. PGA, PGV or S_a) that will occur at the site where a strategic facility of interest is located. It is assumed in this work that *IM* is the predictor on which the decision to take some protective action is based.

10.2.2 Sources of Uncertainty

Uncertainty in the predictor *IM* is a result of the uncertain prediction errors produced by the models M_1 and M_2. The uncertainties for each model are represented in Fig. 10.2, where ε_M, ε_R and ε_{IM} denote the prediction errors for the magnitude and location and the attenuation model, respectively. Uncertainties of each sub-model propagate through the output, so each uncertainty plays an important role in the definition of the final quality of the intensity measure, *IM*.

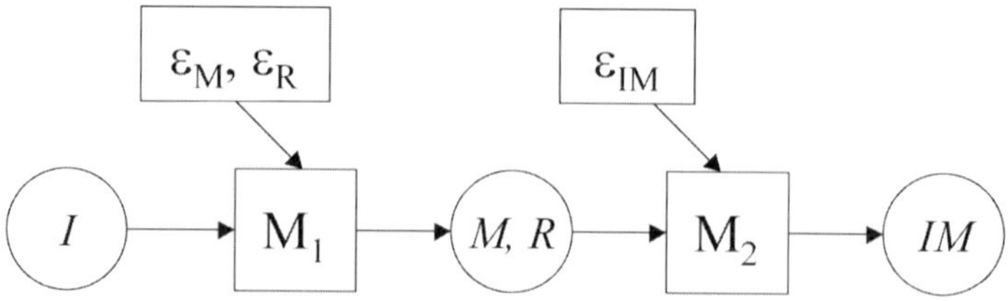

Fig. 10.2 The multi-component model for EWS uncertainty propagation.

A Gaussian probability distribution model is chosen for ε_M and ε_{IM} to model the magnitude and attenuation model uncertainties. The uncertainty in the predicted magnitude can be well modeled as a Gaussian distribution, as shown by Cua and Heaton (2004), with standard deviation dependent on the prediction model. The magnitude error has zero-mean and a standard deviation of about 0.4 magnitude units for the Heaton-Cua relation and it decreases with increasing number of data. According to the Allen-Kanamori method, the uncertainty of magnitude prediction is related to the number of stations considered and to the elapsed time, and it assumes a value of 0.7 magnitude units considering only one station, 0.6 for three stations, 0.45 for five stations, and it drops to 0.35 if ten stations are considered (Allen 2003). The errors related to the ground motion parameter conditional on *M* and *R* being given are modeled well by a lognormal distribution, so if the intensity measure *IM* is the logarithm of the ground motion parameter, then *IM* can be assumed to be Gaussian. This assumption is supported by the analysis done by Cua and Heaton (2004) in which the errors were analyzed based on a large number of data from ground motions recorded by the seismic network in Southern California over 4 years.

In this work, the uncertainty $\varepsilon_{\log R}$ in the natural logarithm $\log_e R$ of the epicentral distance prediction is modeled as Gaussian, which means that *R* has a lognormal distribution. It also means that for small prediction errors, $\varepsilon_R \approx \hat{R}\varepsilon_{\log R}$ is approximately Gaussian. It has been suggested that a more complex distribution for ε_R might be appropriate based on the observation that in the case of large-magnitude teleseismic events, the probability of a large prediction error based on the first few seconds of data is likely; in fact, in this case, the network could erroneously locate the epicenter inside the instrumented area (Kanamori and Heaton 2004, personal communication).

10.2.3 Uncertainty propagation

The quality of the predictions for the ground motion parameter of interest is fundamental for optimization of the decision-making process. This quality is influenced by the errors of the sub-models that propagate through to the output and influence the effectiveness of the EWS application. In particular, it is necessary to estimate the total prediction error in order to quantify the performance of the EWS prediction process in terms of the probability of false and missed alarms. For this purpose, all the prediction errors influencing the intensity measure have to be considered, as shown in Fig. 10.2, along with their corresponding probability distributions.

The total prediction error is given by comparing the predicted intensity measure, $\hat{IM}$, to the actual *IM*:

$$\varepsilon_{tot} = IM - \hat{IM} \tag{10.3}$$

where $\hat{IM}$ is a function of the predicted values $\hat{M}$ and $\hat{R}$, which can be expressed in terms of the actual M and R and the uncertain prediction errors ε_M and $\varepsilon_{\log R}$:

$$\hat{IM} = f(\hat{M}, \log_e \hat{R}) = f(M - \varepsilon_M, \log_e R - \varepsilon_{\log R}) \tag{10.4}$$

where the function f represents the ground motion attenuation model as follows:

$$f(M, \log_e R) = \alpha + \beta M + \gamma \log_e R. \tag{10.5}$$

Most published attenuation models have this form if *IM* denotes $\log_e$ PGA or $\log_e S_a$ (Seismological Research Letters 1997); in the examples later, *IM* is taken to be $\log_e$ PGA. The actual intensity measure, *IM*, is represented by:

$$IM = f(M, \log_e R) + \varepsilon_{IM} \tag{10.6}$$

where ε_{IM} is the prediction error in the ground motion attenuation model, given M and R.

Under previous assumptions about ε_M, $\varepsilon_{\log R}$ and ε_{IM}, the total prediction error ε_{tot} also follows a Gaussian distribution with a mean and variance that depends on the means and variances of these contributing prediction errors, since:

$$\begin{aligned}\varepsilon_{tot} &= IM - \hat{IM} = f(M, \log_e R) + \varepsilon_{IM} - f(M - \varepsilon_M, \log_e R - \varepsilon_{\log R}) \\ &= \beta\varepsilon_M + \gamma\varepsilon_{\log R} + \varepsilon_{IM} \\ &\approx \beta\varepsilon_M + \frac{\gamma}{\hat{R}}\varepsilon_R + \varepsilon_{IM}\end{aligned} \quad (10.7)$$

Under the assumption of independence of errors, the variance of ε_{tot} is:

$$\begin{aligned}\sigma_{tot} &= \sqrt{\beta^2\sigma_M^2 + \gamma^2\sigma_{\log R}^2 + \sigma_{IM}^2} \\ &\approx \sqrt{\beta^2\sigma_M^2 + \frac{\gamma^2}{\hat{R}^2}\sigma_R^2 + \sigma_{IM}^2}\end{aligned} \quad (10.8)$$

and the mean of ε_{tot} is equal to:

$$\mu_{tot} = \beta\mu_M + \gamma\mu_{\log R} + \mu_{IM}\,. \quad (10.9)$$

If the empirically-derived models M_1 and M_2 (Figs. 10.1 and 10.2) are unbiased, then the mean $\mu_{tot} = 0$. In fact, μ_{tot} does have a value close to zero in the Virtual Seismologist method (Cua and Heaton 2004). If more complex attenuation models, or more complex probability models for ε_M, $\varepsilon_{\log R}$ and ε_{IM}, are used, this analytical approach may be not applicable and then a Monte Carlo method is suggested to quantify the uncertainty in ε_{tot} (Grasso et al. 2005a).

10.3 Probability of Wrong Decisions: Pre-installation Analysis

10.3.1 Probabilities of False and Missed Alarms: Pre-installation Analysis

When examining the feasibility of installing an EWS for a facility, it is important to have a mechanism to control the probabilities of false and missed alarms. Since the decision to activate the alarm is based on a predictor, $\hat{IM}$, false alarms can be caused by the predictor exceeding the warning threshold even though the actual intensity measure, *IM*, that occurs at the site does not reach the critical value. Similarly, missed alarms can be caused by the predictor not exceeding the warning threshold even though the actual intensity measure reaches its critical value.

For a given facility, the critical threshold, a, of IM must be chosen by the user based on a vulnerability analysis of the system to be protected; for example, it could be chosen as the value of IM for which there is a high probability that damage (or significant economic losses) will occur. To control the probability of wrong decisions, the warning threshold is chosen as the product of the critical threshold a and a parameter, c, to be specified during the design process. The critical threshold depends on the facility, structure or equipment to be protected, but the warning threshold $c \cdot a$ depends on the design process chosen to optimize the automated alarm activation system. The design parameter c provides a mechanism to control the incidence of false and missed alarms. It is not possible to simultaneously reduce both of these but the design parameter c can be used to control the trade-off between them.

A false alarm occurs when the EWS predicts a value, $\hat{IM}$, that exceeds the warning threshold, $c \cdot a$, while the actual value, IM, of the intensity measure at the site turns out to be less than the value of the critical threshold, a. For a pre-installation analysis, the probability of a false alarm is therefore given by:

$$P_{fa} = \mathrm{P}[IM \leq a \mid \hat{IM} > c \cdot a]. \tag{10.10}$$

Similarly, the probability of a missed alarm is given by:

$$P_{ma} = \mathrm{P}[IM > a \mid \hat{IM} \leq c \cdot a]. \tag{10.11}$$

The values of the probabilities of false and missed alarms, P_{fa} and P_{ma}, are an important tool for the decision-making process during pre-installation design and during operation in a seismic event.

During design, the anticipated rate of missed and false alarms represents a guideline for EWS feasibility. The realization of EWS could be feasible or not, depending on whether the requirements in terms of warning time available and the probability of wrong decisions can be met. A useful tool to evaluate an EWS may be constructed by using a seismic hazard map to provide the probability of exceedance of ground shaking intensity, given a site and time interval of interest, to produce a map of probability of wrong decisions. Such a map would help when performing a territorial feasibility assessment of EWS applications.

During a seismic event, the (automated) decision to activate protective measures may be done either by comparing the requirements in terms of warning time needed and the tolerable level of P_{fa} (or P_{ma}) based on the information made available by the EWS, or by monitoring a time-varying threshold $c(t)\cdot a$. This case is examined later in Sect. 10.5.

The main reason for evaluating the probability of wrong decisions using a pre-installation analysis is to design the EWS application, which can be based on the probabilities of false and missed alarms that are tolerable to the owner or manager of the facility to be protected. By estimating the probability of false and missed alarms during design of the EWS application, we are primarily trying to answer to the question: How would an EWS perform during earthquakes that might occur in the area of interest, in terms of a false and missed alarm rates?

The probabilities of false and missed alarms in a pre-installation analysis are evaluated as a time-independent variable based on the seismicity of the area of interest. Time dependence can be neglected as a first approximation. A more refined analysis is presented later in Sect. 10.5 for the operation of an EWS application during a seismic event, where the changes in estimated magnitude and location coming from the EWS are taken into account. In any case, the prediction uncertainty stabilizes after some time following the first triggering. In some special cases, such as Mexico City, the location of the fault area and the configuration of the seismic stations lead to stabilization of the uncertainty of the prediction after the first few seconds.

10.3.2 Prior Information: Hazard Function

Prior information can be expressed by using the hazard (rate) function (Kramer 1996) that gives the mean annual rate of events with intensity measure exceeding a certain critical value, given a site of interest. The hazard function for a site comes from a PSHA (probabilistic seismic hazard analysis). It directly provides information related to the expected frequency of occurrence of the intensity measure *IM* for a given site and time period of interest; on the other hand, the EWS gives the predicted intensity measure, $\hat{IM}$, based on data from triggered stations and this will differ from the actual *IM* by the prediction error, ε_{tot}. This error depends on the errors related to the prediction of magnitude, *M*, and epicentral distance, *R*, and the prediction error of the ground motion attenuation model. Figure 10.3 shows the relation between *IM* and $\hat{IM}$ given by ε_{tot}, which depends on ε_M, ε_R and ε_{IM} as in Eq. 10.7.

The goal of the analysis that follows is to describe the EWS behaviour in terms of the probabilities of having false and missed alarms, based on a given seismic hazard environment represented by the intensity measure hazard function, $\lambda(IM)$. Considering all the possible events, the probability of a wrong decision is determined for a given site and a given period of time (e.g. one year). The key to forecasting the behaviour of the EWS is the prior knowledge about the errors that are committed when predicting the intensity measure. This prior knowledge comes from an analysis of the uncertainty propagation of the errors, as described in Sect. 10.2.

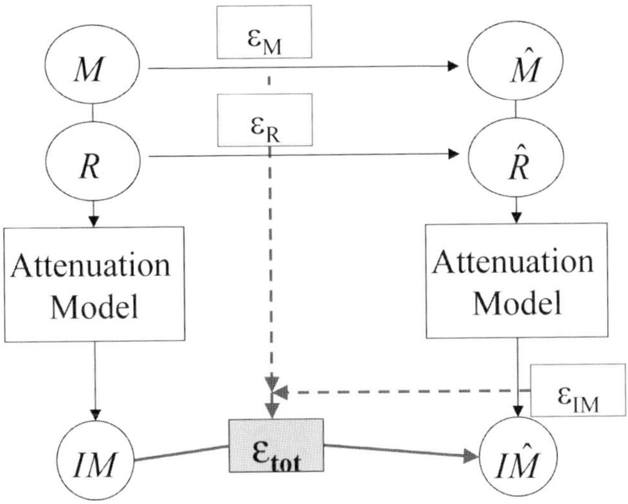

Fig. 10.3 Simulation of the EWS process of prediction in a pre-installation analysis.

The hazard function defines the mean annual rate of exceedance of a critical value of the ground motion intensity measure; from this mean rate, the probability of exceedance of a critical value, given that an earthquake of interest has occurred, can be determined as follows, which is based on a Poisson process model for the temporal occurrence of earthquakes:

$$\lambda(IM_c) = \lambda(IM_0) \cdot P[IM > IM_c \mid IM > IM_0] \tag{10.12}$$

where IM_0 is the minimum value of the intensity measure that is of interest (cut-off value) and it is used to define the earthquakes of interest.
An exponential model is assumed for the hazard function for a site (but recall that the choice of IM used in the examples is $\log_e$ PGA so this corresponds to a power law on PGA):

$$\lambda(IM) = k_0 10^{-k_1 IM} \tag{10.13}$$

where k_0 and k_1 can be obtained by fitting the hazard function from a PSHA for the site. This model implies from Eq. 10.12 that:

$$P\left[IM > IM_c \mid IM > IM_0\right] = 10^{-k_1(IM_c - IM_0)}. \tag{10.14}$$

The cumulative distribution function is then:

$$P\left[IM \leq IM_c \mid IM > IM_0\right] = 1 - 10^{-k_1(IM_c - IM_0)} \tag{10.15}$$

and the expression for the PDF (probability density function) is derived by differentiating this cumulative distribution function:

$$p\left(IM \mid IM > IM_0\right) = k_1 \cdot \log_e 10 \cdot 10^{-k_1(IM - IM_0)} = \overline{c} \cdot 10^{-k_1 IM} \tag{10.16}$$

where $\overline{c} = k_1 \cdot \log_e 10 \cdot 10^{k_1 IM_0}$.

In this work, the parameter k_1 is estimated from a hazard function for the site of interest by using a minimum entropy criterion in which the relative entropy E is minimized with respect to k_1:

$$E = \sum_i p_i \log\left(\frac{p_i}{q_i}\right) \tag{10.17}$$

where p_i represents the discrete probability distribution function derived from $p\left(IM \mid IM > IM_0\right)$ and q_i is the discrete probability distribution function derived from the given hazard function $\lambda(IM)$ using Eq. 10.12, so q_i is obtained by numerically differentiating the cumulative distribution function in the same way as $p\left(IM \mid IM > IM_0\right)$ was derived above. Therefore, p_i is a function of the parameter k_1 but q_i is not. By minimizing the relative entropy, we determine the parameter k_1 so that the model PDF is the best fit in an information-theoretic sense to the PDF implied by the hazard function for the site.

10.3.3 Probability of False Alarm: Pre-installation Analysis

The probability of false alarms, as defined in Eq. 10.10, can be expressed using Bayes rule as:

$$P_{fa} = P\left[IM \le a \mid \hat{IM} > c \cdot a, IM > IM_0 \right] = \frac{P\left[IM \le a \cap \hat{IM} > c \cdot a \mid IM > IM_0 \right]}{P\left[\hat{IM} > c \cdot a \mid IM > IM_0 \right]} \tag{10.18}$$

where it is assumed that an earthquake of interest, i.e. $IM > IM_0$, has occurred and that $a > IM_0$.

Eq. 10.18 can be expressed as (Grasso et al. 2005a):

$$P_{fa} = \frac{\int_{IM_0}^{a} \int_{ca}^{\infty} p(\hat{IM} \mid IM) \cdot p(IM \mid IM > IM_0) d\hat{IM} dIM}{\int_{IM_0}^{\infty} \int_{ca}^{\infty} p(\hat{IM} \mid IM) \cdot p(IM \mid IM > IM_0) d\hat{IM} dIM} \tag{10.19}$$

where $p\left(IM \mid IM > IM_0\right)$ is given by Eq. 10.16 and $p(\hat{IM} \mid IM)$ is a Gaussian distribution with *IM* representing the mean value (if ε_{tot} has zero mean) and standard deviation σ_{tot} given by Eq. 10.8. In the case that the prediction, $\hat{IM}$, is affected by a significant bias error (i.e. mean μ_{tot} of ε_{tot} is not close to zero), then the mean of the Gaussian distribution $p(\hat{IM} \mid IM)$ is (IM - μ_{tot}) (see Eq. 10.3).

The integral of the Gaussian distribution over $\hat{IM}$ can be expressed in terms of the standard Gaussian cumulative distribution function Φ, so Eq. 10.19 can be cast in a simpler form:

$$P_{fa} = \frac{\int_{IM_0}^{a} \Phi\left(-\frac{ca - IM}{\sigma_{tot}} \right) \cdot 10^{-k_1 IM} dIM}{\int_{IM_0}^{\infty} \Phi\left(-\frac{ca - IM}{\sigma_{tot}} \right) \cdot 10^{-k_1 IM} dIM}. \tag{10.20}$$

The integrals in the denominator and numerator here can be evaluated numerically for different values of *c*, given the value of *a*; then curves of P_{fa} versus *c* can be plotted for different critical thresholds *a*. Examples are given later in Sect. 10.6.

10.3.4 Probability of missed alarm: pre-installation analysis

The probability of missed alarms, as defined in Eq. 10.11, can be written using Bayes rule as:

$$P_{ma} = P\left[IM > a \mid \hat{IM} \le c \cdot a, IM > IM_0\right] = \frac{P\left[IM > a \cap \hat{IM} \le c \cdot a \mid IM > IM_0\right]}{P\left[\hat{IM} \le c \cdot a \mid IM > IM_0\right]} \tag{10.21}$$

where once again it is assumed that an earthquake of interest has occurred and that $a > IM_0$. Eq. 10.21 can be expressed as (Grasso et al. 2005a):

$$P_{ma} = \frac{\int_{a}^{\infty}\int_{-\infty}^{ca} p(\hat{IM} \mid IM) \cdot p(IM \mid IM > IM_0) d\hat{IM} dIM}{\int_{IM_0}^{\infty}\int_{-\infty}^{ca} p(\hat{IM} \mid IM) \cdot p(IM \mid IM > IM_0) d\hat{IM} dIM} \tag{10.22}$$

where $p(IM \mid IM > IM_0)$ is given by Eq. 10.16 and $p(\hat{IM} \mid IM)$ is a Gaussian distribution as before (see after Eq. 10.19). Substituting in Eq. 10.22 and expressing the integral of the Gaussian distribution over $\hat{IM}$ in terms of the standard Gaussian cumulative distribution function Φ, gives:

$$P_{ma} = \frac{\int_{a}^{\infty} \Phi\left(\frac{ca - IM}{\sigma_{tot}}\right) \cdot 10^{-k_1 IM} dIM}{\int_{IM_0}^{\infty} \Phi\left(\frac{ca - IM}{\sigma_{tot}}\right) \cdot 10^{-k_1 IM} dIM}. \tag{10.23}$$

10.4 Threshold Design Based on Cost-Benefit Considerations

Instead of directly specifying tolerable probabilities of wrong decisions to represent the design parameters for setting the threshold, it may be more natural to derive them by examining the consequences of wrong decisions using a cost-benefit analysis. In this case, the decision criterion may be taken to be the minimization of the expected consequences over the two possible actions of raising the alarm or doing nothing. A cost benefit analysis is based on the details shown in Table 10.1.

Table 10.1 Cost benefit analysis for threshold design

Action	Cost for case: IM<a	Cost for case: IM>a
Raise Alarm	False alarm: C_{fa}	Good Alarm: C_{ga}
No Alarm	Good Missed Alarm: C_{gm}	Missed Alarm: C_{ma}

where:

$$C_{ga} = C_{eq} - C_{save} \qquad C_{ma} = C_{eq} \qquad C_{gm} \simeq 0$$
$$P_{ga} = 1 - P_{fa} \qquad P_{gm} = 1 - P_{ma} \tag{10.24}$$

Here, C_{eq} represents the expected costs due to the earthquake, C_{fa} is the cost of a false alarm and C_{save} is the expected savings as a consequence of the activation of the protective measure. If the alarm is raised, the expected cost is given by:

$$\begin{aligned} E[\text{cost} \mid \text{alarm}] &= C_{fa} \cdot P_{fa} + C_{ga} \cdot P_{ga} \\ &= C_{fa} \cdot P_{fa} + \left(C_{eq} - C_{save}\right) \cdot \left(1 - P_{fa}\right) \end{aligned} \tag{10.25}$$

On the other hand, if the alarm is not raised, the expected cost is given by:

$$\begin{aligned} E[\text{cost} \mid \text{no-alarm}] &= C_{gm} \cdot P_{gm} + C_{ma} \cdot P_{ma} \\ &= C_{ma} \cdot P_{ma} \end{aligned} \tag{10.26}$$

The decision criterion for deciding between the options, raising the alarm or not, is represented by the minimum cost rule: Raise the alarm if and only if

$$E[\text{cost} \mid \text{no-alarm}] \geq E[\text{cost} \mid \text{alarm}] \tag{10.27}$$

that is,

$$\begin{aligned} C_{eq} \cdot P_{ma} &\geq C_{fa} \cdot P_{fa} + (C_{eq} - C_{save}) \cdot (1 - P_{fa}) \\ &= (C_{save} + C_{fa} - C_{eq}) \cdot P_{fa} + (C_{eq} - C_{save}) \end{aligned} \tag{10.28}$$

Since the probabilities of false and missed alarms, P_{fa} and P_{ma}, are given by Eqs. 10.20 and 10.23, respectively, and so may be evaluated as a function of the warning threshold parameter c, Eq. 10.28 can be taken as an equality to select an appropriate value of c. The tolerable value of P_{fa} and P_{ma} may then be determined from Eq. 10.20 and 23 for this value of c.

The above approach assumes that σ_{tot} in Eqs. 10.20 and 10.23 is time invariant so that the alarm threshold value $c \cdot a$ for $\hat{IM}$ can be set prior to operation of the EWS, where the value of critical threshold, a, is specified by the user. During a seismic event, however, more and more information becomes available to the EWS and so σ_{tot} will decrease with time. A refined analysis using a time-dependent warning threshold, $c(t)a$, would then be more appropriate. Alternatively, the probability of false and missed alarms could be monitored as a function of time and then the alarm would be raised when the tolerable level of probability of $P_{fa}(t)$ or $P_{ma}(t)$ is exceeded.

In the next section, we consider a refined decision-making procedure that is appropriate during a seismic event and which takes into account that the quality of the *IM* prediction improves as more and more information is obtained by the EWS. It is shown in Sect. 10.5.1 that in this case: $P_{fa}(t) + P_{ma}(t) = 1$, so Eq. 10.28 implies that the probability of a false alarm is tolerable if and only if:

$$P_{fa}(t) \leq \beta = \frac{C_{save}}{C_{fa} + C_{save}}. \tag{10.29}$$

Similarly, since the alarm is not raised if and only if:

$$C_{eq} \cdot P_{ma} < (C_{save} + C_{fa} - C_{eq}) \cdot P_{fa} + (C_{eq} - C_{save}) \tag{10.30}$$

it follows that the probability of a missed alarm is tolerable if and only if:

$$P_{ma}(t) < \alpha = \frac{C_{fa}}{C_{fa} + C_{save}}. \tag{10.31}$$

It is clear from Eqs. 10.29 and 10.31 that in this case, where the decision criterion is based on a cost-benefit analysis, $\alpha + \beta = 1$, which directly exhibits the trade-off between the threshold probabilities that are tolerable for false and missed alarms. If the threshold β is reduced to make false alarms less likely, then the threshold α for missed alarms becomes correspondingly larger.

10.5 Decision Making in EWS during a Seismic Event

10.5.1 Real-time Uncertainty Analysis during an Event

During a seismic event, the probability of false and missed alarms will be updated with time as more stations are triggered by the seismic waves and more data comes in from those that have already been triggered. This increase in data available will produce a decrease with time in the uncertainty in the predicted earthquake location and magnitude. Therefore, the prediction of the intensity measure can be updated with time and the characterization of its uncertainty, $\sigma_{tot}(t)$, will vary with time. As a consequence, it is important to update the probability of false and missed alarms as the seismic event evolves.

Recall that the predicted intensity measure, $\hat{IM}$, is estimated from the attenuation model based on the predictions of earthquake magnitude and location provided by EWS and so it can be updated as a function of time during the event. On the other hand, the actual intensity measure value, *IM*, that will occur at the site is unknown. The predicted and actual values of the intensity measure differ by $\varepsilon_{tot}(t)$ as in Eq. 10.3:

$$IM = \hat{IM}(t) + \varepsilon_{tot}(t). \tag{10.32}$$

As shown in Fig. 10.4, the total error $\varepsilon_{tot}(t)$ is related to $\varepsilon_M(t)$, $\varepsilon_R(t)$ and ε_{IM}, which can be continually updated as additional information becomes available by using Bayesian updating, as in the Virtual Seismologist method (Cua and Heaton 2004). Given the predicted value of $\hat{IM}$ at time *t*, the uncertainty in *IM* can be modeled by a Gaussian distribution with mean equal to the prediction $\hat{IM}(t)$ and standard deviation equal to $\sigma_{tot}(t)$, which can be evaluated by an analysis of the uncertainty propagation of the error, as was done to derive Eq. 10.8.

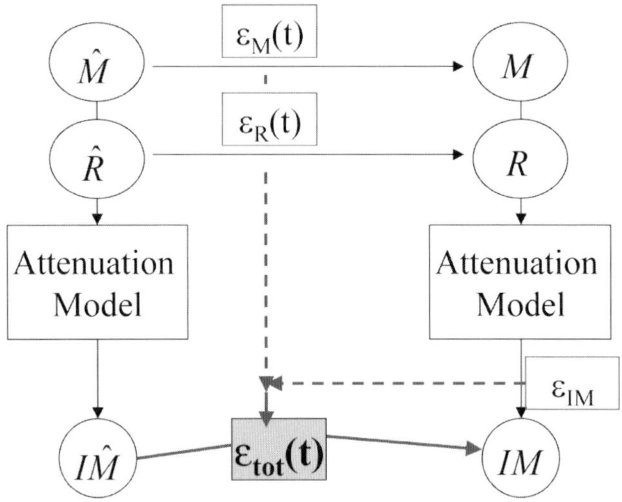

Fig. 10.4 Simulation of the EWS prediction process, during the seismic event.

The potential probability of a false alarm is estimated as the probability of *IM* being less than the critical threshold, *a*, given the predicted value $\hat{IM}(t)$ (if the alarm is raised, it becomes an actual probability of false alarm):

$$P_{fa}(t) = \mathrm{P}[IM \leq a \mid \hat{IM}(t)]. \tag{10.33}$$

Since the uncertainty in *IM* is modeled as a Gaussian distribution with mean equal to the predicted $\hat{IM}(t)$ (if there is a known bias in the prediction, it should be added to this mean) and with standard deviation σ_{tot}(t), evaluated as a function of the updated uncertainties for the earthquake magnitude and location, it follows that:

$$P_{fa}(t) = \int_{-\infty}^{a} \frac{1}{\sigma_{tot}(t)\sqrt{2\pi}} \exp\left[\frac{-(IM - \hat{IM}(t))^2}{2\sigma_{tot}(t)^2}\right] dIM = \Phi\left(\frac{a - \hat{IM}(t)}{\sigma_{tot}(t)}\right) \tag{10.34}$$

where Φ is the standard Gaussian cumulative distribution function. The potential probability of a missed alarm is equal to the probability of *IM* being greater than the critical threshold (if the alarm is not raised, it becomes the probability of a missed alarm):

$$P_{ma}(t) = \mathrm{P}[IM > a \mid \hat{IM}(t)] \tag{10.35}$$

and

$$P_{ma}(t) = \int_a^\infty \frac{1}{\sigma_{tot}(t)\sqrt{2\pi}} \exp\left[\frac{-(IM - \hat{IM}(t))^2}{2\sigma_{tot}(t)^2}\right] dIM = 1 - \Phi\left(\frac{a - \hat{IM}(t)}{\sigma_{tot}(t)}\right). \tag{10.36}$$

Since the two conditions ($IM \le a$ and $IM > a$) are mutually exclusive and exhaustive, the probabilities $P_{fa}(t)$ and $P_{ma}(t)$ always sum to one.

10.5.2 Decision Making during the Seismic Event

The potential probabilities of false and missed alarms in Eqs. 10.34 and 10.36 provide fundamental guidelines for the user's decision making during the seismic event, since they quantify the reliability of the information provided by the EWS. While the event is occurring, the decision to raise the alarm can be based on real-time monitoring of the probability of wrong decisions, focusing on the situation (false alarm or missed alarm) that the user is more concerned about. It will be demonstrated in this section that this can be done by monitoring whether the predicted intensity measure exceeds a time-varying warning threshold, $c(t)\ a$.

The proposed procedure is as follows. Using Eqs. 10.34 and 10.36, the value of the probabilities $P_{fa}(t)$ and $P_{ma}(t)$ are evaluated as time goes by during an event and they are compared to the tolerable values, which may be established based on cost-benefit considerations (see Sect. 10.4). The alarm is raised when either $P_{fa}(t)$ or $P_{ma}(t)$ reaches its tolerable value, β and α, respectively. This assumes that the time available is sufficient for activation of the protective measures. If the time available reaches the minimum time necessary for activation of the protective measure, the alarm will be raised only if the probability of a wrong decision, evaluated at that time, can be accepted.

For the case of a missed alarm, the condition $P_{ma} > \alpha$ and Eq. 10.36 give the time-varying expression for the warning threshold as follows:

$$P_{ma}(t) > \alpha \Leftrightarrow \hat{IM}(t) > a\left[1 - \frac{\sigma_{tot}(t)\Phi^{-1}(1-\alpha)}{a}\right] = c_{ma}(t) \cdot a. \tag{10.37}$$

Therefore, the setting of the alarm based on the probability of a missed alarm becoming unacceptable occurs if $\hat{IM}(t) > c_{ma}(t) \cdot a$ where:

$$c_{ma}(t) = 1 - \frac{\sigma_{tot}(t) \cdot \Phi^{-1}(1-\alpha)}{a}. \tag{10.38}$$

The alarm is also set if the probability of a false alarm falls below the tolerable level β and based on Eq. 10.34:

$$P_{fa}(t) < \beta \Leftrightarrow I\hat{M}(t) > a\left[1\text{-}\frac{\sigma_{tot}(t)\Phi^{-1}(\beta)}{a}\right] = c_{fa}(t)\cdot a \tag{10.39}$$

that is, the alarm is set if $I\hat{M}(t) > c_{fa}(t)\cdot a$ where:

$$c_{fa}(t) = 1 - \frac{\sigma_{tot}(t)\cdot\Phi^{-1}(\beta)}{a}. \tag{10.40}$$

Notice that if β<1-α, then $c_{ma}(t) < c_{fa}(t)$ and so the concern about missing an alarm will control the setting of the alarm; on the other hand, if β>1-α, then concern about causing a false alarm will control the setting of the alarm. Of course, making an alarm decision based on the exceedance of the predictor above the time-varying warning threshold is equivalent to monitoring the probability of $P_{fa}(t)$ and $P_{ma}(t)$ and raising the alarm based on exceedance of the tolerable level β and α, respectively.

It was pointed out in Sect. 10.4 that when the tolerable probabilities β and α to use during operation are based on cost-benefit considerations, they are related by: $\beta = 1 - \alpha$. Therefore, since both the alarm probabilities and their tolerable values sum up to one, the alarm probabilities will reach their critical thresholds at the same time, so one can choose to monitor either $P_{fa}(t)$ and $P_{ma}(t)$. Similarly, if the predictor $I\hat{M}(t)$ is monitored, the critical thresholds, $c_{ma}(t)a$ and $c_{fa}(t)a$, are equal and so are reached at the same time.

10.6 Application of EWS

10.6.1 Pre-installation Analysis: Southern California

We suppose the future realization of a seismic EWS for the protection of facilities in Southern California and examine the question of the feasibility of applications of interest to the potential end-user. In particular, we address the question: How would an EWS perform during earthquakes that might occur in the area, viewed in terms of false and missed alarms? We can quantify the effectiveness of the EWS application by providing to the end-user the probabilities of wrong decisions to see whether they are acceptable and we can also set the warning threshold in order to match the user's requirements.

To illustrate the process, we choose a hazard function appropriate for the Los Angeles area to evaluate the probabilities of wrong decisions as a function of the warning threshold *c.a* and demonstrate that it can be set based on the tolerable level of probabilities of wrong decisions that come from a cost-benefit analysis. We assume that the critical threshold, *a*, has been selected by the user based on the EWS application of interest.

The Virtual Seismologist method (Cua and Heaton 2004; Cua 2004) is chosen to give the earthquake predictive model (M_1 in Fig. 10.1) and the attenuation model (M_2 in Fig. 10.1) which in the VS method (Cua and Heaton 2004) is defined as:

$$IM = \log_e PGA = aM - b[R_1 + C(M)] - d\log_e[R_1 + C(M)] + e + \varepsilon \qquad (10.41)$$

where M is the magnitude; R_1 depends on R, which is the epicentral distance; $C(M)$ is a correction factor depending on magnitude; the residual term ε is a zero-mean error term representing the prediction uncertainty and e is a constant error which includes station corrections; the parameters a, b, d, e were estimated by Cua and Heaton during the calibration of the model from data for different soil types and for rock, they are: $a = 0.779$, $b = 2.55 \cdot 10^{-3}$, $d = 1.352$, $e = -0.645$; and ε is Gaussian (0, 0.243), i.e. with zero mean and standard deviation 0.243.

The intensity measure *IM* is the peak ground acceleration (PGA) on a $\log_{10}$ scale. The chosen hazard function for the Los Angeles area is shown in Fig. 10.5 and it represents the mean rate of exceedance versus PGA in g units. The hazard function is fitted, as described previously, by minimizing the relative entropy in order to estimate the coefficient k_l to describe the probability distribution for *IM* and a value of k_l=1.06 is obtained.

To simulate the behaviour of the EWS from *IM* to the predicted value, $\hat{IM}$, we need to know the total error associated with this process, as defined in the earlier uncertainty propagation analysis. It is assumed that the errors associated with the magnitude and location estimation and the attenuation model (Cua and Heaton 2004) are described by:

- ε_M : Gaussian (0, 0.5)
- ε_R : ignored at this stage
- ε_{IM} : Gaussian (0, 0.243)

Therefore, the total error associated with the predicted $\hat{IM}$, as given by Eqs. 10.8 and 10.9, is:

- ε_{tot} : Gaussian (0, 0.44)

The probabilities of false and missed alarms are evaluated based on Eqs. 10.20 and 10.23 and are shown in Figs. 10.6 and 10.7 as a function of the warning threshold factor c for different values of the critical threshold a. Notice that for P_{fa} in the approximate range of 0.05 to 0.4, the choice of c is insensitive to a. If it is false alarms that the user is more concerned about, the warning threshold can be set so that the tolerable level of probability of a false alarm is not exceeded. For example, if this tolerable level is P_{fa} = 0.4, then the warning threshold ca will be set equal to 2.22, as shown in Fig. 10.8. Based on this warning threshold, the probability of a missed alarm that has to be accepted can then be evaluated as P_{ma} = 0.05, as shown in Fig. 10.9 by the point on the curve corresponding to ca=2.22.

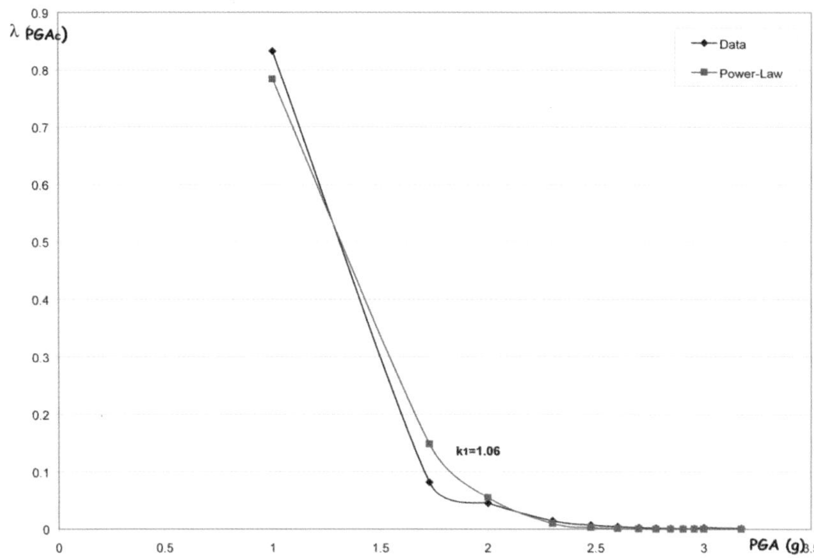

Fig. 10.5 Fitting the hazard function for Los Angeles area with a power-law in PGA

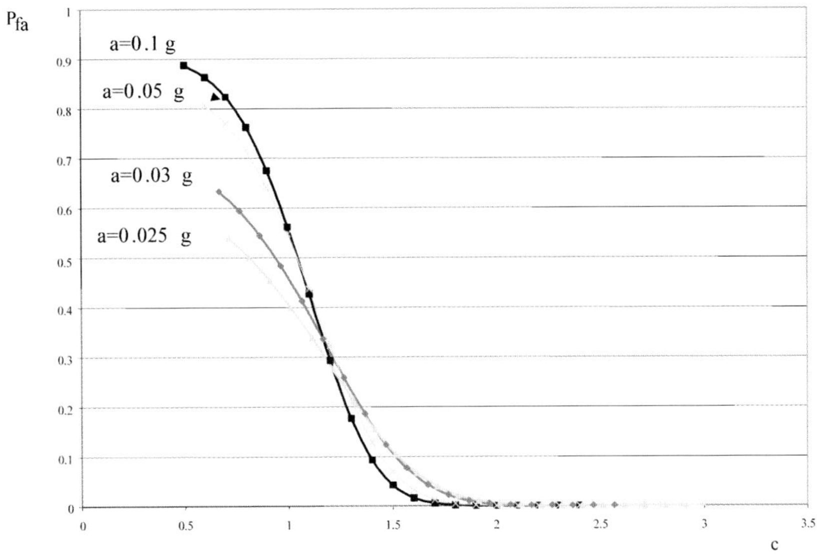

Fig. 10.6 Probability of false alarm as a function of warning threshold factor *c* for different critical thresholds, *a*, expressed in g's.

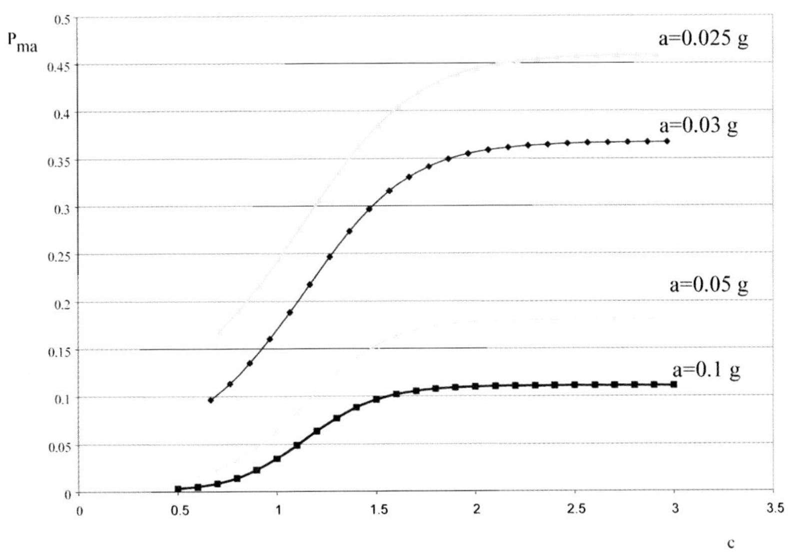

Fig. 10.7 Probability of missed alarm as a function of warning threshold factor *c* for different critical thresholds, *a*, expressed in g's.

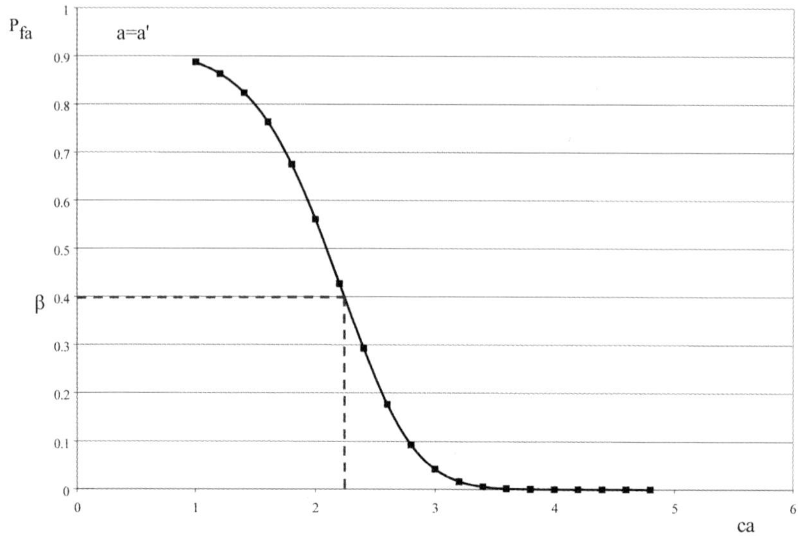

Fig. 10.8 Probability of false alarm for a given critical threshold a′ as a function of the warning threshold.

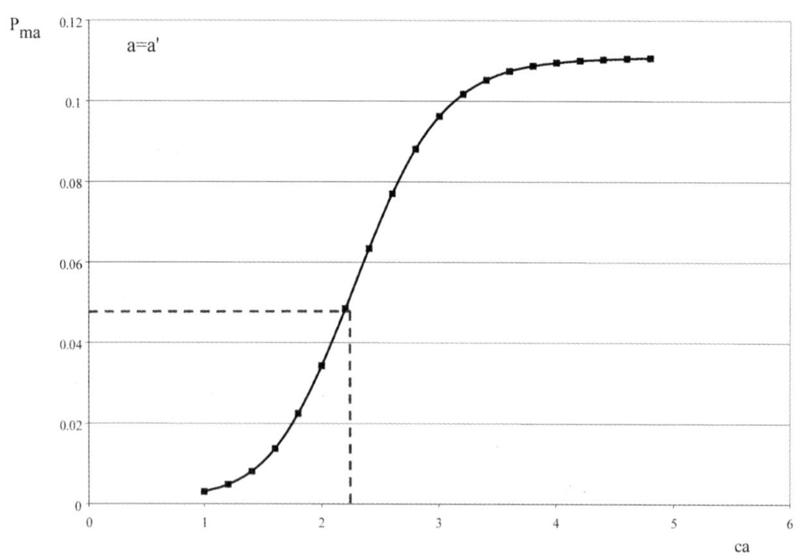

Fig. 10.9 Probability of missed alarm for a given critical threshold a′ as a function of the warning threshold.

10.6.2 Yorba Linda Earthquake: *M*=4.75

On 3 September 2002, the Yorba Linda earthquake occurred in Orange County, California, with magnitude 4.75. The epicenter has been located at 33.9173^0 N and 117.7758^0 W with a depth of 12.92 Km. The area of the main shock is densely instrumented and the first station that was triggered by the event was the Serrano station at 9.9 Km from the epicenter. To simulate how the decision process would operate during a seismic event, the Virtual Seismologist magnitude and location estimates are used (Cua and Heaton 2004; Cua 2004).

The Virtual Seismologist method is based on continual Bayesian updating of the predictions of magnitude and location as an increasing amount of data arrives from newly triggered stations and from continued recording of data from the stations already triggered. Different prior information can be considered in the method, including the Gutenberg-Richter law for a prior probability density (PDF) on magnitude, Voronoi cells to define the most likely locations, and recently observed seismicity to take into account any foreshocks that may have occurred in the area of interest in the 24 hours before the main shock.

As the first station is triggered, the Voronoi cell related to this station gives an area of possible locations since the cell is defined as all epicentral points that would give an event that is first recorded at the station. As a consequence, it is possible to define the prior PDF of most likely locations; the successive P arrivals that trigger other stations then provide additional information for the Bayesian updating of the prediction of epicentral distance. The most likely predictions of magnitude and location are those that maximize the joint posterior PDF given by:

$$p\left(M,R\,|\,\text{data}\right)\propto p(\text{data}\,|\,M,R)\cdot p(M,R)\,. \tag{10.42}$$

The predictions from the Virtual Seismologist method for the magnitude and epicentral distance are available at 5, 10, 15, 20 and 50 seconds after the first station was triggered by the Yorba Linda earthquake. The epicentral distance prediction is considered time invariant, but the magnitude uncertainty is assumed to decrease as $1/\sqrt{N}$ where N is the number of stations contributing information (Cua and Heaton 2004). The updating of magnitude uncertainty is continued because the error associated with magnitude is more influential in the *IM* prediction process than the error associated with epicentral distance.

Using the temporal sequence of updated magnitude and location predictions given by Cua and Heaton (2004) for the Yorba Linda earthquake, we estimate the peak ground acceleration in $\log_{10}$ scale based on their attenuation relationship. The total uncertainty is updated every second as a function of the magnitude uncertainty and attenuation model uncertainty using Eq. 10.8 (neglecting location uncertainty, as mentioned above, because it is not as influential as the other two sources of uncertainty). The potential probabilities of wrong decisions are then evaluated at each second from Eqs. 10.34 and 10.36. The results as a function of time are presented in Fig. 10.10, which shows the PGA prediction $\hat{IM}(t)$ (most probable value of $IM = \log_{10}$PGA at time t), the standard deviation σ_{tot} of the total error associated with the prediction and the potential probabilities of wrong decisions that were evaluated assuming the critical threshold for PGA is a=0.025g (corresponding to a=1.4 in $\log_{10}$ scale in cm\s\s). These potential probabilities of wrong decisions may be used to make a decision during the event of whether to raise the alarm or do nothing, based on tolerable values of P_{fa} or P_{ma} derived from a cost-benefit analysis. For example, Fig. 10.11 shows the plot of P_{fa} as a function of time and at about 28 seconds after the first station is triggered by the event, the tolerable value, $\beta = 0.4$, is reached and so the alarm would have then been raised (assuming that the minimum warning time for activation of the protective measure had not been reached before then).

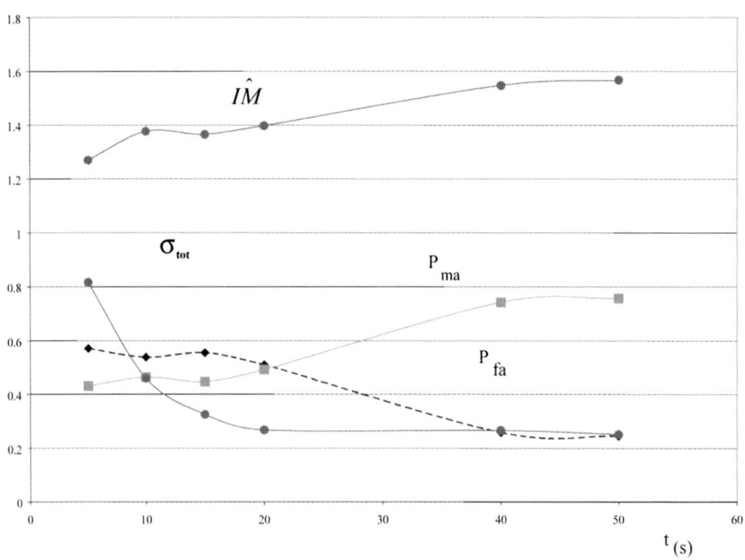

Fig. 10.10 Yorba Linda 2002: Evolution of the prediction of IM, the standard deviation of magnitude prediction and the probabilities of wrong decisions (false and missed alarms) with time.

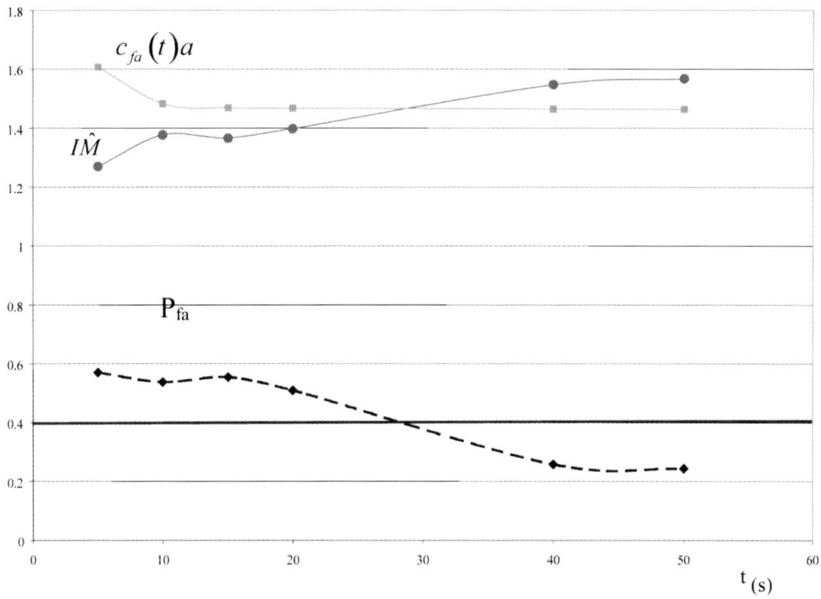

Fig. 10.11 Yorba Linda 2002: Decision making based on tolerable probability of false alarm.

Alternatively, the threshold level $c_{fa}(t)$ in Eq. 10.40 can be determined and the alarm raised when $I\hat{M}(t)$ exceeds $c_{fa}(t)a$, which, of course, also occurs at about 28 seconds.

10.7 Concluding Remarks

The theory presented here using the ground motion intensity as the predictor could be readily extended to consider other quantities of interest that more closely represent the consequences of concern to the user, such as in terms of structural response (for example, interstory drift), non-structural and structural damage, safety or economic losses. These consequences can be predicted based on facility-specific loss estimation methods, such as the performance-based earthquake engineering (PBEE) framework (Porter et al. 2002, 2004).

The setting of the alarm during real-time operation of the EWS could then be based on monitoring the probabilities that specified engineering or economic parameters of interest would exceed their critical thresholds. The analysis is a natural extension of the previous theory; for example, to extend the probability of a missed alarm based on *IM* to that for an engineering demand parameter (*EDP*):

$$P_{ma}(t) = \mathrm{P}[EDP > e \mid I\hat{M}(t)] \qquad (10.43)$$
$$= \int_{-\infty}^{\infty} \mathrm{P}\left(EDP > e \mid IM\right)\mathrm{p}\left(IM \mid I\hat{M}(t)\right) dIM$$

where e is the critical threshold value of the *EDP*, $\mathrm{p}(IM \mid I\hat{M}(t))$, is a Gaussian PDF as before (see the integrand in Eq. 10.36) and $\mathrm{P}(EDP > e \mid IM)$ comes from a seismic vulnerability analysis for the facility (Porter et al. 2002, 2004). Similarly, the probability of a missed alarm can be extended to damage or loss estimation:

$$P_{ma}(t) = \mathrm{P}[DV > d \mid I\hat{M}(t)] \qquad (10.44)$$
$$= \int_{-\infty}^{\infty} \int_{0}^{\infty} \mathrm{P}\left(DV > d \mid EDP\right)\mathrm{p}\left(EDP \mid IM\right)\mathrm{p}\left(IM \mid I\hat{M}(t)\right) dIMdEDP$$

where DV is a decision variable quantifying the damage or loss and d is the critical threshold value.

References

Allen RM (2004) Rapid Magnitude Determination for Earthquake Early Warning. In: Proceedings of Workshop on Multidisciplinary Approach to Seismic Risk Problems, Sant'Angelo dei Lombardi, September 22, 2003

Allen RM, Kanamori H (2003) The Potential for Earthquake Early Warning in Southern California. Science 300:786-789

Barroso LR, Winterstein S (2002) Probabilistic Seismic Demand Analysis of Controlled Steel Moment-resisting Frame Structures. Earthquake Engineering and Structural Dynamics 31:2049-2066

Bates S, Cullen A, Raftery A (2003) Bayesian Uncertainty Assessment in Multicompartment Deterministic Simulation Models for Environmental Risk Assessment. Environmetrics 14:355-371

Cua G (2004) Creating the Virtual Seismologist: developments in ground motion characterization and seismic early warning. PhD Thesis in Civil Engineering, California Institute of Technology, Pasadena, December 2004, http://resolver.caltech.edu/CaltechETD:etd-02092005-125601

Cua G, Heaton T (2004) Characterizing Average Properties of Southern California Ground Motion Envelopes. In: 2004 SCEC Annual Meeting Proceedings and Abstracts, vol XIV

Cua G, Heaton T (2004) Illustrating the Virtual Seismologist (VS) Method for Seismic Early Warning on the 3 September 2002 M=4.75 Yorba Linda, California Earthquake. In: 2004 SCEC Annual Meeting Proceedings and Abstracts, vol XIV

Grasso VF, Beck JL, Manfredi G (2005a) Seismic Early Warning Systems: Procedure for Automated Decision Making. Earthquake Engineering Research Laboratory, California Institute of Technology, Report Number EERL 2005-02

Grasso VF, Iervolino I, Occhiuzzi A, Manfredi G (2005b) Critical Issues of Seismic Early Warning Systems for Structural Control. In: Proceedings of 9th International Conference on Structural Safety and Reliability, Rome, Italy, June 2005

Kanda K, Kobori T, Ikeda Y, Koshida H (1994) The Development of a Pre-arrival Transmission System for Earthquake Information Applied to Seismic Response Controlled Structures. In; Proceedings of 1st World Conference on Structural Control, 2, IASC, California, USA, 1994

Kramer SL (1996) Geotechnical Earthquake Engineering. Prentice-Hall

Lee WHK, Espinosa-Aranda JM (1998) Earthquake Early Warning Systems: Current Status and Perspectives. In: Proceedings of International Conference on Early Warning Systems for Natural Disaster Reduction, pp 409-423

Occhiuzzi A, Grasso VF, Manfredi G (2004) Early Warning Systems from a Structural Control Perspective. In: Proceedings 3rd European Conference on Structural Control, Vienna University of Technology, Austria, July 2004

Paté-Cornell E (1986) Warning Systems in Risk Managment. Risk Analysis 6(2):223-234

Porter KA, Beck JL, Shaikhutdinov RV (2002) Sensitivity of Building Loss Estimates to Major Uncertain Variables. Earthquake Spectra 18:719-743

Porter KA, Beck JL, Shaikhutdinov RV, Au SK, Mizukoshi K, Miyamura M, Ishida H, Moroi T, Tsukada Y, Masuda M (2004) Effect of Seismic Risk on Lifetime Property Values. Earthquake Spectra 20:1211-1237

Saita J, Nakamura Y (1998) UrEDAS: The Early Warning System for Mitigation of Disasters caused by Earthquakes and Tsunamis. In: Proceedings of International Conference on Early Warning Systems for Natural Disaster Reduction, EWC98, pp 453-460

Seismological Research Letters (1997) Issue on Ground Motion Attenuation Models. 68(1)

Wald A (1947) Sequential Analysis. J. Wiley & Sons, New York, and Chapman & Hall, London

Wieland M (2001) Earthquake Alarm, Rapid Response and Early Warning Systems: Low Cost Systems for Seismic Risk Reduction. In: Proceedings of International Workshop on Disaster Reduction, Reston, Virginia, U.S., August 2001

Wieland M, Griesser L, Kuendig C (2000) Seismic Early Warning System for a Nuclear Power Plant. In: Proceedings of 12th World Conference on Earthquake Engineering, Auckland, New Zealand, 2000

11 The Crywolf Issue in Earthquake Early Warning Applications for the Campania Region

Iunio Iervolino[1], Vincenzo Convertito[2], Massimiliano Giorgio[3], Gaetano Manfredi[1], Aldo Zollo[4]

[1] Dipartimento di Ingegneria Strutturale, Università di Napoli Federico II, Napoli, Italy
[2] Istituto Nazionale di Geofisica e Vulcanologia, Osservatorio Vesuviano, Napoli, Italy
[3] Dipartimento di Ingegneria Aerospaziale e Meccanica, Seconda Università di Napoli, Aversa, Italy
[4] RISSC-Lab, Dipartimento di Scienze Fisiche, Università di Napoli Federico II, Napoli, Italy

Abstract

Earthquake early warning systems (EEWS), based on real-time prediction of ground motion or structural response measures, may play a role in reducing vulnerability and/or exposure of buildings and lifelines. Indeed, seismologists have recently developed efficient methods for real-time estimation of an event's magnitude and location based on limited information of the P-waves. Therefore, when an event occurs, estimates of magnitude and source-to-site distance are available, and the prediction of the structural demand at the site may be performed by Probabilistic Seismic Hazard Analysis (PSHA) and then by Probabilistic Seismic Demand Analysis (PSDA) depending upon EEWS measures. Such an approach contains a higher level of information with respect to traditional seismic risk analysis and may be used for real-time risk management. However, this kind of prediction is performed in very uncertain conditions which may affect the effectiveness of the system and therefore have to be taken into due account. In the present study the performance of the EWWS under development in the Campania region (southern Italy) is assessed by simulation. The earthquake localization is formulated in a Voronoi cells approach, while a Bayesian method is used for magnitude estimation. Simulation has an empirical basis but requires no recorded signals. Our results, in terms of hazard analysis and false/missed alarm probabilities, lead us to conclude that the PSHA depending upon the EEWS significantly improves seismic risk prediction at the site and is close to what could be produced if magnitude and distance were deterministically known.

11.1 Introduction

Seismic risk management consists of: (1) Risk mitigation by vulnerability or exposure reduction; (2) Emergency rapid response. Emergency preparedness is a near-real-time issue; risk mitigation strategies are typically mid-term (i.e. seismic retrofit of structures and infrastructures) or long-term actions (e.g. land use planning or development of appropriate design standards). An earthquake early warning and rapid response system can provide the critical information needed: (i) to minimize loss of lives and property, and (ii) to direct rescue operations (Wieland 2001). Therefore early warning systems may play a role in both of the risk management issues (Iervolino et al. 2007, this issue). In particular, in near-real-time applications shake maps, which are territorial distributions of ground shaking, are provided by a regional seismic network and already used for emergency management (Wald et al. 1999, Kanamori 2005, Convertito et al. 2007, this issue). On the other hand, seismic early warning systems are now capable of providing, from a few seconds to a few tens of seconds before the arrival of strong ground shaking, a prediction of the ground motion or the seismic demand on structures caused by a large earthquake. Therefore they may be used to take real time action for vulnerability or exposure reduction in the light of seismic risk management.

Earthquake early warning systems (EEWS) may simplistic be classified as *regional* or *site-specific*. Regional EEWS consist of wide seismic networks covering a portion of the area threatened by quake strike. Such systems are designed to provide real-time, or near-real-time information suitable for spreading the alarm to the community or inferring data (i.e. shake maps). Site-specific EEWS also enhance the safety margin of specific critical engineered systems such as nuclear power plants (Wieland 2000) or lifelines. The networks devoted to site specific EEW are much smaller than those of the regional type, only covering the surroundings of the system. The location of the sensors depends on the lead time needed to activate the safety procedures before the arrival of the more energetic seismic phase (i.e. S or superficial waves). Typically the alarm is issued when the ground motion at one or more sensors exceeds a given threshold; uncertainty, in this case, is often neglected since the path between the network and the site is limited.

Due to a large and rapid development of regional networks in recent years worldwide (see SAFER 2005 for example) the question of using EEWS for structure-specific applications is being raised (Iervolino et al. 2005). EEWS predictions may be used for the real time set-up of active or semi-active structural control, in order to achieve a safer structural response to ground motion. The “early” information provided by the regional EEWS in the first seconds of the event can be still used, in case of alarm, to activate different types of security measures, such as the shutdown of critical systems, evacuation of buildings, stopping of high speed trains (Veneziano and Papadimitriou 1998) and shut-off of valves in gas and oil pipelines.

Whether a structure-specific application of a regional EEWS is feasible is the topic of the study presented herein. In this case the ground motion Intensity Measure (IM) or the Engineering Demand Parameter (EDP) of interest has to be estimated far from the sensor network’s recordings and cannot be measured at the site. A scheme of the *hybrid* application of a regional network for structure-specific earthquake early warning is shown in Fig. 11.1.

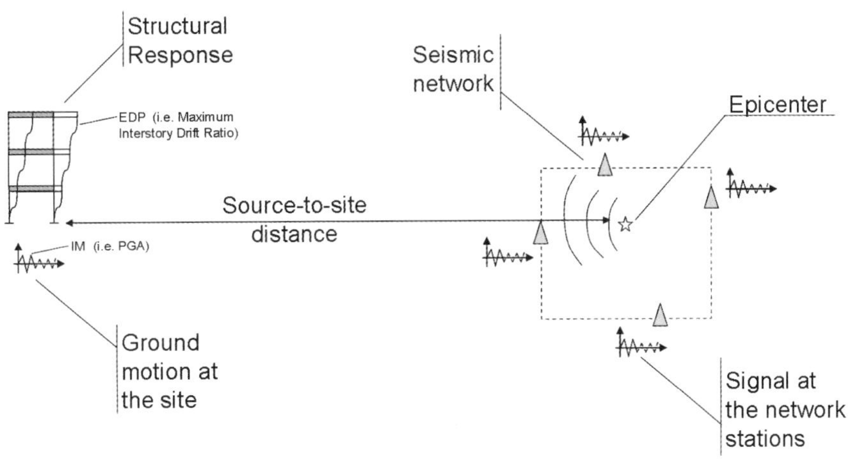

Fig. 11.1 Regional EWWS for structure-specific applications.

While the system captures the earthquake's features and then predicts IM and/or EDP at the site of interest to give additional lead time, this also entails significant uncertainty[1] which may lead to false and missed alarms. Alerting or not alerting both have a cost; in the case of not alerting the loss is associated to an earthquake striking without any countermeasure being taken; in the case of alarm, preparing interventions has a cost (social and/or economic) which may transform into loss if the actual ground motion does not require such action. Therefore, a key issue in assessing EEWS performance is the estimation of missed and false alarm (MA and FA respectively) probabilities associated to the adopted decisional rule (Patè-Cornell 1986). Computation of MA and FA rates of occurrence on an empirical basis should consist of post-event analysis of EEWS predictions and would require a large strong-motion waveforms database both for the network and the site where the structure is located. Since such databases are very rarely available, especially for large earthquakes, the I and II type error probabilities related to MA and FA may be estimated in a simulation framework using appropriate characterizations of the uncertainties involved in the prediction. This approach requires virtually no records other than those used to calibrate the method adopted for the estimation of Magnitude (M) and source-to-site distance (R).

11.2 Seismic Risk Analysis Conditioned to the Earthquake Early Warning System

Recently seismologists have developed several methods to estimate an event's magnitude based on limited information of the P-waves (e.g. first few seconds of velocity recording) for real-time applications (Allen and Kanamori 2003). Similarly, as briefly described below, the source-to-site distance may be predicted by a sequence of network stations triggered by the developing earthquake (Satriano et al. 2007, this issue). Therefore, since it may be assumed that at a given instant estimates of M and R are available, the prediction of the ground motion at the site can be performed in analogy with Probabilistic Seismic Hazard Analysis (PSHA) (Cornell 1968, McGuire 1995). This results in a seismic hazard analysis conditioned (in a probabilistic sense) by the real-time information given by the EEWS. Consequently, the distribution of the structural response may also be computed by Probabilistic Seismic Demand Analysis or PSDA (Carballo and Cornell 2000, Cornell et al. 2002) provided there is an IM-EDP relationship for the structure of interest. It is easy to recognize that the

[1] It is worth noting that the site-specific EEWS reads IM directly while the regional predicts IM/EDP which is a more uncertain process.

probability density function of the structural response at the site when an event is occurring contains the highest level of information available and therefore is the best tool for real-time decision making.

11.2.1 EWWS-conditioned PSHA and PSDA

Let us assume that at a given time t from the earthquake's origin time, the seismic network can provide estimates of M and R. These probability density functions (PDFs) are intrinsically conditioned to a vector of measures, say $\{\tau_1,\tau_2,...,\tau_v\}$ where v is the number of instruments at which the measure of interest is available. Then the PDF of M has to be indicated as $f_{M|\tau_1,\tau_2,...,\tau_v}(m\,|\,\tau_1,\tau_2,...,\tau_v)$; similarly the PDF of R, which for the method used only depends on the sequence of stations triggered, will be referred to as $f_{R|s_1,s_2,...,s_v}(r\,|\,s_1,s_2,...,s_v)$ where $\{s_1,s_2,...,s_v\}$ is such a sequence. Thus it is possible to compute the probabilistic distribution (or hazard curve) of a ground motion Intensity Measure (i.e. Peak Ground Acceleration or PGA) at the site, in analogy with the seismic hazard integral reported in Eq. (11.1), repeating it for several values of IM.

$$f_v(im) = \int_M \int_R f(im\,|\,m,r)\, f_{M|\tau_1,\tau_2,...,\tau_v}(m\,|\,\tau_1,\tau_2,...,\tau_v)\, f_{R|s_1,s_2,...,s_v}(r\,|\,s_1,s_2,...,s_v)\, dr\, dm \quad im \in [0,+\infty] \tag{11.1}$$

where the PDF, $f(im\,|\,m,r)$, is given by an attenuation relationship as in the ordinary PSHA. The subscript v indicates that the computed hazard curve refers to a particular set of triggered stations and changes when a large amount of data is included in the process (e.g. more stations are triggered as time flows).

For structural applications of the EEWS the prediction of the structural response in terms of an Engineering Demand Parameter (EDP), rather than in terms of a ground motion IM, may be of prime concern. This requires a further integration to get the PDF of EDP as reported in Eq. (11.2).

$$f_v(edp) = \int_{IM} f(edp\,|\,im)\, f_v(im)\, dim \quad edp \in [0,+\infty] \tag{11.2}$$

where the PDF, $f(edp\,|\,im)$, is the required probabilistic relationship between IM and EDP. If a Moment Resisting Frame (MRF) structure is concerned, for example, the PSDA procedure allows to obtain the relation expressed in Eq. (11.3) between the Maximum Inter-storey Drift Ratio (MIDR) and $Sa(T_1)$ (first mode spectral acceleration), which are the IM and EDP respectively.

$$MIDR = a\left(Sa(T_1)\right)^b \varepsilon \tag{11.3}$$

where the log of ε is a normal random variable with zero-mean and variance equal to the variance of the logs of MIDR, and the coefficients *a* and *b* are obtained via non-linear incremental dynamic analysis (Vamvakistos and Cornell 2000). Barroso and Winterstein (2002) have proposed a similar relationship for controlled structures.

For the sake of simplicity it will be assumed in the following that the parameter of interest is IM. This keeps the presentation of the method clear and ensures the results of the application are easier to interpret. Since EPD is only a probabilistic transformation of IM this choice does not affect the generality of the discussion.

11.2.2 Magnitude Estimate

The integral given in Eq. (11.1) requires the distribution of magnitude estimated on the basis of data provided by the network at a given time. Allen and Kanamori (2003) provide the relationship between the magnitude of the event and the log of the predominant period $\tau_{P,max}$ (simply τ herein) of the first four seconds of the P-waves for the TriNet network. It has been assumed that the distributions of τ, conditioned to the magnitude of the event $f_{\tau|M}(\tau \mid m)$, are lognormal. The mean of the logs and variance, retrieved from the data in the homoscedasticity hypothesis (Fontanella 2005), are reported in Eq. (11.4).

$$\begin{cases} \mu_{\log(\tau)} = \dfrac{(M-5.9)}{7} \\ \sigma_{\log(\tau)} = 0.16 \end{cases}. \tag{11.4}$$

These distributions enable us to compute the estimation of magnitude, $f_{M|\tau_1,\tau_2,\ldots,\tau_v}(m \mid \tau_1,\tau_2,\ldots,\tau_v)$, using a Bayesian approach. In fact, if at a given time only one station is triggered measuring τ_1 from the first four seconds of the signal, the sought distribution of magnitude, conditioned to such measurement, $f_{M|\tau_1}(m \mid \tau_1)$, is the *posterior* of Eq. (11.5).

$$f_{M|\tau_1}(m \mid \tau_1) = \frac{f_{\tau_1|M}(\tau_1 \mid m) f_M(m)}{\int_{M_{MIN}}^{M_{MAX}} f_{\tau_1|M}(\tau_1 \mid m) f_M(m) dm} \tag{11.5}$$

where $f_M(M)$ is the *a priori* PDF of the magnitude, Eq. (11.6), from the Gutenberg-Richter recurrence relationship and the denominator is the marginal distribution of τ, $f_{\tau_1}(\tau_1)$.

$$f_M(m): \begin{cases} \dfrac{\beta e^{-\beta m}}{e^{-\beta M_{\min}} - e^{-\beta M_{\max}}} & M_{\min} \le m \le M_{\max} \\ 0 & m \notin [M_{\min}, M_{\max}] \end{cases}. \tag{11.6}$$

As time elapses, the number of stations which may be included in the magnitude estimation increases, new data are therefore available, and the posterior distribution may then be updated. At the time when a number ν of stations have measured τ, Eq. (11.5) can be generalized as Eq. (11.7).

$$f_{M|\tau_1,\tau_2,\ldots,\tau_\nu}(m \mid \tau_1,\tau_2,\ldots,\tau_\nu) = \frac{f_{\tau_1,\tau_2,\ldots,\tau_\nu|M}(\tau_1,\tau_2,\ldots,\tau_\nu \mid m) f_M(m)}{\int_{M_{MIN}}^{M_{MAX}} f_{\tau_1,\tau_2,\ldots,\tau_\nu|M}(\tau_1,\tau_2,\ldots,\tau_\nu \mid m) f_M(m) dm}. \tag{11.7}$$

Assuming that, conditionally upon M, the τ measurements are stochastically independent, then $f_{\tau_1,\tau_2,\ldots,\tau_\nu|M}(\tau_1,\tau_2,\ldots,\tau_\nu \mid m) = \prod_{i=1}^{\nu} f_{\tau_i}(\tau_i \mid m)$ which is the product of known terms. Therefore Eq. (11.7) may be rewritten as Eq. (11.8) which, applied for all the values of $m \in [M_{\min}, M_{\max}]$, gives the full magnitude PDF to be plugged into the PSHA integral.

$$f_{M|\tau_1,\tau_2,\ldots,\tau_\nu}(m \mid \tau_1,\tau_2,\ldots,\tau_\nu) = \frac{\left(\prod_{i=1}^{\nu} f_{\tau_i}(\tau_i \mid m)\right) f_M(m)}{\int_{M_{MIN}}^{M_{MAX}} \left(\prod_{i=1}^{\nu} f_{\tau_i}(\tau_i \mid m)\right) f_M(m) dm}. \tag{11.8}$$

It is possible to recognize that the distribution of magnitude (the same applies for distance) is indirectly dependent on time because, if at two different instants two different sets of triggered stations and measurements correspond, they will lead to two distributions of magnitude. Therefore the hazard integral in Eq. (11.1) may be re-computed at every time new stations perform measurements of τ. It will be shown in simulation how the prediction improves with time as the number of triggered stations increases.

11.2.3 Real-time Location and Distance PDF

The real-time location methodology is that of Satriano et al. (2007, this issue) which is based on the equal differential-time formulation (EDT). For a detailed discussion the reader should refer to the above author's paper in this same book and only a brief description of the main features of the procedure are given in this section for readability purposes.

The hypocentral location technique follows an evolutionary and full probabilistic approach. It relies on the stacking of EDT surfaces; this is robust in respect to outlier data (e.g. wrong signal picking in the case of concurrent events). With just one recorded arrival, hypocentral position can be already constrained by the *Voronoi cell* associated to the triggered station. As time flows and more triggers become available, the evolutionary location converges to a standard EDT location.

The algorithm defines a dense grid of points (e.g. 1 km spaced) in the space below the network. At each time step, based only on the information on which stations are triggered and which are not yet triggered, it is possible to assign, to any the of the grid points, the probability of that point being the hypocenter. This leads to the definition of a time-dependent spatial PDF for the location. Therefore, at any time t, the distance estimate in terms of $f_{R|s_1,s_2,...,s_v}(r \mid s_1,s_2,...,s_v)$ may be retrieved by a geometric transformation which associates to any particular distance a probability which is the sum of the probabilities of all points of the grid with the same distance to the site.

11.3 Decisional Rule, False and Missed Alarms

Once the EWWS provides a distribution of the ground motion intensity measure or seismic demand for the structure of interest, a decisional condition has to occur to issue the alarm. Several options are available to formulate a decisional rule, for example: (a) the alarm may be launched if the expected value $(E[IM])$ of the variable exceeds a threshold (IM_C); (b) alternatively, in a more sophisticated way, the alarm may be issued when the probability of the variable exceeding the threshold crosses a reference value (P_C). These decisional rules are given in Eqs. (11.9) and (11.10) respectively.

$$Alarm: E[IM] = \int_0^{+\infty} im\, f_v(im)\, dim > IM_C \tag{11.9}$$

$$Alarm: P[IM > IM_C] = 1 - \int_0^{IM_C} f_v(im)\, dim > P_C . \tag{11.10}$$

It is worth noting that the decisional rule (a) does not require the full computation of the hazard integral, Eq. (11.1). In fact, the expected value of IM may be well approximated by a First Order Second Moment (FOSM) method (Pinto et al. 2004), thereby reducing the computational effort. However, this decisional rule has the disadvantage of not considering the variance of IM nor the shape of its PDF. In the case of option (b) the P_C value has to be set in relation to an appropriate loss function. This second approach is more consistent with a full probabilistic approach to earthquake early warning for seismic risk management.

While performance of the early warning system may be tested to verify whether it correctly predicts the distribution of IM at the site, the efficiency of the decisional rule depends on I and II type errors which are related to the assessment of the false and missed alarm probabilities, P_{FA} and P_{MA} respectively[2]. Referring to Eqs. (11.9) and (11.10) the false alarm occurs when the EEWS issues the alarm while the intensity measure at the site IM_T (T subscript means "true" indicating the realization of the random variable to distinguish it from the prediction of the EEWS) is lower than the threshold IM_C. Probabilities of these events, Eq. (11.11), will be estimated in simulating (e.g. by a Montecarlo approach) the Campania EEWS for the decisional rules considered.

$$\begin{cases} Missed\ Alarm: \{no\ Alarm \cap IM_T > IM_C\} \\ False\ Alarm: \{Alarm \cap IM_T \le IM_C\} \end{cases} . \tag{11.11}$$

It has been discussed how the information and hence the uncertainty involved are dependent on the number of stations triggered at a certain time. Therefore, in principle, the decisional condition may be verified at any time from the triggering of the first station, and consequently the false and missed alarm probabilities are, also indirectly, a function of time. From this point of view the decisional process is again time-dependent, and one

[2] Of course the underlying hypothesis of EEWS is that it is more important to reduce missed alarms rather than false alarms, otherwise the system would be unnecessary.

may decide to alert when the trade-off between the available lead time and the losses related to a missed or false alarm is at its optimum.

11.4 Simulation of the SAMS Earthquake Early Warning System

The Campania early warning system (SAMS - Seismic Alert Management System) is based on the developing seismic network in the Apennines, spanning the regions of Campania and Basilicata (Weber et al. 2007, this issue). This network operates in the seismically most active area for Campania (100 km x 80 km wide) and is designed to acquire non-saturated data for earthquakes larger than 4 [M_w]. In Fig. 4.1 (see Chap. 4) the stations of the EW network (dark squares); the M > 2 events recorded from 1981 to 2002 and the faulting system of the Irpinia 1980 earthquake are given, showing how the network covers the most hazardous area in the region. Light squares represent additional stations which will be used to calibrate local attenuation relationships (Convertito et al. 2007, this issue).

To assess the performance of the EEWS on an empirical basis a large number of recordings should be available. In principle, to simulate the prediction of the IM at the site and comparing it with the actual value experienced by the structure, for any event, a set of recordings in each station and at the site should be available. However, it is possible to compute the false and missed alarm probabilities without data but still on an empirical basis by simulation (e.g. Montecarlo). The procedure has been implemented in a computer code and it takes advantage of empirical methods for the estimation of magnitude and distance calibrated by seismologists off-line.

Each run simulates a specific seismic event occurring in the area of interest and consists of three steps: (1) Simulation of the event's features (e.g. assignment of the event's magnitude; location and true IM at the site); (2) Simulation of the measurements and predictions (e.g. real-time PSHA) made by the network at any instant up to the triggering of all the stations; (3) Verification the decisional condition and of the false/missed alarm; (4) Count of the number of false/missed alarms to compute their frequency of occurrence. The flow chart of the simulation procedure is given in Fig. 11.2.

The site considered in the simulation is assumed to be in the city of Naples which is approximately 110km from the center of the network. In Fig. 11.3a the relative position of the network and the site are given as in the scheme in Fig. 11.1.

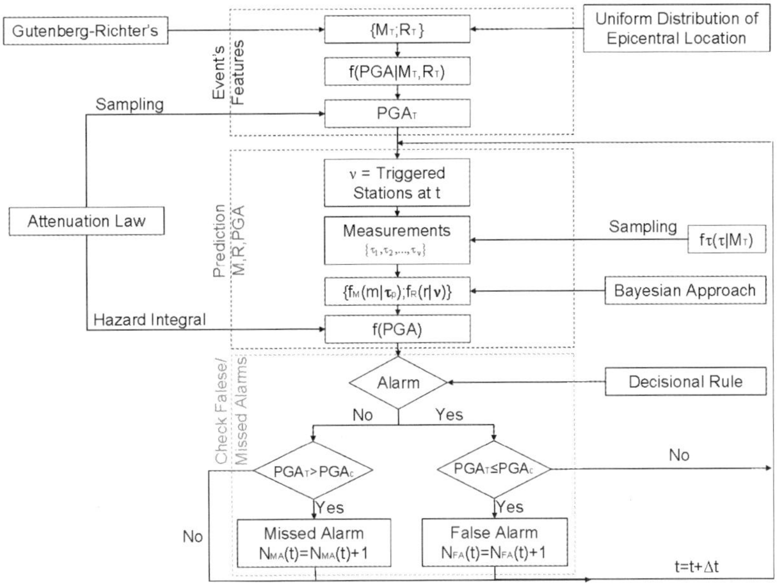

Fig. 11.2 Simulation flow chart.

11.4.1 Event and Ground Motion Feature Generation

Each run in the Montecarlo simulation starts with the generation of the geophysical features the EEWS will try to estimate. These values will completely define the earthquake of that run. In other words, since to compute the PSHA dependent on the EEW information the distributions of M and R are required, we need to establish the true value of those which will be called M_T and R_T (true magnitude and true source-to-site-distance respectively). Moreover, the ground motion intensity measure at the site (IM_T) has to be fixed; it is needed to verify the decisional condition and see whether a false or missed alarm has occurred.

The true magnitude of the event (M_T) may be sampled according to the Gutenberg-Richter recurrence relationship for the Campania region [in Eq. (11.6); $\beta = 1.69$, $M_{min} = 4$, $M_{max} = 7$]. On the other hand, one may be interested in evaluation the EEWS performance with respect to a specific magnitude; hence M_T for the all runs in the simulation has to be set at the same value. This is useful in the light of assessing the EEWS's performance in the case of high magnitude events which are the most threatening. Below, this second option will be followed for sake of clarity and readability of results.

The location of the epicenter is randomly chosen by sampling its coordinates $\left\{x_{epi}, y_{epi}\right\}$ from two s-independent uniform distributions defined in the area covered by the network. Once the epicentral coordinates are set the distance R_T to the site of interest (e.g. Naples) is readily obtained. (In Fig. 11.3b the simulated event locations in 1000 runs are given.) Again, for some purposes one may want to set the location of the epicenter at the same point for all the simulations. Therefore, in this case, the value of R_T is fixed for all the simulations.

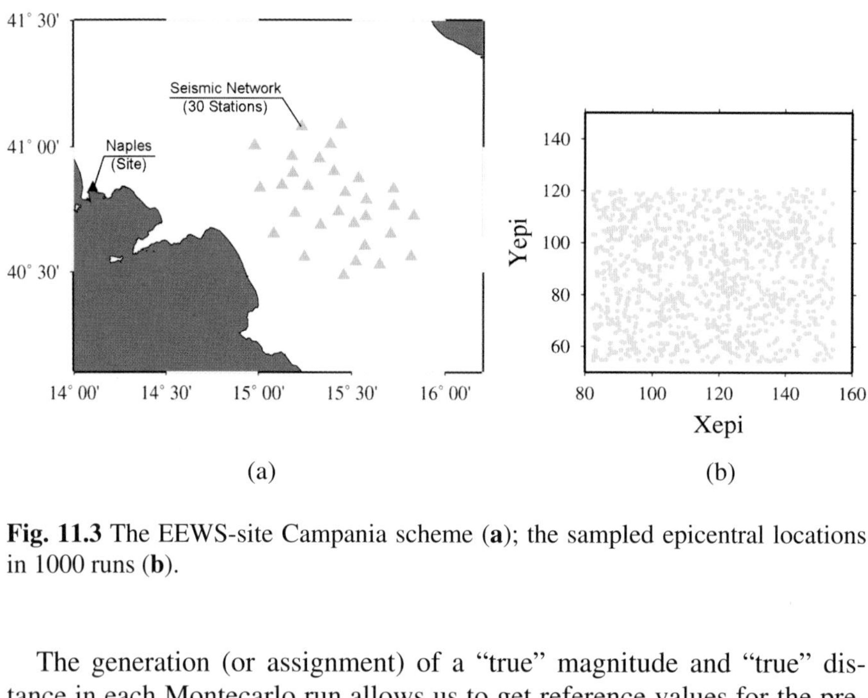

Fig. 11.3 The EEWS-site Campania scheme (**a**); the sampled epicentral locations in 1000 runs (**b**).

The generation (or assignment) of a "true" magnitude and "true" distance in each Montecarlo run allows us to get reference values for the prediction of the EEWS. However, the "true" ground motion at the site (IM_T) should also be set. It is required to verify the decisional condition: for example, it has to be compared to the expected value of IM computed by the EEWS, Eq. (11.9), to establish whether the decision adopted produced, in that run, a missed or false alarm. The value of IM_T at the site, consistent with the values of M_T and R_T, is obtained by sampling the attenuation relationship which, by definition, provides the PDF of the ground motion intensity measure conditioned to $\{M_T, R_T\}$. Herein the Sabetta and Pugliese (1996) attenuation is considered in its epicentral formulation to be consistent with the location estimation method. The considered IM is the PGA; hence, in each run the value of PGA_T is sampled from a completely specified lognormal random variable[3].

Finally the event is completely defined for EEWS purposes since $\{M_T, R_T, PGA_T\}$ of the generic run are set; the next step consists in simulating the measurements at the stations consistently with the event's features.

11.4.2 Station Measurements and M,R Real-time Distributions

In the simulation process, at any given time, the number of stations triggered is computed. This is carried out assuming a homogeneous and isotropic propagation model with P- and S-waves velocities of 5.5km/s (V_P) and 3.5km/s (V_S) respectively. This allows, for any epicentral location, determination of which stations are triggered at any time. Similarly the lead time, defined as the time required for the S-waves to hit the site, may be computed at each instant of time.

Once the event is defined by $\{M_T, R_T, PGA_T\}$, the response and predictions of the seismic network should be simulated (e.g. the measurement of τ) without any recording but consistently with the measures that would be performed in the real case. For example, let first consider the case when only one station is triggered[4]. It is possible to simulate the station's measurement by sampling the empirical distribution of the parameter to be measured conditioned to the true magnitude of the event, $f_{\tau|M}(\tau | M_T)$. Real τ values measured from recorded signals would be distributed as $f_{\tau|M}(\tau | M_T)$ by definition, and therefore such sampling is appropriate in a simulation approach.

To generate τ for more than one station it is assumed that measurements performed by different stations are s-independent conditionally upon the event's magnitude (M_T). Therefore, at a given time t when v stations are triggered, all the v component of the $\{\tau_1, \tau_2, ..., \tau_v\}$ vector are obtained by

[3] If many recorded signals were available at the site for a given magnitude, the empirical distribution of the IM as retrieved by the records should be the same as that provided by the attenuation law.

[4] Again, due to the magnitude estimation method adopted, four seconds have to elapse after the triggering of the station to include it in the estimation process.

sampling ν times the same $f_{\tau|M}(\tau | M_T)$ PDF. Since data by Allen and Kanamori (2003) are based on τ measurements on four seconds of recording, herein the working hypothesis is that any station's measurement is considered in the process if four seconds have elapsed from its triggering. Moreover, no evolution in time of the τ measure is considered.

Once the measurement vector $\{\tau_1, \tau_2, ..., \tau_\nu\}$ is available, the Bayesian method of section 2.2 may be applied to compute the distribution of magnitude. In Fig. 11.4 the resulting magnitude distributions for a simulated M 6 event are given, clearly showing that, when few stations are triggered, the distributions underestimate the magnitude.

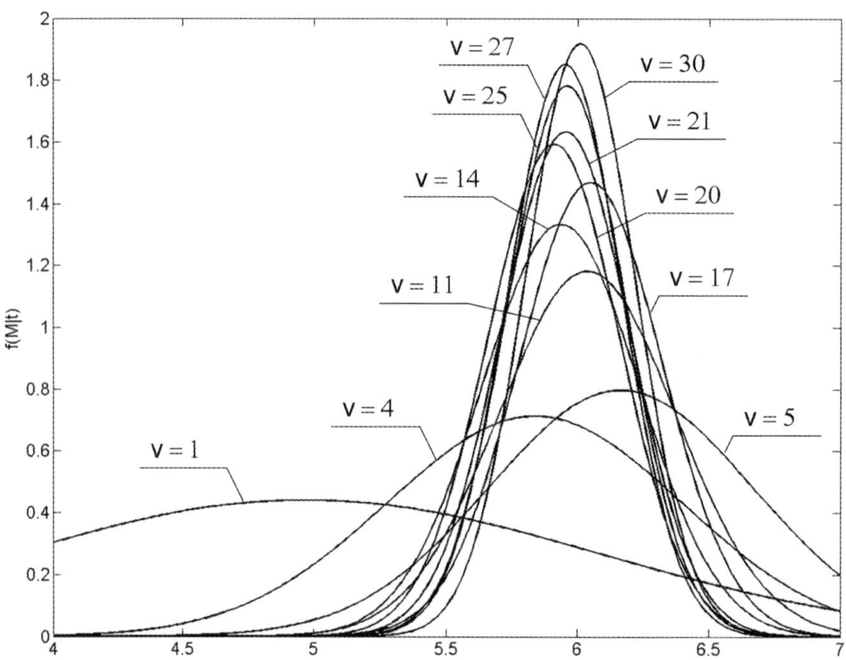

Fig. 11.4 Magnitude distribution as the number of triggered stations increases ($M_T = 6$, $R_T = 91$km).

Indeed, when few data are available, the dominating information is that *a priori* of Eq. (11.6) which naturally tends to give larger occurrence probability to low magnitude events. More precisely, the Bayesian approach will tend to produce overestimates of magnitude when it is below the *a priori* mean and it will tend to underestimate it when it is greater than the mean. This effect is directly proportional to the difference in the expected value of the *a priori* and M_T and inversely proportional to the size of measurement vector. Then, as more measurements became available, the prediction centers on the real value with a relatively small uncertainty. An estimator with these features is said to be biased by classic statisticians and other methods can be considered to obtain an unbiased estimator (i.e. maximum likelihood). However, the Bayesian approach was preferred since, albeit slightly biased, it gives, on average, significantly smaller estimation errors due to the use of the *a priori* information.

A similar observation applies to the distribution of the source-to-site distance. As earthquake location is only dependent on the sequence of stations triggered, no measurements have to be simulated to compute $f_{R|S_1,S_2,...,S_v}(r \mid S_1,S_2,...,S_v)$ once the arrival time ($t_{a,j} = R_j/V_P$ where R_j is the distance of the j-th station from the epicenter) has been computed for all the stations in the network.

The estimation process for the magnitude starts four seconds after the triggering of the first station. At that time it is assumed that the location (and therefore the distance) is known. This is not a limiting hypothesis: simulations show that the localization method after a few seconds (e.g. 3s) reduces the uncertainty on the location to about 1 km, which is negligible in respect to other uncertainties involved in the process.

11.4.3 Seismic Risk Analysis

The estimated distributions of M and R, along with the attenuation law, in the hazard integral allow us to compute the exceeding probability of PGA at the site as the event evolves and the stations trigger. The hazard curves corresponding to the event simulated in Fig. 11.4 are given in Fig. 11.5. It is possible to see the evolution of hazard which stabilizes when a large number of stations provide information about the τ measurements.

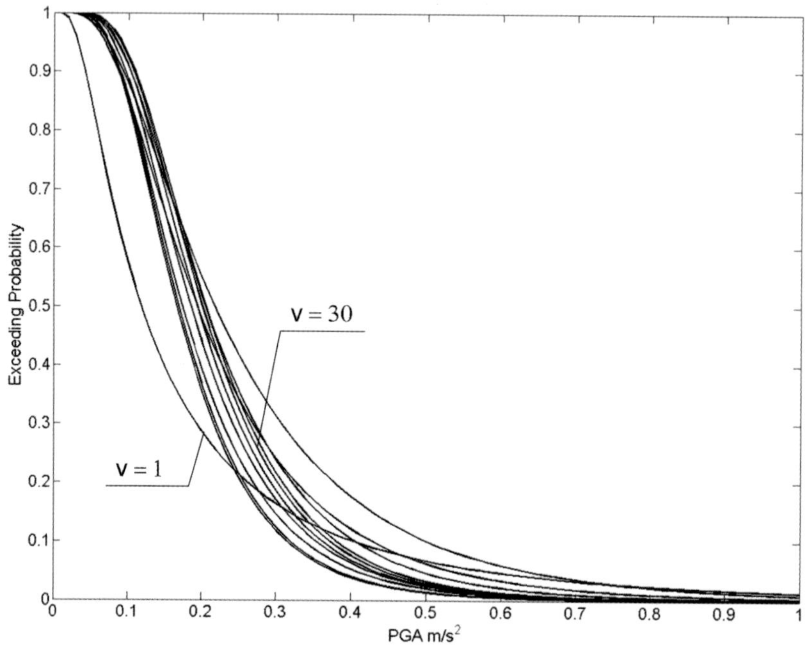

Fig. 11.5 EWWS-conditioned seismic hazard as the number of stations increases ($M_T = 6$, $R_T = 91$km).

To better understand whether the hazard computed with EEWS information is correct, it is worth comparing it to the "maximum knowledge status" of the hazard by adopting the true value of magnitude and distance (as if they were deterministically known). This corresponds to the exceeding probability of PGA when M and R are fixed. This comparison is shown in Fig. 11.6: a thick curve represents the complementary Cumulative Density Function (CDF) for the PGA when M_T (7) and R_T (110km) are known; the black curves are the results of 200 simulations. (In the figure only the hazard curves corresponding to the case when all stations triggered (ν = 30) are reported.) The EEWS hazard may be seen to correctly approximate the maximum knowledge condition.

To reduce the variability of the hazard curves, a strategy would be to increase the number of stations. Indeed, the estimation procedure of the magnitude distribution would benefit from the larger vector of information.

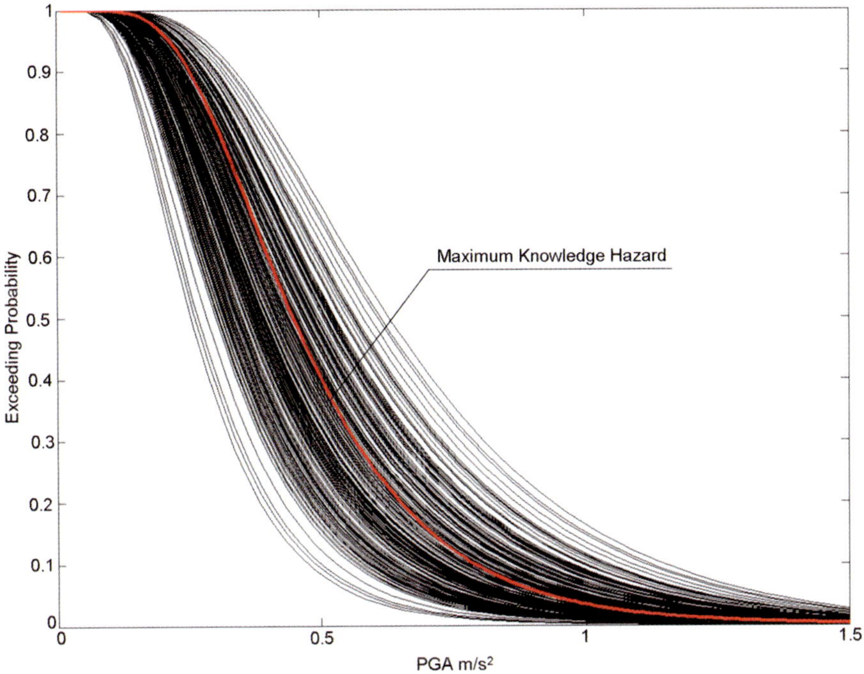

Fig. 11.6 EWWS-conditioned seismic hazard in 200 simulations compared to the maximum knowledge condition ($M_T = 7,\ R_T = 110\text{km}$).

11.4.4 False and Missed Alarm Probabilities

The simulation (Fig. 11.2) also allows the frequency of false and missed alarms, Eq. (11.11), to be computed according to the decisional rule chosen. For example, according to Eq. (11.9) these probabilities are estimated as in Eq. (11.12).

$$\begin{cases} P_{MA} \cong \dfrac{N\left[E[PGA] \le PGA_C \cap PGA_T > PGA_C\right]}{N_{TOT}} \\ P_{FA} \cong \dfrac{N\left[E[PGA] > PGA_C \cap PGA_T \le PGA_C\right]}{N_{TOT}} \end{cases} \tag{11.12}$$

where N is the number of occurrences of MA or FA and N_{TOT} is the number of simulated events. Analogously, for the decisional rule of Eq. (11.10) the probabilities are estimated as in Eq. (11.13).

$$\begin{cases} P_{MA} \cong \dfrac{N\left[P\left[PGA > PGA_C\right] \le P_C \cap PGA_T > PGA_C\right]}{N_{TOT}} \\ P_{FA} \cong \dfrac{N\left[P\left[PGA > PGA_C\right] > P_C \cap PGA_T \le PGA_C\right]}{N_{TOT}} \end{cases} \quad (11.13)$$

In Fig. 11.7 such estimations are given for M 7 (10^4 simulations) events with an epicentral distance of 110km. The PGA_C is arbitrarily set at 0.3m/s^2 and the critical probability of exceedance (P_C) is 0.2.

The real-time PSHA is performed at each second from the nucleation of the event and hence a prediction of the PGA at the site is evolving with time. Consequently; the false and missed alarm occurrence also changes with time, which has implications for the risk management strategy. For example, since the false alarm probability is decreasing for all decisional rules, alerting at a certain time means accepting some greater error probability than if the alarm were issued later, although this implies additional lead time.

To better understand the results of Fig. 11.7 it is useful to discuss the given curves. In particular we will focus on the decisional rule of Eq. (11.10). The critical value of PGA (PGA_C) is 0.3m/s^2, the true value of magnitude and distance are $M_T = 7$ and $R_T = 110\text{km}$ respectively. The chosen attenuation relationship, conditioned by M_T and R_T, gives $P\left[PGA > PGA_C\right] = 0.81$. Hence if P_C is equal to 0.2, the right decision would be to alert at every run. As a consequence, the probability of a missed alarm is zero because the alarm should always be issued and the probability of a false alarm is $P\left[PGA \le PGA_C\right]$ or 1-0.81 = 0.19. These probabilities are intrinsic to the decisional rule and the thresholds set. However, as discussed, the EEWS cannot perfectly estimate the hazard curve with M_T and R_T known (thick curve). In fact, due to the variability in the estimated hazard, the value $P\left[PGA > PGA_C\right]$ is sometimes underestimated and sometimes overestimated. For example the underestimation of $P\left[PGA > PGA_C\right]$ leads to the alarm not being given even if required and therefore the missed alarm curve is not zero. In particular, when there are few triggered stations this underestimation effect is strong and the missed alarm probability is relatively high because when the alarm is not launched (incorrectly) it will most likely result is a missed alarm. As time elapses, the estimation improves, $P\left[PGA > PGA_C\right]$ tends to its correct value (0.81) and the missed alarm probability also tends to its correct value (0). On the other hand the false alarm tends to 0.19. This means that, when all stations are triggered, the systems will work according to what has been designed.

The shape of the curves depends on both the chosen values of PGA_C and P_C and may thus be very different from those discussed in this example if other values of the thresholds are concerned. Nevertheless, given the missed and false alarm reference values calculated by means of the hazard conditioned by M_T and R_T, the system may be calibrated by setting PGA_C and P_C appropriately.

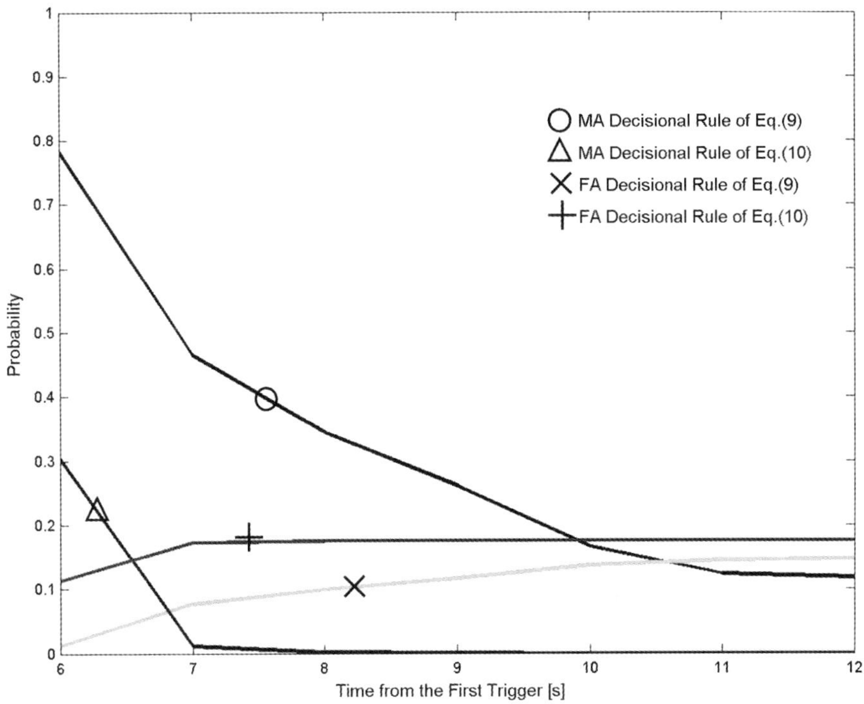

Fig. 11.7 False and missed alarm probabilities for 10^4 events ($M_T = 7$, $R_T = 110\text{km}$, $PGA_C = 0.3\text{m}/\text{s}^2$).

11.5 Conclusions

The discussed method aims to assess whether it is possible to use real-time information provided by an EEWS to estimate the seismic performance of a structural or infrastructural system of interest. Magnitude and source-to-site distance probabilistic distributions are plugged into the hazard analysis, which may be further processed to obtain a prediction of the structural response for the event occurring. Real-time seismic risk analysis seems the way to use all the information provided by the earthquake early warning system for real-time decision making. However, since the site is most likely far from the network, the uncertainty related to the prediction cannot be neglected.

The approach was tested by simulating the Campania early warning system. Results indicate that the PSHA, conditioned to the EEWS measures, correctly approximates the hazard computed if the magnitude and distance were deterministically known, which is the maximum level of knowledge possible. A significant reduction in the dispersion of the hazard curves would be obtained by increasing the number of sensors in the area.

The approach is also used to test possible decisional rules to issue the alarm. Decisional rules and alarm thresholds have intrinsic (by design) missed and false alarm probabilities which may be changed according to appropriate loss functions. Simulation shows how the missed and false alarm probabilities estimated by the EEWS are evolving with time, approaching their design values as the number of stations increases. Such curves may be used for risk management, optimizing the trade-off between the probability of wrong decisions and the available lead time for risk reduction action.

11.6 Acknowledgements

The study presented in this paper was developed in the framework of the P.O.R. Campania 2000 - 2006, Misura 1, founded by the Campania Regional Authority. The authors are also grateful to Professor Daniele Veneziano (Massachusetts Institute of Technology) for his helpful discussions on the topic.

References

Allen RM, Kanamori H (2003) The Potential for Earthquake Early Warning in Southern California. Science 300:786-789

Barroso LR, Winterstein S (2002) Probabilistic Seismic demand analysis of controlled steel moment-resisting frame structures. Earthquake Engineering and Structural Dynamics 31:2049-2066

Carballo JE, Cornell CA (2000) Probabilistic Seismic Demand Analysis: Spectrum Matching and Design. Department of Civil and Environmental Engineering, Stanford University. Report No. RMS-41

Convertito V, De Matteis R, Romeo A, Zollo A, Iannaccone G (2007) A strong motion attenuation relation for early-warning application in the Campania region (Southern Apennines). In: Gasparini P, Manfredi G, Zschau J (eds) Earthquake Early Warning Systems. Springer

Cornell CA (1968) Engineering seismic risk analysis. Bull Seismol Soc Am 58:1583-1606

Cornell CA, Jalayer F, Hamburger RO, Foutch DA (2002) The Probabilistic Basis for the 2000 SAC/FEMA Steel Moment Frame Guidelines. J of Struct Eng 128(4):526-533

Fontanella N (2005) Gestione del Rischio Sismico nella Regione Campania: Formulazione e Calibrazione del Simulatore del Sistema di Early Warning Sismico per il Progetto SAMS, MSc. Thesis, University of Naples Federico II. Advisors: M. Giorgio, I. Iervolino, V. Convertito (in Italian)

Iervolino I, Convertito V, Manfredi G, Zollo A, Giorgio M, Pulcini G (2005) Ongoing Development of a Seismic Alert Management System for the Campanian Region. Part II: The cry wolf issue in seismic early warning applications, Earthquake Early Warning Workshop, Caltech, Pasadena, CA [http://www.seismolab.caltech.edu/early.html]

Iervolino I, Manfredi G, Cosenza E (2007) Earthquake early warning and engineering application prospects. In: Gasparini P, Manfredi G, Zschau J (eds) Earthquake Early Warning Systems. Springer

Kanamori H (2005) Real-time seismology and earthquake damage mitigation. Annual Review of Earth and Planetary Sciences 33:5.1-5.20

McGuire RK (1995) Probabilistic seismic hazard analysis and design earthquakes: Closing the loop. Bulletin of the Seismological Society of America 85:1275-1284

Pate-Cornell ME (1986) Warning systems in risk management. Risk Management 6:223-234

Weber E, Iannaccone G, Zollo A, Bobbio A, Cantore L, Corciulo M, Convertito V, Di Crosta M, Elia L, Emolo A, Martino C, Romeo A, Satriano C (2007) Development and testing of an advanced monitoring infrastructure (ISNet) for seismic early-warning applications in the Campania region of southern Italy. In: Gasparini P, Manfredi G, Zschau J (eds) Earthquake Early Warning Systems. Springer

Pinto PE, Giannini R, Franchin P (2004) Seismic reliability analysis of structures. IUSSPress, Pavia, Italy

Sabetta F, Pugliese A (1996) Estimation of response spectra and simulation of nonstationarity earthquake ground motion. Bulletin of the Seismological Society of America 86:337-352

Satriano C, Lomax A, Zollo A (2007) Optimal, real-time earthquake location for early warning. In: Gasparini P, Manfredi G, Zschau J (eds) Earthquake Early Warning Systems. Springer

Seismic eArly warning For EuRope – Safer (2005) Sixth Framework Programme Call: Fp6-2005-Global-4 Sustainable Development, Global Change and Ecosystem Priority 6.3.IV.2.1: Reduction of seismic risks

Vamvatsikos D, Cornell CA (2002) Incremental Dynamic Analysis. Earthquake Engineering and Structural Dynamics 31(3):491-514

Veneziano D, Papadimitriou AG (1998) Optimization of the Seismic Early Warning System for the Tohoku Shinkansen. 11th European Conference on Earthquake Engineering. Paris, France

Wieland M (2001) Earthquake Alarm, Rapid Response, and Early Warning Systems: Low Cost Systems for Seismic Risk Reduction. Electrowatt Engineering Ltd. Zurich, Switzerland

Wald DJ, Quitoriano V, Heaton TH, Kanamori H, Scrivner CW, Worden BC (1999) TriNet "ShakeMaps": Rapid Generation of Peak Ground Motion and Intensity Maps for Earthquake in Southern California. Earthquake Spectra 15:537-555

Wieland M, Griesser M, Kuendig C (2000) Seismic Early Warning System for a Nuclear Power Plant. Proc. of 12th World Conference on Earthquake Engineering. Auckland, New Zealand

12 Earthquake Early Warning and Engineering Application Prospects

Iunio Iervolino, Gaetano Manfredi, Edoardo Cosenza

Dipartimento di Ingegneria Strutturale, Università di Napoli Federico II, Napoli, Italy

Abstract

The foreseeable future of Earthquake Early Warning Systems (EEWS) is their use as a tool for real-time seismic risk management and mitigation. The applicability potential of EEWS seems to be more related to the immediate activation of safety measures for critical systems rather than as a massive alert to the public. Evacuation of buildings requires warning times which are unlikely to be available in many urbanized areas threatened by seismic hazard, whereas the protection of critical systems may still significantly help to reduce the losses subsequent to a catastrophic event and to increase the resiliency of communities to earthquakes.

Real-Time Seismology (RTS), which consists of methods and procedures for the rapid estimation of earthquake and ensuing ground motion features based on measurements made on the first few seconds of the P-waves, is the focus of a great deal of research. In principle, it may boost the potential of regional seismic sensor networks for site-specific applications, in other words: *hybrid* EEW. Thus the next challenge of early warning and earthquake engineering is geographically distributed seismic networks for the protection of several critical systems and lifelines at the same time. The key issue is related to uncertainty in the estimation of the event's features. Therefore, the performance target and feasibility factor of such an EEWS is no longer only to maximize the warning time but also calibrate, in a full probabilistic approach, the alarm thresholds and the decisional rules in order to maximize loss reduction following the decision. This paper reviews and discusses some issues raised for hybrid EEWS in the light of performance-based earthquake engineering (PBEE) for risk reduction applications.

12.1 Specific vs. Regional EEWS

The basic elements of an Earthquake Early Warning System (EEWS) are a network of seismic instruments, a station (local or central) processing the data measured by the sensors and a transmission infrastructure spreading the alarm to end users (Heaton 1985) to initiate personal or automatic security measures. An EEWS is considered to be an attractive and moderately costly solution for risk mitigation, the attractiveness being related to the reduction in total losses produced in a large region or for very critical facilities.

EEWS may be distinguished by the configuration of their seismic network as *regional* or *site-specific* (Kanamori 2005). Regional EEWS consist of wide seismic networks covering a portion of the area which is likely to be the source of a catastrophic earthquake and/or the urbanized area exposed to the strike. Data recorded by the seismic instruments are further processed to retrieve information such as magnitude and/or location, faulting mechanism or spectral response. This information may be used to estimate the level of shaking in the affected area. Such processing may require significant time and, in a possibly large portion of the region, called the *blind zone* (Kanamori 2005), the alarm may rarely be issued before the ground motion hits. Regional systems are mainly devoted to applications such as *shake maps* (Wald et al. 1999), which are territorial distributions of ground shaking available immediately after the event for emergency management for example, aiding in directing rescue teams in the zones which are expected to be subjected to the largest shaking and are therefore expected to suffer the highest losses. In this case the system works in near-real-time as a *Rapid Response System* (Wieland 2001) introducing another classification of the EEWS by operating time-scale.

When the system can spread the alarm during the event, before the ground motion hits some sites of interest, it is operating in real-time for seismic alert purposes. In only a few cases will regional systems have enough time to process the data and spread the evacuation alarm. This is the case for the early warning system of Mexico City where the seismic source zone is clearly known and sufficiently far away, such that large segments of the population can be warned by the media. In Mexico City, public schools and government agencies are directly connected with the alarm system. The seismic alert system is an EEWS for large earthquakes which have their source in the subduction zone of the Pacific coast. The seismic sensors network consists of 12 digital strong motion field stations located along a 300 km stretch of the Guerrero coast. Each field station includes a computer that continually processes seismic activity which occurs within a 100 km radial area around each station. The transmission infrastructure consists of a central radio relay station and three relay stations located between the coast and Mexico City. Two seconds are required for the information on an event to reach Mexico City. Data received are processed automatically to estimate the magnitude of the event and to issue a public alert (4.4 million people are potentially covered by the system). The system disseminates the warnings to the public and to specific entities via commercial radio stations and audio alerting mechanisms via specially designed receivers.

While regional systems directly improve the resiliency of communities to earthquakes, site-specific EEWS are devoted to enhancing in real-time the safety margin of specific critical engineered systems such as nuclear power plants, lifelines or transportation infrastructures, mitigating the seismic risk by reducing the exposure of the facility by automated safety actions. The networks for specific EEW are much smaller than those of the regional type, only covering the surroundings of the system creating like a barrier for the seismic waves. The location of the sensors depends on the lead time needed to activate the safety procedures before the arrival of the more energetic seismic phase at the site. In these *Seismic Alert Systems* the alarm is typically issued when the S-phase ground motion at one or more sensors exceeds a given threshold and there is no attempt to estimate the event's features. Although, the knowledge of the seismic parameters is desirable it is not essential for issuing the alarm in critical facilities. This is because the latter is time-consuming and also because the uncertainty related to the propagation of seismic waves is generally moderate since the path between the network and the site is limited. Errors in the alerting decision are not considered much of an issue since the risk of system failure is always assumed to be greater than the losses related to a false alarm. Incidentally, it is worth recalling that if an EEWS is required intrinsically this means that the missed alarms are more important than the false alarms.

Among site-specific systems a paradigmatic example is that of the Ignalina nuclear power plant in Lithuania (Wieland et al. 2000). The system is designed to detect potentially damaging earthquakes and to provide an alarm before the arrival of the shear waves at the reactor. The seismic network is made up of six stations that are installed at a distance of 30 km from the power plant (Fig. 12.1). An earthquake with an epicenter outside the fence of stations may trigger an alarm about 4 to 8 seconds before the ground motion reaches the reactor. As the required time to insert the control rods is 2 seconds, the reactor could be secured before the earthquake arrives.

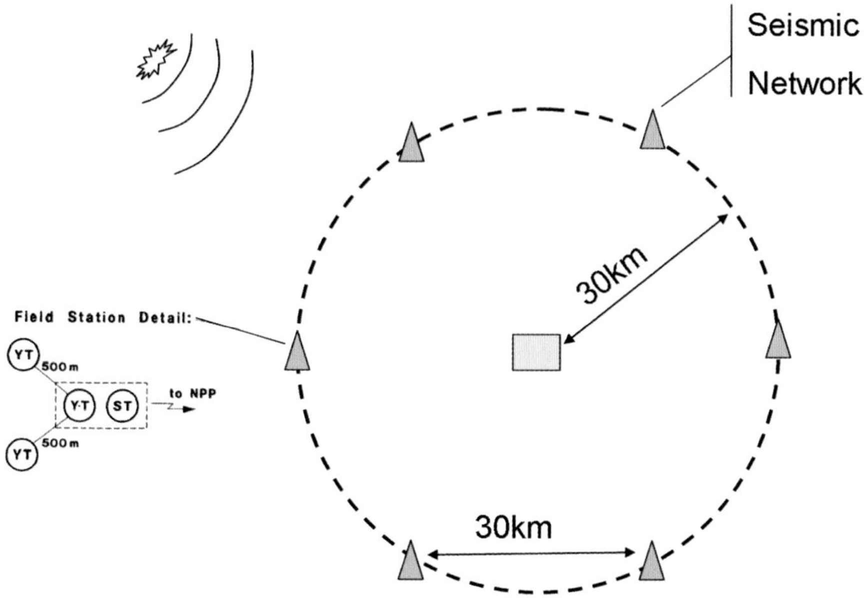

Fig. 12.1 Ignalina EEWS schematics (Wieland et al. 2000).

The alarm threshold is set at an acceleration value of 0.025 g. The decision to alert is taken by a "two out of three" logic which in statistical terms is a *partial parallel* system giving the same protection level against missed and false alarms. As discussed below, any threshold carries intrinsic false and missed alarm rates, which have to be assessed for calibration of the threshold before the seismic alert system can be used to initiate safety procedures such as control rod activation. Aspects to be considered in this respect are the acceptable losses related to both possible decisional errors. For the case of the false alarm, for example, they may be associated to the downtime of the facility.

Another example of a specific EEWS is that protecting the Thoku Shinkanzen high speed train in Japan. The fence of seismic stations is placed along the coast to protect the systems from off-shore events (Fig. 12.2). A second set of instruments, located along the track, protects the trains from inland earthquakes.

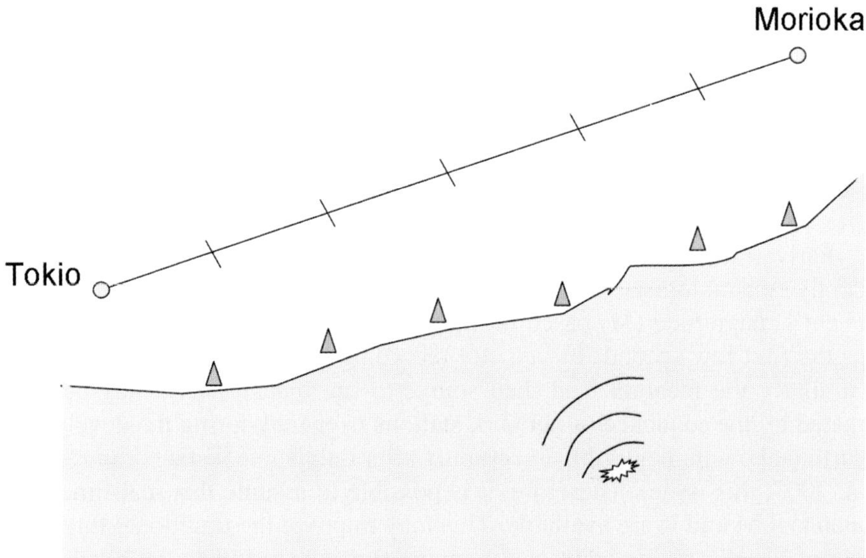

Fig. 12.2 Tohoku Shinkanzen schematics (Veneziano and Papadimitriou 1998).

The system prevents the train from running on viaducts or in tunnels potentially damaged by the earthquake, which may cause catastrophic derailment. Originally the system was designed, as for the Ignalina powerplant, to issue the alarm when the S-waves acceleration recorded at the coastal stations exceeded a threshold; the train can then be stopped and, eventually, the railway inspected for damage. The available lead time is about 20 seconds.

The EEWS for the Tokio-Morioka Shinkanzen required extensive study for its optimization (Veneziano and Papadimitriou 1998). The original system caused frequent delays and train cancellations due to false alarms. The study shows how an engineering approach can improve the performance of the system: optimizing the alarm thresholds by considering the seismic fragility of the track which may suffer damage during an earthquake could reduce the annual rate of false alarms by several orders of magnitude.

12.2 Real-Time Seismology and Hybrid Systems

State-of-the-art site-specific EEWS, working as seismic alarm systems, require a dedicated seismic network around the facility to protect, while regional networks monitoring a potential seismic source zone, due to the computational effort needed, are mostly used as rapid response systems, producing shake maps for emergency preparedness. Due to the considerable development of regional networks worldwide in recent years the question of using EEWS for site-specific applications is rising. A major step for the EEWS may be the use of regional networks to protect multiple critical systems and/or the community and then a *hybrid* use of regional and on-site warning methods (Kanamori 2005).

Early warning is the current focus of considerable research effort. Recently seismologists have developed several methods to estimate the event's magnitude (M) based on limited information of the P-waves, such as the first few seconds of velocity recording (Allen and Kanamori 2003). Similarly the location, and then source-to-site distance (R), may be estimated by the sequence of network stations triggered during the developing earthquake with negligible uncertainty after only some instants (Satriano et al. 2007, this issue). Therefore, it is possible to assume that real-time estimates of M and R are available. This may improve the traditional functioning of EEWS, giving additional warning time and reducing the blind zone. However, this information may also be used to design EEW engineering applications. For example, the M and R estimates can provide a prediction of the ground motion at the site, which can be performed in analogy with common Probabilistic Seismic Hazard Analysis (PSHA). It results in seismic hazard analysis reflecting real-time information.

Computing the seismic hazard *conditionally* (in a probabilistic sense) upon the EEWS allows it to account properly for all uncertainties related to both the estimates of the seismic networks and also the propagation of the seismic waves from the source zone to the sites of interest by an appropriate attenuation law. Consequently, the performance or even the losses related to a structure or engineered system of interest may be computed. Most such types of analyses may be optimized so as not to require significant additional processing time (Iervolino et al. 2007, this issue). A scheme of the hybrid application of a regional network for structure-specific earthquake early warning is shown in Fig. 12.3.

Employing *Real-Time Seismology* in an earthquake engineering framework means updating the knowledge of the seismic hazard from the data gathered by the network. This allows re-evaluation of the seismic risk conditioned to the measures of the network for risk management purposes. Indeed, virtually all the knowledge and decision-making approaches developed in the framework of Performance-Based Earthquake Engineering (PEER 2004) may be applied to early warning, helping to design such systems on a quantitative and consistent basis.

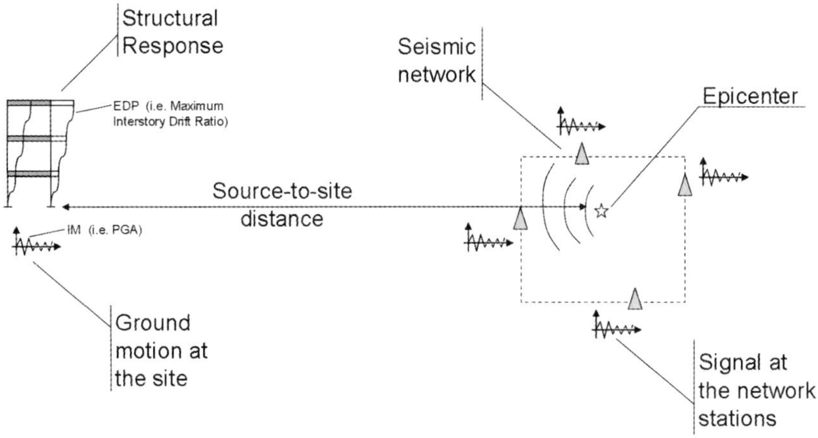

Fig. 12.3 Regional EWWS for structure-specific applications.

Obviously, the probability density function of the structural response and/or consequent expected losses, conditioned to the measures of the seismic instruments, when an event is occurring contains the highest level of information available. Real-time risk analysis in the case of hybrid EEWS, in principle, allows application of regional networks to multiple specific systems at the same time as critical systems and lifelines and possibly gives a quantitative basis for automated decision making. Contrasting with the current EEWS calibration approach, adjusting the alarm threshold by predicting the consequences is more consistent with an engineering approach to the seismic risk management.

Hybrid systems designed in this way may álso overcome an intrinsic limit of existing EEWS. The latter currently help reduce the loss related to exposure (e.g. casualties in case of evacuation) but they do not help to decrease economic losses due to structural damage in buildings, infrastructures and other engineered systems. Now it seems possible to take real-time action to reduce the structural vulnerability of specific systems (to follow). For example, if real-time hazard analysis allows the response spectra at the site to be estimated before the ground motion hits, semi-active control devices, which need milliseconds to seconds to set, may change the vibrating characteristics of the structure accordingly.

12.3 Applicability Potential of EEWS

Seismic risk may be defined, whilst separating causes and effects, as the combination of: hazard, vulnerability and exposure. Risk management consists of: (1) Risk mitigation by vulnerability or exposure reduction; and (2) Emergency preparedness (Fig. 12.4). The latter is a near-real time issue; the former consists of strategies which are typically mid-term (i.e. seismic retrofit of structures and infrastructures) or long-term actions (i.e. urban land use planning or development of appropriate design standards). From the brief review given in the previous sections, it is clear that EEWS may play a role in both policies, whether in minimizing loss of lives and property or directing rescue operations (Wieland 2001).

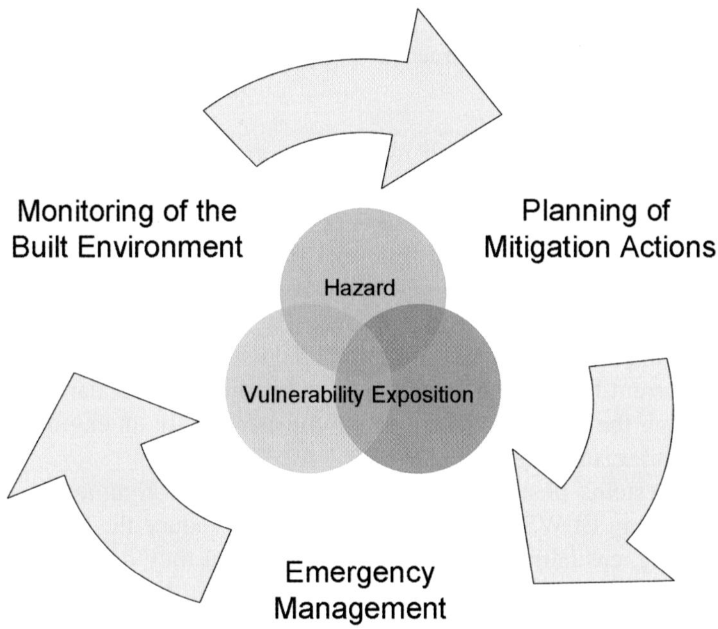

Fig. 12.4 Risk factors and risk mitigation strategies.

The traditional approach to risk mitigation by EEW deals with those facilities and processes where rapid response can contribute to the reduction in the *value* exposed. For example, operations of critical facilities and processes are stopped, trains are slowed down, traffic lights are switched to red on critical infrastructure tracks such as bridges, valves in gas and oil pipelines in hazardous industrial facilities are closed, and power plants are secured. Personal protective measures are undertaken at home and in the workplace, including getting under desks and moving away from dangerous equipment or materials. All the listed actions following an EEW reduce losses following damage (exposure) of the engineered systems but do not prevent such damage (vulnerability). However, EEWS are now capable of providing, from a few seconds to a few tens of seconds before the arrival of strong ground shaking, a prediction of the ground motion or the seismic demand on structures caused by a large earthquake in a fully probabilistic way. Therefore the question raised now is how such early warning information may be used to take real-time action for vulnerability reduction in the light of real-time seismic risk management and effective damage mitigation.

Several studies discussed, as an engineering application of EEWS, the semi-active control of structures (Grasso et al. 2005) such that the building can change its dynamic properties within a few seconds to better withstand the predicted ground motion features. A semi-active control device is a passive system which has controllable properties which may change the dynamic properties of the structures it is applied to. For example, the magneto-rheological dampers use fluids which contain micro-sized iron particles that, if a magnetic field is applied, form chains increasing the viscosity. The intensity of the magnetic field may be regulated to adjust the viscosity; this may change the structural damping. How to change the damping in the semi-active control strategies is dependent on the response spectra which reflect the hazard at the site. Then, although EEWS development for such applications will require a dedicated and reliable infrastructure that can utilize the information and operate very rapidly and automatically, integration with an EEWS in the light of real-time structural control now seems feasible since it is possible to have the expected, or the full probabilistic distribution, of the spectral ordinates in real time. On the other hand, it seems harder to integrate the EEW with active control strategies requiring the full waveform prediction to operate.

An application of semi-active control implementing some of these concepts, albeit for a traditional site-specific EEWS, is that under development by ENEA in Italy to protect Michelangelo's David displayed at the Uffizi museum (Florence). The system is made of a seismically isolated stand, which in non-earthquake conditions is tightly secured to the floor to prevent accidental movement, while in the case of an early warning alarm the locks are automatically released, isolating the statue from the floor motion.

Although interaction of EEW and semi-active control is a real-time application and quite innovative, there is another engineering application which has received considerable attention, namely integration with structural monitoring systems. Structural health monitoring is used to represent the evolution of structural conditions during service life. By contrast, performance-based earthquake engineering methodologies employ structural-response information to estimate probabilistic losses related to seismic performance. It seems quite straightforward to combine these capabilities to automatically estimate in near-real-time the probabilistic performance of an instrumented building after the hit of the strong motion (Porter et al. 2004). This application of EEW may enhance the potential of the system in the light of the rapid response to catastrophic events. Indeed, in the case of critical systems, which have to be operable for emergency management purposes such as hospitals, fire stations or even lifelines, rapid damage assessment may provide a useful picture of the situation of such important resources available during the emergency.

Finally, another possible evolution of near-real-time applications of EEW for the implementation of a rapid response system is the evolution of shake maps into *damage maps*. Research is being conducted on the fragility functions for classes of buildings, which are probabilistic distributions of structural damage conditioned to a seismic intensity measure (spectral acceleration for instance) retrieved on an analytical or empirical basis. Then, if the spatial distribution of the inventory of any category of buildings is available, it is possible to retrieve in near-real-time maps of the structural damage, which is more informative for emergency management rather than the distribution of the shaking level. This holds particularly for those countries where the building stock is very heterogeneous and structures in the same region may be old masonry constructions, reinforced concrete frames whether seismically designed or pre-code, pre-cast and even steel structures. Thid happens, for example, in Europe and in Mediterranean countries, where the shaking maps are not the best proxy for damage since the listed categories are very differently sensitive to the ground motion level.

12.4 Beyond the False Alarms: the Loss Estimation Approach to Early Warning

In hybrid EEWS the warning time is not the only parameter to optimize: estimation of event features by real time seismology is a process based on empirical relationships and carries significant uncertainty. Moreover, ground motion prediction, structural response, damage and loss relationships further introduce uncertainty in the prediction at the site. The uncertainty may lead to errors in alerting decisions. Alerting or not alerting both have a cost: in the case of not alerting the loss is associated to an earthquake striking without any countermeasure taken; in the case of alerting, preparedness interventions have a cost (social and/or economic) which may transform into loss if the actual ground motion does not require such action. As discussed by Goltz (2002), false alarms are important for alerts to the community because they can result in a reduction of credibility (the "cry wolf") with even legal liability. In automated decision making for engineering applications costly risk reduction measures must not be taken if not strictly needed. For example, the downtime of lifelines may be costly and has to be limited. It has to be kept in mind that the real-time actions featuring the larger risk mitigation potential often also require the larger warning time. Also, the cry wolf phenomenon does not have the same importance for every intervention category: its impact depends on the extent of the systems the alarm affects, and on the cost of downtime (Fig. 12.5).

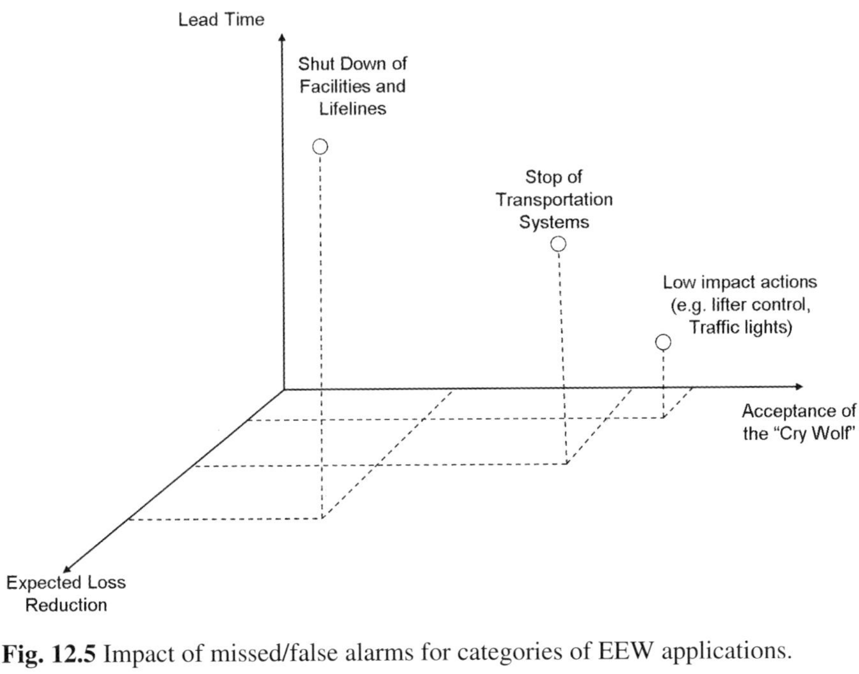

Fig. 12.5 Impact of missed/false alarms for categories of EEW applications.

Any decisional rule and alarm threshold have intrinsic false and missed alarm probabilities which constitute a trade-off. Indeed, the reduction in false alarms by adopting high warning thresholds is dangerous since it would intrinsically increase the chance of a missed alarm. These error rates may be effectively adjusted only by improving the estimation method of the parameters the decision is based on (acceleration, for example). Moreover, the false and missed alarm rates also change with time, therefore the decision of issuing the alarm may be taken in advance with some probability of errors which may change if more data are available. This is another crucial trade-off in the design of seismic early warning applications because the uncertainty only decreases when more information is collected by the seismic instruments, and then only when the available lead-time is reduced. Estimation of missed and false alarm probabilities associated to an alarm threshold is one way to understand the implications of a decision based on that threshold. Computation on an empirical basis should consist of post-event analysis of EEWS predictions and would require a large strong-motion waveform database both for the network and the site where the structure is located. Since such databases are very rarely available, especially for large earthquakes, the I and II type errors may be approximated in a simulation framework using appropriate characterizations of the uncertainties involved in the prediction.

Estimation of false missed alarm rates is a first approach to test performance of a hybrid early warning system. A more sophisticated way to calibrate design of an EEWS for automated decision making may be based on the minimization of expected losses. Let us say that two actions are possible, based on the data from the seismic instruments: (1) alarm; (2) no alarm. Consider that the decision to alert should be made if a statistic of the measures (decision variable) made by the network exceeds a given threshold. To establish which threshold should be set, the expected losses following action (1) or (2) have to be computed conditionally upon any value of the decision variable. The decision associated to the lower expected cost indicates which action should be taken for that value of the decision variable. This approach also leads to the definition of the threshold value.

For example, in the case of the methodology suggested by Allen and Kanamori (2003) the alarm threshold could be set at the predominant period of the first few seconds of the P-waves (τ_c) because this parameter is correlated to the magnitude of the event and, together with an estimation of source location, it may be used to predict ground motion, structural performance or economic losses for a system of interest. Considering no risk mitigation action, it is possible to compute the expected value of losses in the case of not issuing the alarm. Similarly, in the case of a protective action being taken, the expected losses may be computed if the alarm is issued. Estimating these two losses conditionally upon any possible value of τ_c will lead to sub-divide the space of τ_c into two regions: (1) the region where the expected losses in the case of no alarm is lower than the expected losses if any action is taken; (2) the region where there are expected losses if the alarm is issued are lower than if the alarm is not issued. The limiting value of τ_c separating these two regions is the optimal threshold to set. This approach overcomes the false and missed alarm approaches to establishing the alarm threshold since the decision always minimizes the expected loss (Iervolino et al. 2006).

12.5 Concluding Remarks: Future Prospects of Seismic Early Warning Engineering

EEWS may be regional or site-specific. Currently the regional warning method using a network of stations can provide more detailed but less rapid information about the ground motion. In contrast, the on-site method provides a more rapid warning, but the information coming from the on-site warning is limited to relatively simple parameters. A hybrid use of a regional and on-site warning may enhance the usefulness and reliability of an EEWS.

Real-Time Seismology may help to overcome some of the limitations of the EEWS developed or implemented so far which only provide warnings regarding the severity of impending earthquakes. Now information regarding the characteristics of the ground motion, at least on the response spectrum, may be given from the first few seconds of the event. The integration of real-time seismology with performance-based earthquake engineering allows the EEWS to be capable of providing the real-time predictions of those information which are useful for design of engineering applications with a quantification of the uncertainty related as a function of time.

Current EEW projects worldwide are chiefly of the regional type, since they rely on the development of national or regional seismic networks. Hence the question of using such EEWS for engineering applications has arisen. Indeed, several countries are developing regional seismic networks aiming to have RTS capabilities. For example, Japan, Turkey, Romania, Greece, United States (California) and Italy have several earthquake early warning projects (see also www.seismolab.caltech.edu/early.html). However, few of them are ready to implement a prototypal system of real-time earthquake engineering, even though all the projects set this as a major goal.

All the projects and current EEWS tend to reduce exposure of critical systems. However, it seems a natural development of EEWS to reduce damage, i.e. to mitigate seismic risk by reducing structural vulnerability. It seems possible with the application of real-time seismology as an input to semi-active structural control. To this aim the interaction of real-time engineering seismology and real-time earthquake engineering is required to develop design guidelines for engineering EEWS applications.

This is the foreseeable future of EEWS. Whether this kind of application is feasible depends on the lead time provided but also on the failure rate of the prediction. There is extensive discussion on early warning systems as to how the alarm threshold should be set. In many EEWS, typically site-specific but also regional, in some cases, this threshold is set on a ground motion level, i.e. the acceleration recorded by the seismic network. In general, the threshold should be set on the parameter the seismic stations record as a proxy for the features of the event. Although false/missed alarm rates may now be estimated, calibration of the seismic early warning system and set-up of an alarm threshold should be done in a loss estimation approach, i.e. the action to be taken to reduce the risk is that which minimizes the expected losses.

Among its requisites, EEWS for real-time engineering applications should have a network able to measure parameters of use for RTS, rapid processing capabilities and a reliable, then redundant, dedicated transmission infrastructure. In the case of real-time vulnerability reduction, as for semi-active structural control, the system to protect should also be supplied with an automatic system able to operate the devices or initiate any other security measure in case of an alarm. These kinds of applications most likely will require the development of new technologies specific to real-time earthquake engineering, which may be a critical issue for the development of hybrid EEWS.

References

Allen RM, Kanamori H (2003) The Potential for Earthquake Early Warning in Southern California. Science 300:786-789

Goltz JD (2002) Introducing Earthquake Early Warning in California: A Summary of Social Science and Public Policy Issues, A Report to OES and the Operational Areas. California Governor's Office for Emergency Services

Grasso VF, Iervolino I, Occhiuzzi A, Manfredi G (2005) Critical Issues of Early Warning Systems for Active Structural Control, ICOSSAR05, 9th Conference on Structural Safety and Reliability, Rome, Italy

Heaton TH (1985) A model for a seismic computerized alert network. Science 228:987-90

Iervolino I, Convertito V, Giorgio M, Manfredi G, Zollo A (2007) The crywolf issue in earthquake early warning applications for the Campania region. In: Gasparini P, Manfredi G, Zschau J (eds) Earthquake Early Warning Systems. Springer

Iervolino I, Giorgio M, Manfredi G (2006) Expected Loss-Based Alarm Threshold Set for Earthquake Early Warning Systems. Submitted for publication

Kanamori H (2005) Real-time seismology and earthquake damage mitigation. Annual Review of Earth and Planetary Sciences 33:5.1-5.20

PEER 2004/05 (2004) Performance-Based Seismic Design Concepts and Implementation Proceedings of the International Workshop Bled, Slovenia, June 28-July 1 2004. Fajfar P, Krawinkler H (eds) Pacific Earthquake Engineering Research Center, Richmond, CA, USA

Porter KA, Beck JL, Ching JY, Mitrani-Reiser J, Miyamura M, Kusaka A, Kudo T, Ikkatai K, Hyodo Y (2004) Real-time Loss Estimation for Instrumented Buildings. Technical Report: CaltechEERL:EERL-2004-08. Earthquake Engineering Research Laboratory, Pasadena, CA

Satriano C, Lomax A, Zollo A (2007) Optimal, real-time earthquake location for early warning. In: Gasparini P, Manfredi G, Zschau J (eds) Earthquake Early Warning Systems. Springer

Veneziano D, Papadimitriou AG (1998) Optimization of the Seismic Early Warning System for the Tohoku Shinkansen. 11th European Conference on Earthquake Engineering. Paris, France

Wald DJ, Quitoriano V, Heaton TH, Kanamori H, Scrivner CW, Orden BC (1999) TriNet "ShakeMaps": Rapid Generation of Peak Ground Motion and Intensity Maps for Earthquake in Southern California. Earthquake Spectra 15:537-555

Wieland M (2001) Earthquake Alarm, Rapid Response, and Early Warning Systems: Low Cost Systems for Seismic Risk Reduction. Electrowatt Engineering Ltd. Zurich, Switzerland

Wieland M, Griesser M, Kuendig C (2000) Seismic Early Warning System for a Nuclear Power Plant. 12th World Conference on Earthquake Engineering. Auckland, New Zealand

13 UrEDAS, the Earthquake Warning System: Today and Tomorrow

Yutaka Nakamura, Jun Saita

System and Data Research Co. Ltd.

Abstract

The most important countermeasure against earthquake risk is to have all structures vulnerable enough for the possible earthquake load. In this regard, an early warning system should be installed to reduce the possibility of earthquake disaster. An early warning system is required mainly to issue an alarm to have a time margin for evacuating or shutting down key facilities, and not to determine exact earthquake parameters. Thus the early warning system must be realized independently and the government and other public authorities must release accurate earthquake information immediately.

The necessary qualities for early warning systems may be summarized as follows:

- **Fully Automated:** As the time margin is limited, the facility should be directly controlled without human judgment.
- **Quick and Reliable:** As there is limited time to respond to earthquake motion, this kind of system is required to be quick and reliable.
- **Small and Cheap:** To install easily, the system must be small and cheap.
- **Independence:** To issue fail-safe alarms, the system must be independent of other systems.
- **Easy to Connect Network:** To deliver the earthquake information, the system must be easy to connect network.
- **Accuracy is Better:** For the alarm, accuracy of the information is not such a serious problem.

UrEDAS, Urgent Earthquake Detection and Alarm System, is the first real-time P-wave alarm system in practical use in the world. It is able to process digitized waveforms step by step without storing the waveform data. As the amount of processing does not differ whether or not an earthquake occurs, system failure due to overload will not occur.

Here, the present condition of the P-wave early detection system UrEDAS is viewed under working conditions and the results of test observations where faults exist are also reported. Then a new real-time data processing system for new generation will be considered.

13.1 Introduction

The most important countermeasure against earthquake risk is to have all structures vulnerable enough for the possible earthquake load. In this regard, an early warning system should be installed to reduce the possibility of earthquake disaster. An early warning system is required mainly to issue an alarm to have a time margin for evacuating or shutting down key facilities, and not to determine exact earthquake parameters. Thus the early warning system must be realized independently and the government and other public authorities must release accurate earthquake information immediately.

It is considered that an early warning system should have the following functions:

1. Rapid earthquake detection. Installing seismometers far from the target (e.g. urban area) is the easiest way to create enough time to escape. The time is caused by the difference in velocity between telecommunications (300,000 km/sec) and the seismic wave (8 km/sec). This type of early warning is called a "Front-detection system". Moreover if the system can detect P waves and determine the earthquake parameters or estimate the dangerousness of the earthquake motion, the time margin will be greater. In the event of P-wave detection, even near the target the warning system can obtain a time margin based on the difference between the P wave and S wave. This type of early warning system is called an "On-site system".
2. Automatic management. All procedures for early warning and alarm must be carried out automatically, because human judgment may need time and can cause misjudgment.
3. Education and training in the system. It is necessary to educate the public with regard the meaning of the information or alarm from the early warning system. Also it is important to train staff how to behave in case of early warning and to promote manuals for the earthquake countermeasures.
4. That the possibility of false alarm and information error be recognized. Since there is always a possibility of issuing a false alarm, organizations that use the alarm system should understand taking risk. Obviously, attempts should be made to reduce the possibility of false alarms.

UrEDAS, Urgent Earthquake Detection and Alarm System, is the first real-time P-wave alarm system in practical use in the world. It is able to process digitized waveforms step by step without storing the waveform data. As the amount of processing does not differ whether or not an earthquake occurs, system failure due to overload will not occur.

Here, the present condition of the P-wave early detection system UrEDAS is viewed under working conditions and the results of test observations where faults exist are also reported. Then a new real-time data processing system for new generation will be considered. Table 13.1 shows the practice of early warning systems.

Table 13.1 Practice of early warning systems.

	P-wave Alarm	S-wave Alarm
On-Site Detection/Alarm	Shinkansen lines since 1998 and Tokyo Metro since 2001 using Compact UrEDAS with One Second Alarm after P-wave Detection	Tokaido Shinkansen since 1964 using Mechanical Detector
	FREQL: Sub-Seconds Alarm after P-wave Detection as On-Site system for Hyper Rescue Teams since 2005 and for Tokyo Metropolitan Subway since 2007	AcCo: more than 200 Users at the end of 2005, mainly for On-Site S-wave Alarm
Front Detection/Alarm	One Second Alarm/ Information after P-wave Detection as Front system for Nuclear Power Plant since 2006	
	Tokaido Shinkansen since 1992, Sanyo Shinkansen since 1996, for Wakayama Tsunami Warning System since 2001 using UrEDAS with Three Seconds Alarm/ Information after P-wave Detection as Front system	Coast Line Detection System for Tohoku Shinkansen since 1982

13.2 The History of Early Warning

13.2.1 The First Concept of Early Warning

The main concept of the early warning system was introduced by Dr. Cooper in the San Francisco Daily Evening Bulletin of 3rd November 1868 (see Appendix). This report explained the concept as follows:

"A very simple mechanical contrivance can be arranged at various points from 10 to 100 miles from San Francisco, by which a wave of the earth high enough to do damage, will start an electric current over the wires now radiating from this city, and almost instantaneously ring an alarm bell, ... This bell should be very large, of peculiar sound, and known to everybody as the earthquake bell. Of course nothing but the distant undulation of the surface of the earth should ring it. This machinery would be self-acting and not dependent on the telegraph operators."

At that time, no system could realize this idea. Figure 13.1 shows the concept of Dr. Cooper's idea of the front detection system.

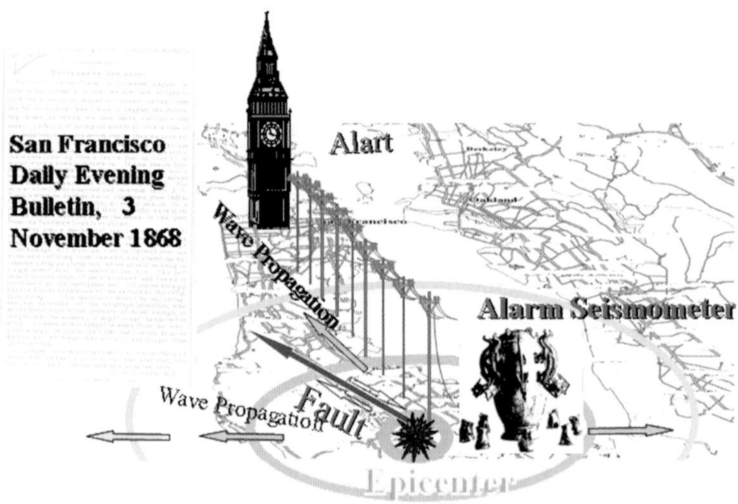

Fig. 13.1 The concept of the first detection system by Dr. Cooper.

13.2.2 Earthquake Alarm for Railways

Earthquake detectors for the railway system were developed and spread from the second half of 1950s in Japan. It was started by the strong motion observation with SMAC, the first strong-motion seismograph in Japan developed in 1953. The 1964 Niigata earthquake (M7.5) triggered debate on an earthquake warning system for Shinkansen under construction.

However, in April 1965, the year after starting operation of Tokaido Shinkansen (bullet train), an M6.1 earthquake occurred in Shizuoka prefecture and some structures in Shinkansen were damaged. Subsequently, JNR, Japanese National Railways, decided to construct a new earthquake warning system with ordinary alarm seismometers and waveform recording seismometers. These seismometers were installed every 20 to 25 km along the Shinkansen line to issue an alarm if the acceleration of horizontal ground motion exceeds 40 Gal (= cm/sec^2). This preset level, 40 Gal, was determined as a level to detect earthquakes exactly, so as not to issue alarms in the event of small earthquakes, and not to issue errors with passing trains or other environmental noise. Figure 13.2 shows the examples of these historical earthquake detectors.

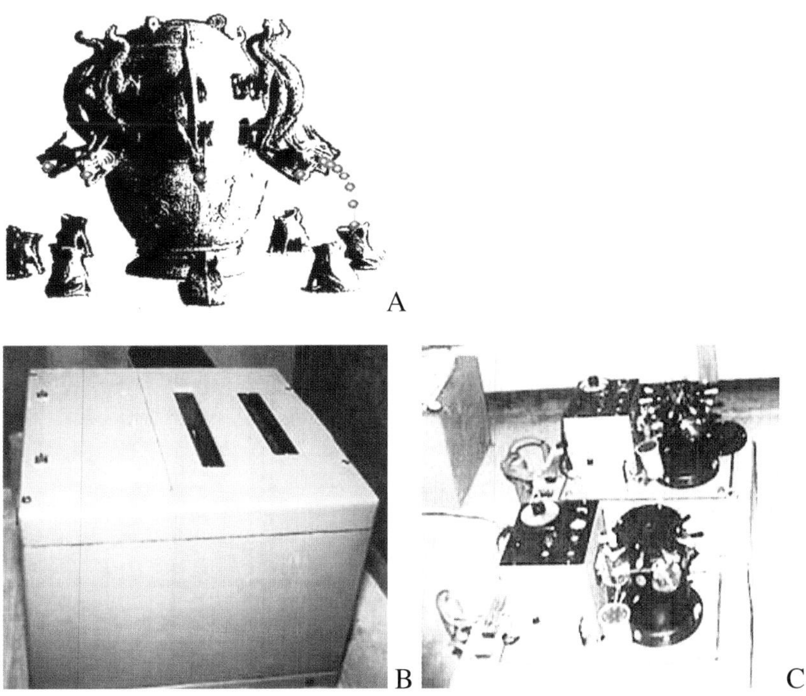

Fig. 13.2 Historical earthquake detectors. (A) The first earthquake detector in ancient China (model). (B) SMAC type strong motion detector (appearance). (C) Earthquake detector for Shinkansen (inside).

In 1972, researchers in earthquake disaster prevention in Japan advocated the "Strong earthquake alarm system 10 seconds before". Although this was an idea similar to Dr. Cooper's front detection system in 1868, nobody had put it into practical use until then. The world's first front detection system, "a coastline detection system" with the simple triggering seismograph, began operations at the time of the completion of the Tohoku Shinkansen in 1982. This is the first example which carried through Dr. Cooper's idea into practice, followed by the Mexico-based SAS (Sistema de Alerta Sísmica) installed in 1991 similar to this coastline detection system. Figure 13.3 shows the coast line detection systems for the Tohoku Shinkansen line and so on.

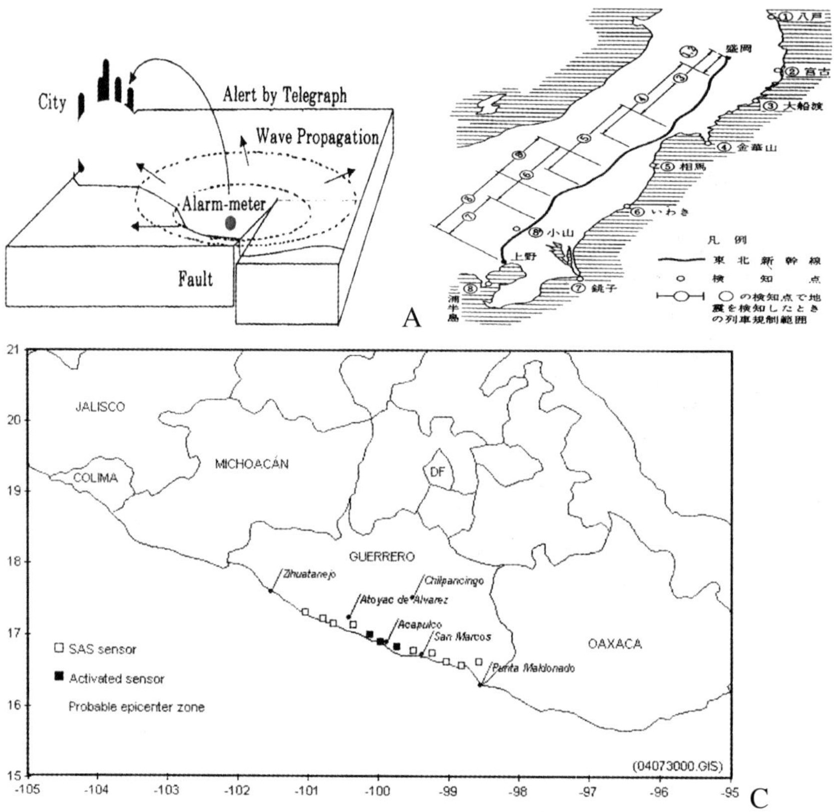

Fig. 13.3 Coastline detection systems. (A) Dr. Cooper's idea (1868). (B) Coastline detection system for Tohoku Shinkansen (1982). (C) SAS, Sistema de Alerta Sísmica, of Mexico (1991).

13.2.3 Birth of UrEDAS

Since earthquake detection with a simple trigger is required to set the threshold level high, earthquake detection tends to be late after S-wave arrival. If it is possible to detect a P wave, the duration of the preliminary tremor can be used for countermeasures. If a front detection system with P-wave detection can be created, then the time margin may be expected to increase. Thus P-wave detection and an alarm system were developed as an indispensable system for a high-speed train. This system was completed as a UrEDAS which can estimate the earthquake parameters and issue an alarm in three seconds after P-wave detection. The prototype UrEDAS started observations in 1984. Since 1988, UrEDAS has been used in practice as part of an earthquake disaster prevention system for the Seikan undersea tunnel (see Fig. 13.4). For the Shinkansen lines, after trial operation started in 1990, UrEDAS started operating with 14 stations for the Tokaido Shinkansen line in 1992. This is the first front P-wave detection alarm system in practical use as an automatic train controlling system.

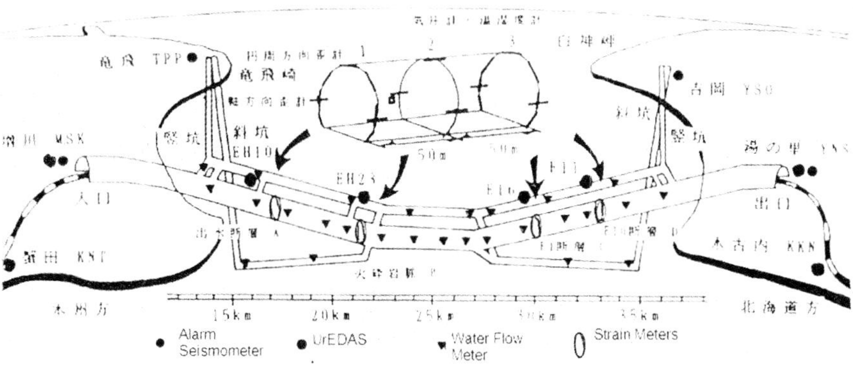

Fig. 13.4 Earthquake disaster detection system for Seikan undersea tunnel.

13.2.4 After the Kobe earthquake

The 1995 Kobe Earthquake (M7.2) caused extensive and severe damage to viaducts and other structures. This time along-the-railroad alarm seismometers issued the alarm immediately since the station was close to the epicenter. This earthquake led to a plan to install the Sanyo Shinkansen UrEDAS. The system was completed and operation was started with five stations in 1996. In 2000, a UrEDAS was installed in Kushimoto, the southern end of Honshu Main Island, as a tsunami warning system by Wakayama prefecture, and observations have continued since then.

The Kobe Earthquake also provided the motivation for Compact-UrEDAS development. In the VTR shown in Fig. 13.5, initial P-wave motion was detected as something happening, and then severe motion started. In an interview with victims, although there were only a few seconds between detection of something happening to earthquake recognition, there was anxiety and fear because they could not understand what was happening during this period and felt relieved after recognition of earthquake occurrence. To counter this kind of feeling, earlier earthquake alarm was required: Compact UrEDAS was developed to issue the alarm within one second of P-wave arrival.

Fig. 13.5 VTR at the time of the Kobe earthquake.

After the Kobe earthquake, there was a plan to update and renew the alarm system for the Shinkansen lines of East Japan Railways in the northern part of Japan. A compact UrEDAS was chosen for this plan, which can issue the alarm immediately - at almost one second after P-wave detection - based on the dangerousness of the detected earthquake motion, and also issue the alarm with an S wave. 56 sets of Compact UrEDAS were installed for Shinkansen lines of East Japan Railways and started operation only with an S-wave alarm in 1997. After adjustment for the P-wave alarm, this system started full operations in 1998 as an along-the-railroad on-site P-wave detection system.

For the subway network in the Tokyo metropolitan area, six sets of Compact UrEDAS were installed in 1998 and started operation immediately with only S-wave alarms. In 2001, this subway Compact UrEDAS was in practical use as an automatic train control system with a P-wave alarm system. This subway Compact UrEDAS was replaced to FREQL, the next generation of UrEDAS and Compact UrEDAS, in 2007.

Figure 13.6 shows the distribution map of the stations for UrEDAS and Compact UrEDAS after the Kobe earthquake.

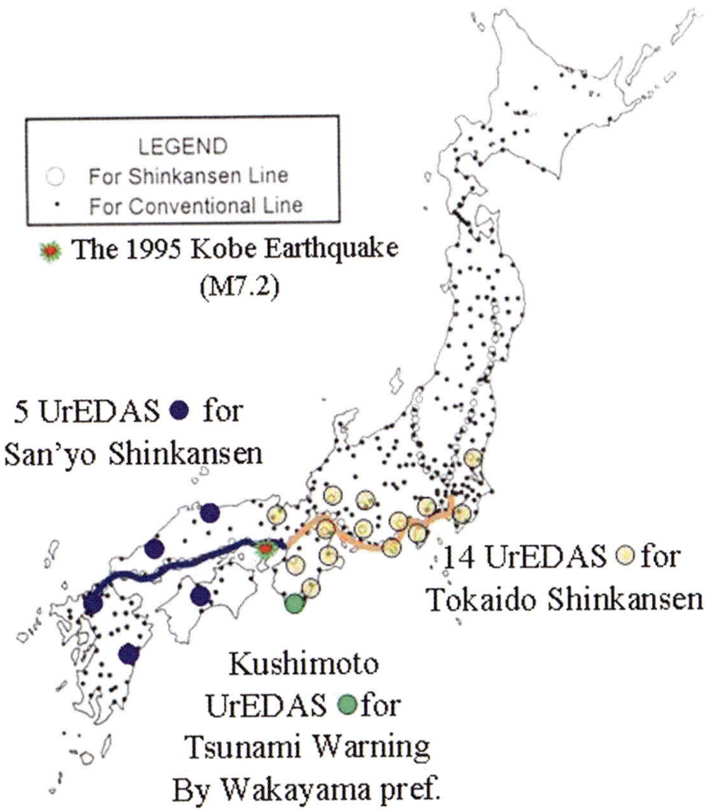

Fig. 13.6 Distribution of UrEDAS and Compact UrEDAS.

13.3 UrEDAS

13.3.1 UrEDAS Functions

Main UrEDAS functions are estimation of magnitude and location, vulnerability assessment and warning within a few seconds of initial P wave motion at a single station. Unlike the existing automatic seismic observation systems, UrEDAS does not have to transmit the observed waveform in real time to a remote processing or centralized system and thus the system can be considerably simplified.

UrEDAS calculates parameters such as back azimuth, predominant frequency and vertical to horizontal ratio, using amplitude level for each sampling in real time. These calculations are basically processed in real time without storing waveform data. UrEDAS calculates these values continuously regardless of whether or not an earthquake occurs, and calculates just like filtering, so the number of procedures is not increased in the event of an earthquake. UrEDAS can detect earthquakes in P-wave triggering with the amplitude level, and then estimates earthquake parameters such as magnitude, epicentral and hypocentral distance, depth and back azimuth from the result of real-time calculation in a fixed period. Moreover UrEDAS can support restarting operation based on the detailed earthquake parameters.

13.3.2 Estimation of Magnitude and Location

(1) P-wave Recognition and Estimation of Azimuth Using Single Station Information

Figure 13.7 shows the methods for P-wave recognition and estimation of azimuth using single station information, seismic wave recognition and epicentral azimuth estimation using three components of the single station. If the vertical component is larger than the horizontal component, the wave will be a P wave. Figure 13.8 shows a comparison between the initial motion periods automatically read by the UrEDAS and the magnitudes given by JMA. The magnitude estimated by UrEDAS seems to be within almost ±0.5 of its JMA counterpart.

(2) Magnitude

The magnitude of an earthquake relates to the size of a seismic fault: the larger the fault, the greater the magnitude. Also the duration and the predominant period of the motion are proportional to earthquake magnitude. Therefore, the magnitude of an earthquake can be predicted from the predominant period of the initial motion as in Fig. 13.9.

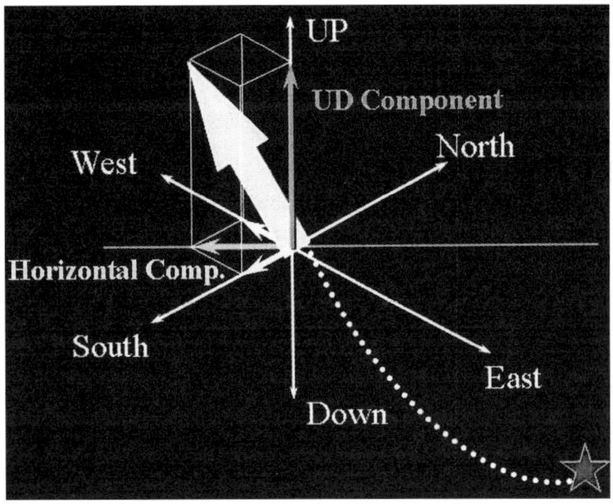

Fig. 13.7 Estimation of P wave recognition.

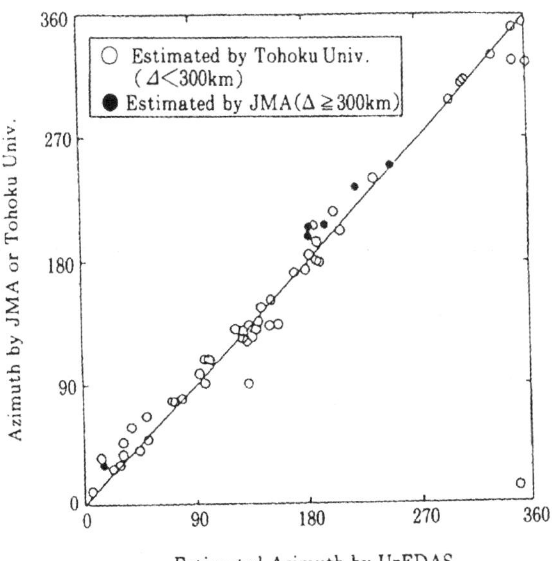

Fig. 13.8 Estimation of back azimuth.

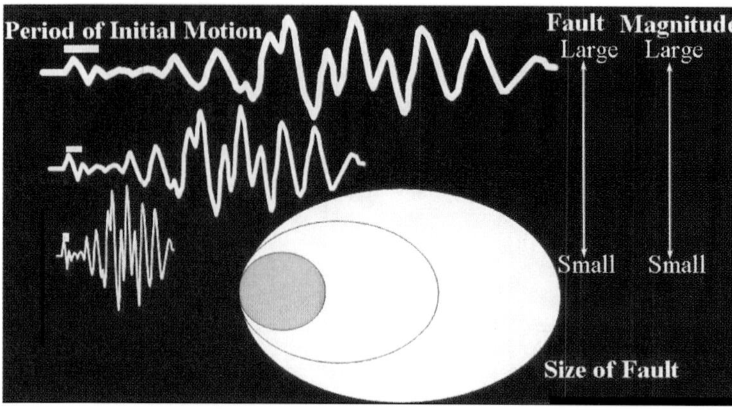

Fig. 13.9 Relationship between the period of initial motion and magnitude.

UrEDAS uses both acceleration and velocity of the earthquake motion for continuous estimation of the predominant frequency in real time. Amplitude level of the motion is calculated continuously with exponential smoothing and then the predominant frequency is estimated from the ratio of acceleration level to velocity level. This corresponds to the frequency of the gravity center of velocity spectrum. Figure 13.10 shows the change in the predominant frequency of actual earthquake motion, and shows it is possible to grasp the change of the predominant frequency.

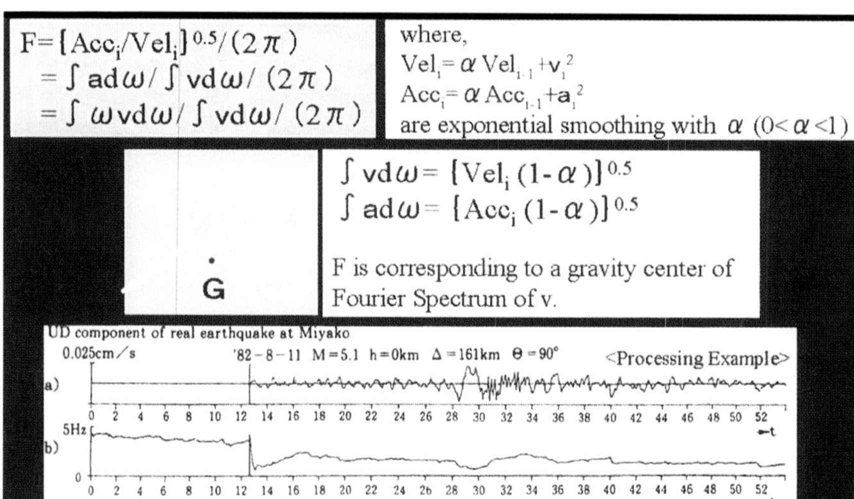

Fig. 13.10 Estimation of the predominant frequency.

Figure 13.11 shows the performance of continuous estimation of the predominant frequency based on the proposed technique. The first example shows the result of inputting a sine curve with various amplitudes and frequencies. Amplitude does not affect the estimation result but the change in predominant frequency is immediately detected. In the case of real earthquake motion, since amplitude increases drastically with P-wave arrival, detection of the change in predominant frequency is extremely rapid.

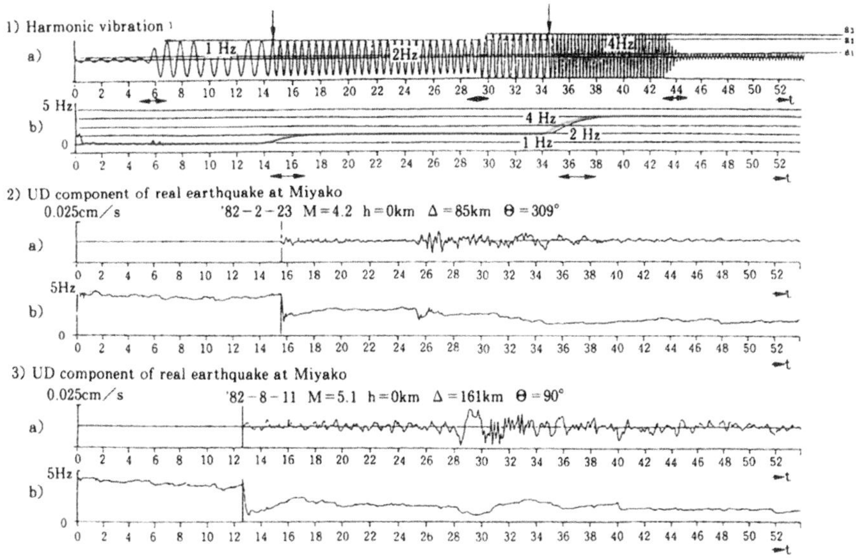

Fig. 13.11 Performance of continuous estimation of predominant frequency.

Figure 13.12 shows a comparison between the initial motion periods automatically read by the UrEDAS and the magnitudes given by JMA. The magnitude estimated by UrEDAS seems to be almost within ±0.5 that of JMA. Figure 13.13 shows the comparison between the official magnitude of JMA and the UrEDAS magnitude estimated in real time with the initial P wave. Estimated magnitude is a little larger but the scatter is in ±1. It is not a serious problem for the alarm system to estimate slightly large because of the view of the safety side.

Additionally magnitude of distant and deep earthquakes are estimated smaller but this constitutes no problem in terms of alarm, as in general deep earthquakes do not cause serious damage.

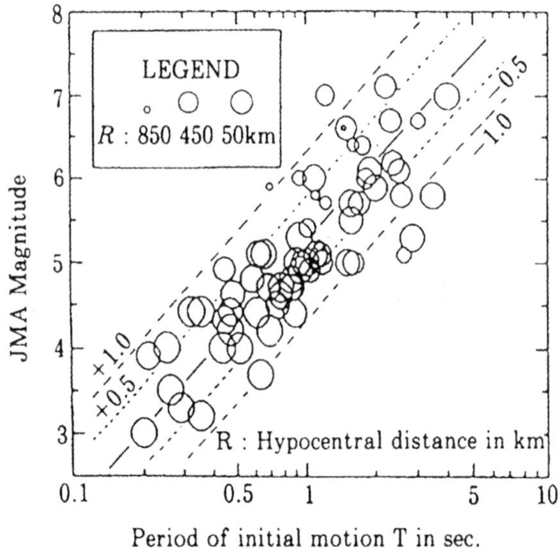

Fig. 13.12 Comparison between the period of the initial motion and magnitude.

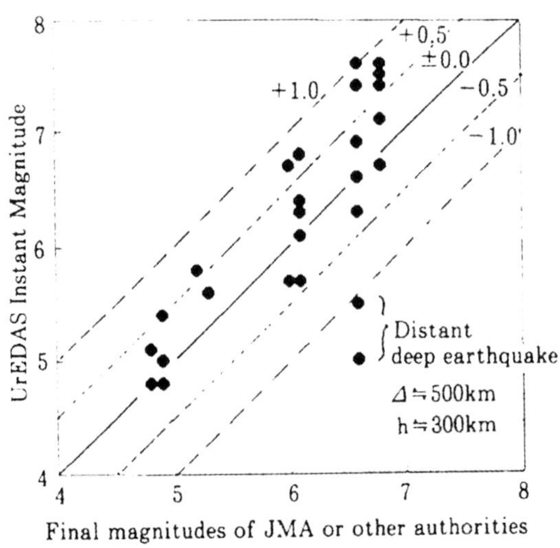

Fig. 13.13 Estimated magnitude and JMA magnitude.

(3) Epicentral Distance and Depth

In general, the magnitude of an earthquake is predicted from the amplitude of the initial motion and the distance from the epicenter. As the amplitude of an initial motion can be automatically measured and the magnitude of the earthquake can be estimated from the period of the initial motion, the epicentral distance can be estimated from this information. The depth can be estimated statistically using a parameter, the ratio of the vertical initial motion to the horizontal initial motion as a kind of the angle of seismic wave incidence. Although the accuracy of estimating the epicentral distance in this way is not so high (half to twice the distance), more accurate estimation by using the duration of preliminary tremors is possible after the arrival of the principal motion.

13.3.3 Vulnerability Assessment and Warning Based on M-Δ Relation

Past earthquake damage for railway structures is plotted in this M-Δ diagram, with earthquake magnitude on the x axis and epicentral distance on the y axis as in Fig. 13.14.

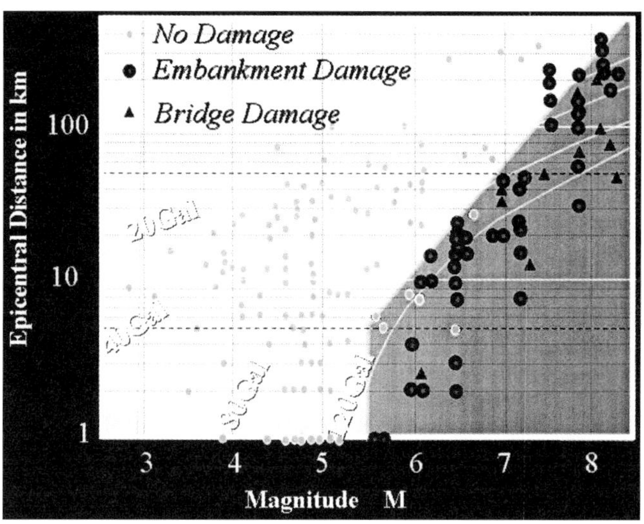

Fig. 13.14 M-Δ diagram

It clearly shows damage occurs for M>5.5, and damaged areas are confined within a certain distance range around the epicenter. For example, earthquakes with M6, M7 and M8 give damage 12 km, 60 km and 300 km within epicentral distances, respectively. If we could rapidly estimate the earthquake magnitude and epicenter location with depth, the area to be alerted is clearly shown by the M-Δ diagram and a reasonable alarm can be issued immediately after earthquake detection. This new alarm is referred to as an M-Δ Alarm.

13.4 Compact UrEDAS

13.4.1 Assessment Index for Vulnerability of Strong Motion

Compact UrEDAS estimates the expected destructiveness of the earthquake immediately from the earthquake motion directly, not from the earthquake parameters as UrEDAS, and then issues the alarm if needed. To estimate earthquake dangerousness, the power of the earthquake motion is calculated from the inner product of acceleration vector and velocity vector, but this value will be large. Hence Destructive Intensity (DI) is defined as the logarithm of absolute value of this inner product as in Fig. 13.15.

Figure 13.16 shows the change in DI as a function of time. When the P wave arrives, DI increases drastically. PI value is defined as maximum DI within t seconds after P-wave detection. This value is suggested to be used for P-wave alarm. Subsequently, DI continues to increase slowly until the S-wave arrival, after which it reaches its maximum value which is called the DI value. This value relates to earthquake damage and is similar to the Instrumental Intensity scale of JMA or Modified Mercalli Intensity (MMI).

Instrumental JMA seismic intensity can be determined only after the earthquake has terminated. On the other hand, DI has a very important practical advantage, because it can be calculated in real time soon after the P-wave arrival with physical meaning. In other words, with the continuous observations of DI, an earthquake alarm can be issued efficiently and damage can be estimated precisely.

Figure 13.17 shows the relationship between the DI value, PI value and maximum acceleration. Common alarm seismometers are triggered 40Gal and 80Gal of acceleration. These trigger levels correspond to 2.0 and 2.5 of PI value, respectively. The actual Compact UrEDAS are set 2.5 to 3.0 of PI value.

Fig. 13.15 Overview of Destructive Intensity, DI.

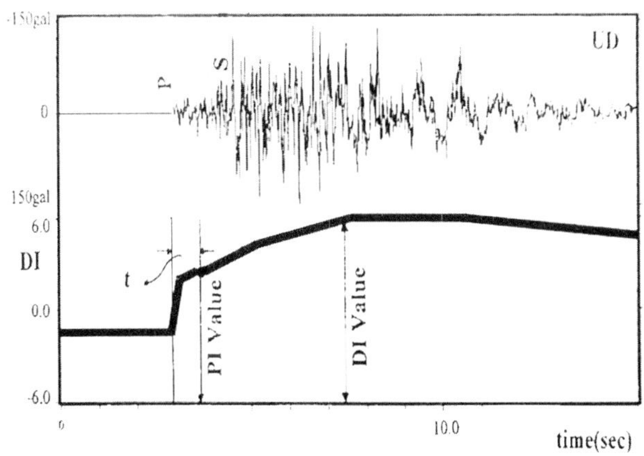

Fig. 13.16 Change of DI.

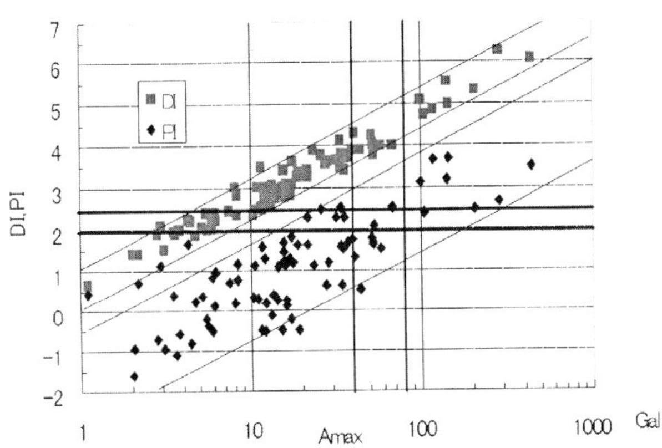

Fig. 13.17 Relationship between DI value, PI value and maximum acceleration.

13.4.2 Alarms of Compact UrEDAS Based on Destructive Intensity and Acceleration Level

Figure 13.18 is a schematic diagram for alarms of Compact UrEDAS. Compact UrEDAS issues alarms not only on P-wave arrival but also S-wave arrival. Thus Compact UrEDAS achieves both rapidity and reliability by combining P-wave and S-wave alarms.

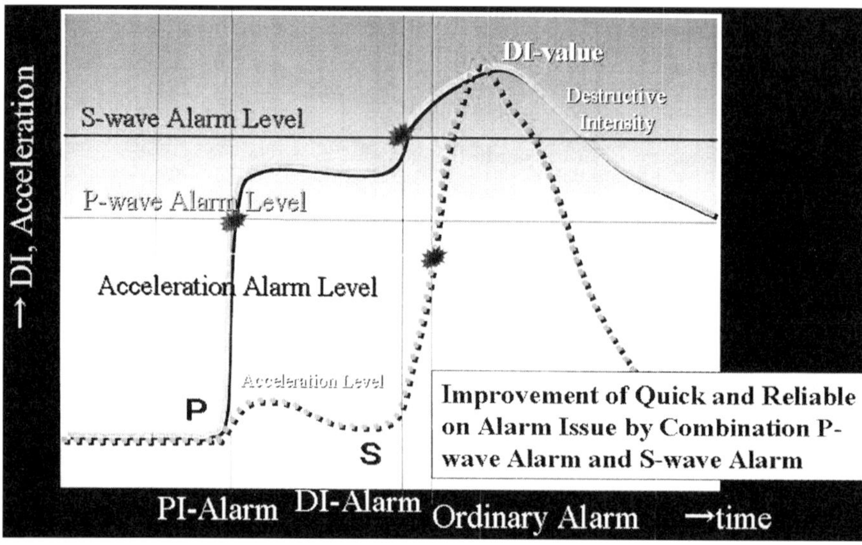

Fig. 13.18 Schematic diagram for Compact UrEDAS alarm.

13.5 Operating Conditions

13.5.1 Overview of the Operating Conditions

Table 13.2 shows the result of UrEDAS operations on Tokaido Shinkansen for approximately 14 years. Initially, the UrEDAS alarm threshold level was set at M4.5 for adjustment. As a result, the UrEDAS alarm was issued more than the ordinary alarm. But after adjustment, the number of UrEDAS alarms decreased, which shows the system has achieved a rational alarm.

Table 13.2 An example of UrEDAS operation result.

Period	Total Num.	UrEDAS Alarm			Common Alarm	Ordinary Alarm (5HzPGA> 40Gal)
		M>5.5	EN[*1]	DBE[*2]		
3-12, 1992	11	7[*3]	2	1	0	1
1-12, 1993	7	2	0	2	0	3
1-12, 1994	4	1	0	0	1	4
1-12, 1995	7	2	0	0	1[*4]	6
1-12, 1996	3	0	0	0	0	3
1-12, 1997	5	1	0	0	1	5
1-12, 1998	2	1	0	0	0	1
1-12, 1999	1	0	0	0	0	1
1-12, 2000	5	2	0	1	0	2
1-12, 2001	4	0	0	0	0	4
1-12, 2002	0	0	0	0	0	0
1-12, 2003	1	0	0	0	0	1
1-12, 2004	2	2	0	0	2	2
1-12, 2005	2	1	0	0	1	2

*1 EN: Electrical Noise.

*2 DBE: Distance Big Earthquake.

*3 M>4.5.

*4 The 1995 Kobe earthquake (Mjma7.2, Mw6.8).

13.5.2 Practical Use

(1) The 1994 Northridge earthquake

During the first 24 hours after the Northridge earthquake, UrEDAS detected about 700 aftershocks whose magnitude and hypocenter locations were automatically estimated as in Fig. 13.19.

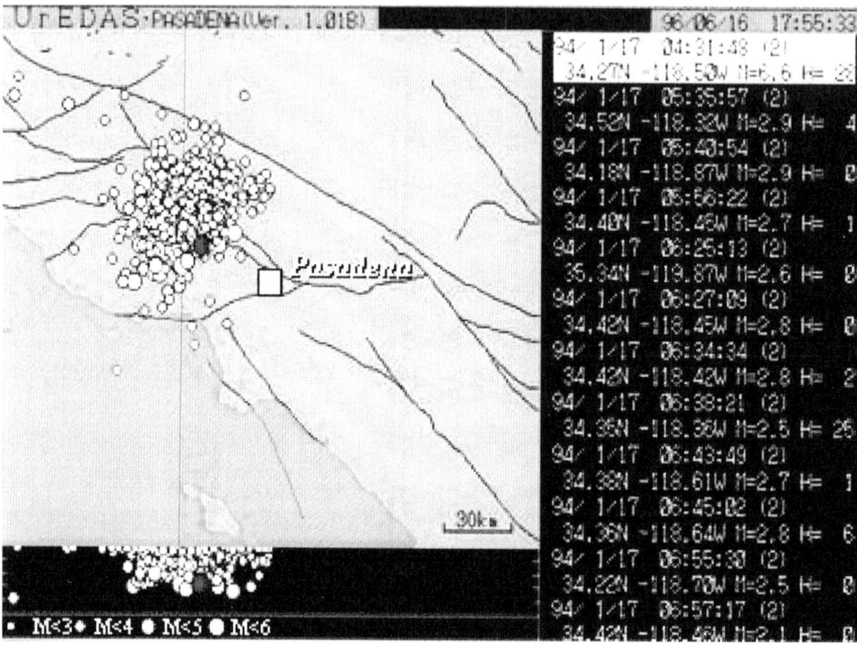

Fig. 13.19 The 1994 Northridge earthquake and after shocks observed by Caltech UrEDAS.

(2) The 1995 Kobe earthquake

Figure 13.20 shows the results of monitoring the Kobe Earthquake sequence during the two weeks after the main shock. Although these figures are based on the data from only one UrEDAS station, aftershock activity was almost correctly traced. At the time of the earthquake, an onsite alarm seismometer issued an alarm instantaneously due to the distance from the epicenter, but also UrEDAS issued an alarm at the same time or a little later. Unfortunately, the alarm did not arrive at the target area due to the transmission system failure. This is a typical example showing the difficulty of controlling the remote target.

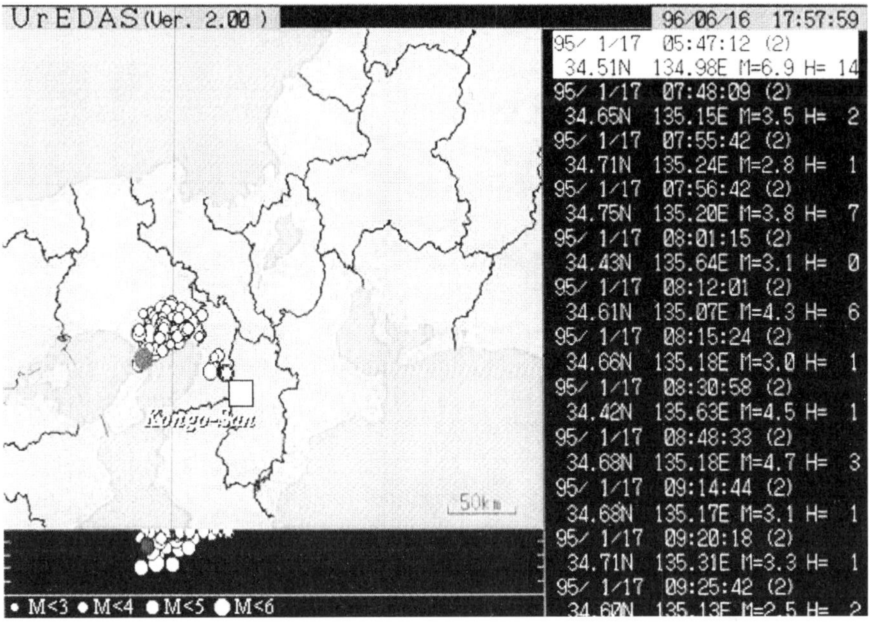

Fig. 13.20 The result of monitoring the 1995 Kobe earthquake.

(3) The 2003 Miyagiken-Oki earthquake

At the time of the Miyagiken-Oki Earthquake, Kitagami plain, situated in the northern part of Honshu Island, was hit by large earthquake motion and damage was caused. The Tohoku Shinkansen line runs along the eastern edge of this plain and suffered damage to the columns of viaducts. The Coastline Compact UrEDAS for front detection issued a P-wave alarm for the damaged section three seconds after P-wave detection, and the alarm reached there before the P-wave arrival at the section. The on-site Compact UrEDAS then issued the P-wave alarm one second after P-wave detection. After that, the on-site Compact UrEDAS re-issued the 40 Gal alarm before S-wave arrival.

Figure 13.21 is a schematic illustration showing the relationship between the wave propagation and the P-wave alarm. The times shown in Fig. 13.21 are based on the origin time. Observed acceleration along the Shinkansen line was high, ranging from 300 Gal to 600 Gal.

At the time of this earthquake, there were trains stopping at stations or running at low speeds near the station at the early warning section. Only one train (Hayate #26) was running at full speed on the new section outside the early warning section. Although Hayate #26 failed to notice the earthquake due to the train's speed, around the train about 300 Gal was observed on the ground.

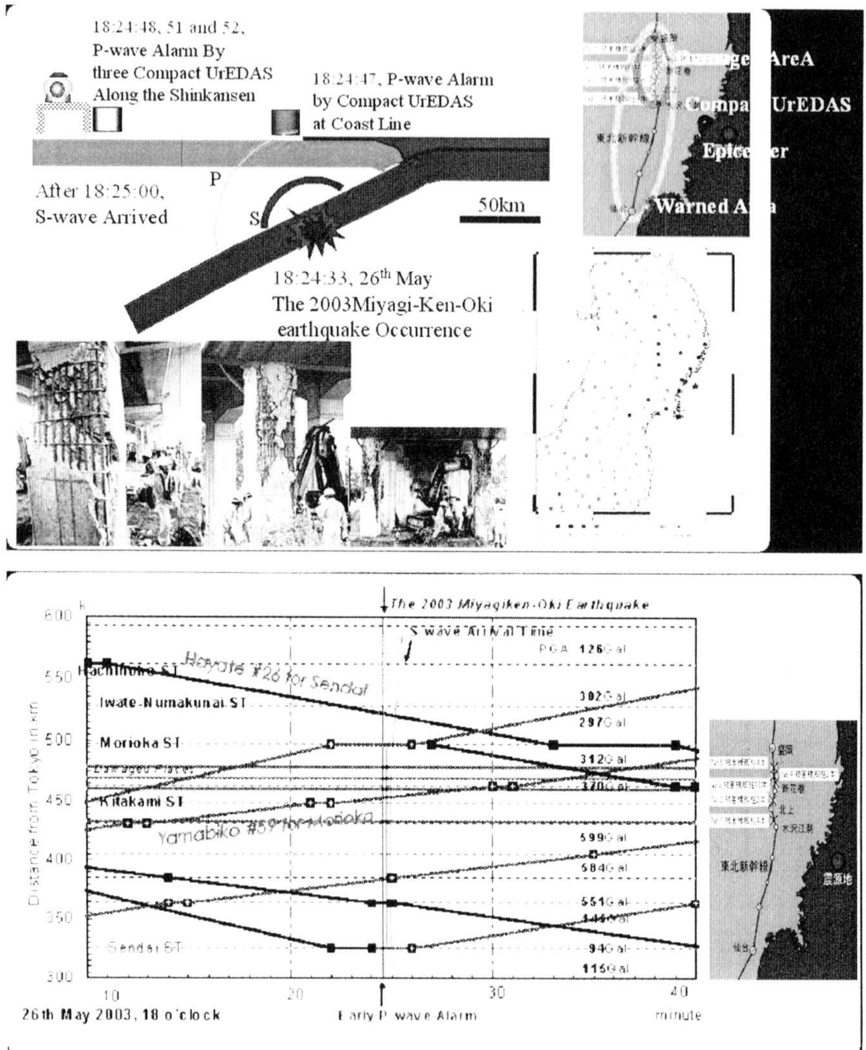

Fig. 13.21 Train operation at the time of the 2003 Miyagiken-Oki earthquake.

On this occasion, there was no train near the damaged area. If the earthquake had occurred a few minutes after or before, the train might have derailed as it was traveling along the section displaced by earthquake damage. On this section, viaducts were actually damaged and it was just a fortunate nothing more than to lead the hazard because of no train.

The first P-wave alarm was issued from the Coastline Compact UrEDAS, and P-wave alarms were then issued from three Compact UrEDAS along the Shinkansen line. This earthquake caused damage to the Shinkansen viaducts. The damaged area is located in the first warned area and the alarm was issued prior to the arrival of the destructive strong motion. Three Compact UrEDAS issued P-wave alarms located near the damaged area. The Compact UrEDAS showed very good performance as the typical merit of the early warning system during this earthquake.

(4) The 2004 Niigataken Chuetsu earthquake
At the time of the Niigataken Chuetsu Earthquake, there were four trains running in the focal area. There are four observatories called Oshikiri SP, Nagaoka SSP, Kawaguchi SS and Muikamachi SP, from north to south. Of these stations, Kawaguchi and Nagaoka issued both P-wave and S-wave alarms, and the others issued only an S-wave alarm. Every station issued an alarm for the section to the next station (see Fig. 13.22). At first Kawaguchi and then Nagaoka issued the P-wave alarm. Subsequently, Oshikiri and Muikamachi issued a 40 Gal alarm. As a result, trains Toki #325 and #332 received the alarm 3.6 seconds after the earthquake, Toki #406 4.5 seconds after and Toki #361 11.2 seconds. The section damaged was between Muikamachi and Nagaoka. Trains traveling on this section received the alarm immediately, proving that the alarm system settings were appropriate.

The UD component of earthquake motion predominate the high frequency more than 10 Hz. The Shinkansen line runs from north to south and the EW component seems to effect derailment. In the case of the EW component, there is a peak at 1.5 Hz and the range of 1 to 2.5 Hz predominates. The natural frequency of the Shinkansen vehicle is included this frequency range.

The Kawaguchi observatory detected the P wave 2.9 seconds after the earthquake occurred, and one second after that, or 3.9 seconds after the event, issued a P-wave alarm. When the derailed train, Toki #325, encountered the earthquake motion when traveling at 75 m from the Takitani tunnel end, it was three seconds after earthquake occurrence. 3.9 seconds after the earthquake, the train received the alarm from the Compact UrEDAS and the power supply was interrupted. The Shinkansen train situated automatically to apply the break immediately at the interruption of power supply. The driver put on the emergency brake after recognizing the Compact UrEDAS alarm. The S-wave hit the train 2.5 seconds after the alarm and more one second later a large shock hit the train, which continued for about five seconds. Figure 13.23 shows the schematic diagram for this earthquake.

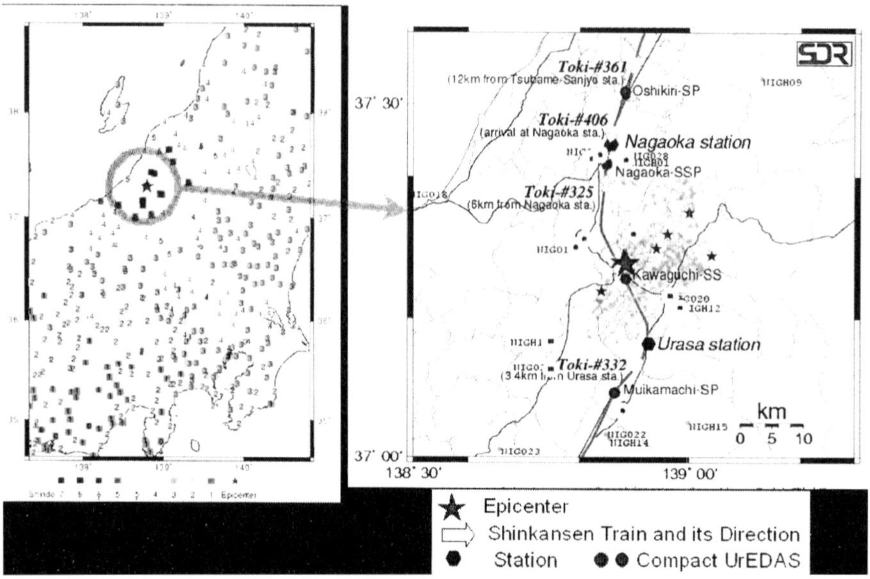

Fig. 13.22 Overview of the 2004 Niigataken Chuetsu earthquake.

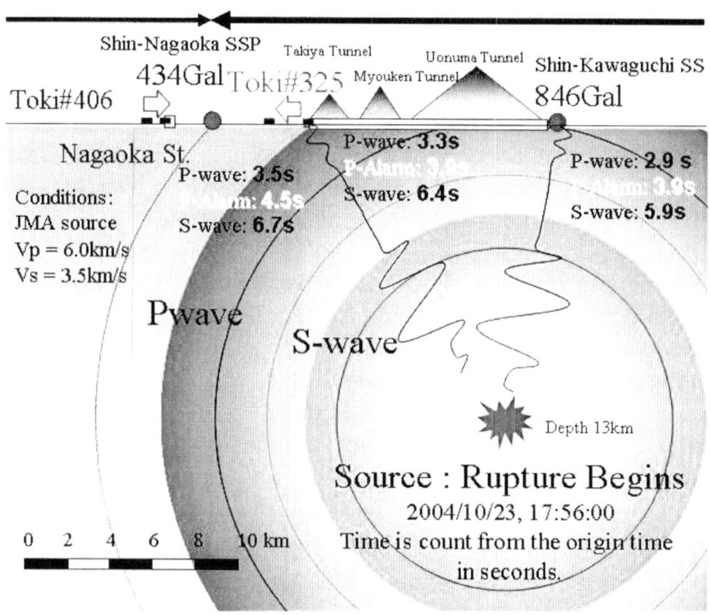

Fig. 13.23 Schematic diagram of the 2004 Niigataken Chuetsu earthquake.

As the result of simulation using the strong-motion records at Kawaguchi and Nagaoka, real-time intensity (RI) rose sharply with the earthquake motion arrival and immediately reached the P-wave alarm level. This RI is a real-time value and the maximum value fits the instrumental intensity of JMA. Because FREQL, the new generation of Compact UrEDAS, improves the reliability of P-wave distinction, FREQL can issue the alarm immediately after the P-wave alarm threshold is exceeded. If FREQL had been installed instead of Compact UrEDAS, both Kawaguchi and Nagaoka observatory would issued the P-wave alarm 0.2 and 0.6 seconds after P-wave detection, respectively. Table 13.3 summarizes the simulation results. In this case, the P-wave alarm reached the derailed section before P-wave arrival. Accordingly, FREQL minimizes the process time for alarm.

Table 13.3 The result of the simulation for Niigataken Chuetsu Earthquake.

Alarm and Accident Site	Kawaguchi	Tunnel Exit	Nagaoka
5HzPGA(Gal)	864		434
RImax(MMI)	6.6 (10.9)		5.8 (9.6)
Origin Time	17:56:00	17:56:00	17:56:00
Recorded Detection Time	03		04
Estimated P wave Arrival Time	2.9	3.3	3.5
Time of RI>2	3.1		4.1
P wave Alarm Time	3.9	3.9	4.5
Time of Acc>10Gal	3.4		4.7
Time of Acc>40Gal	4.2		5.9
Time of Max 5Hz PGA	7.7		9.4
Time of Max RI	8.1		9.5

Figure 13.24 shows the details of the derailment. The derailed train, Toki #325, consisted of 10 cars, from car 10 to car 1 along the traveling direction. The number of derailed axles is 22 out of a total of 40 axles. The last car, #1, fell down the drain besides the track and tilted by about 30 degrees. The open circle indicates the location of broken window glass. The quantity of broken grass appears greater on the left due to the something bounce from the sound barrier, and tends to break one or two cars after the derailed car. The amount of broken glass from car 1 is exceeded by that of car #2.

If it is assumed that the glass broken of car #2 was caused by the derailment of cars #4 and #3, the paucity of broken glasses from car #1 suggests that car #2 did not derail during the earthquake motion. It is estimated that the frictional heat between the vehicle and the rails caused elongation and large displacement at the joints, and car #1 derailed, making car #2 derail.

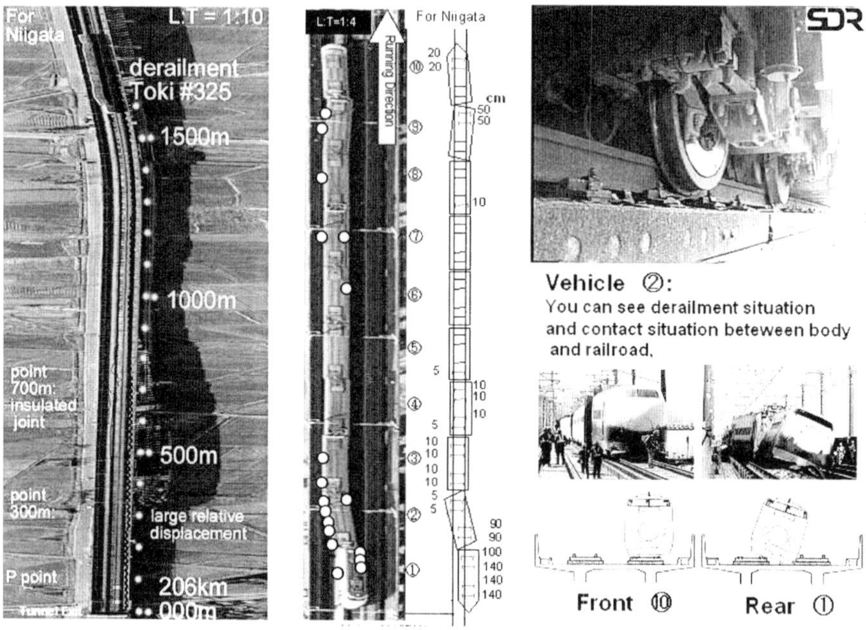

Fig. 13.24 Detailed situation of the derailment.

Deformation performance of viaducts is specified within one cm from the seismic design code. Although the designed natural frequency corresponding to the deformation performance is 2.5 Hz, in practice it is 3.5 Hz. The viaduct may thus be considered to behave statically against the earthquake motion less than around 1.5 Hz. Figure 13.25 shows the relative deformation derived from the dimension of the viaduct columns. The heavy line shows the averaged deformation for each viaduct block, and it is estimated that the relative large occurred at the area farther from the tunnel exit. Taking into account the timing of earthquake occurrence, this is the point of derailment.

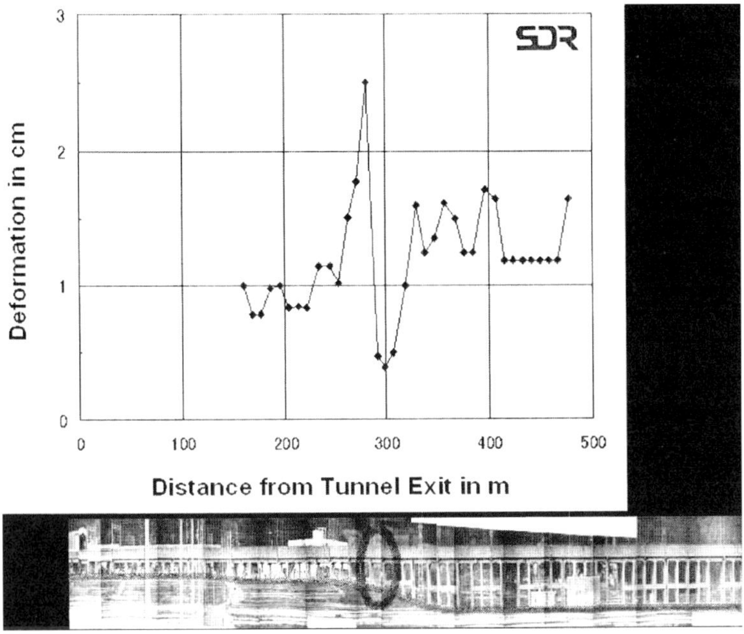

Fig. 13.25 Performance of the deformation (photo is mirror image).

Figure 13.26 outlines the circumstances of the derailment. It seems that the derailed cars were on the large displacement section by chance. The later the alarm reached, the more the number of derailed car, because of the risk of running the large displacement section. As a result, if the friction heat release value were higher, the derailment situation were more severe. On the other hand, the early warning slows the train down, which means that the main shock hits the train before the large displacement section and decreases while the train travels the section. The number of the derailed cars is thus expected to decrease and the derailment damage must be minor. In this regard, the P-wave alarm of the Compact UrEDAS demonstrates its effectiveness at making the derailment non-catastrophic.

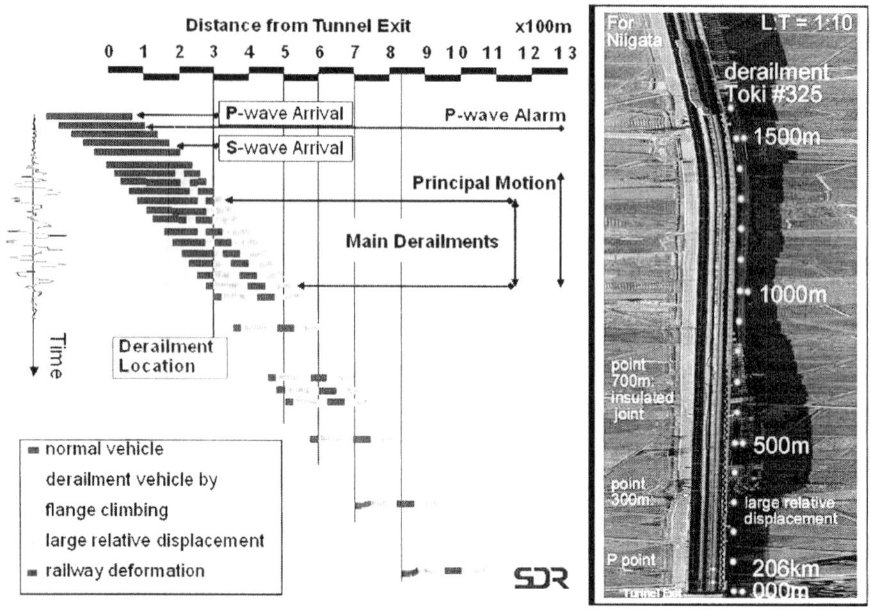

Fig. 13.26 Outline of the situation of the derailment.

13.5.3 Research UrEDAS Worldwide

Other than UrEDAS in practical use, more than five sets of UrEDAS have been installed for our research on high accuracy and shortening estimation times. Figure 13.27 shows the distribution map of the systems. Currently, Istanbul and Mexico City UrEDAS have been stopped but both the Caltech and UC Berkeley systems are working, and send an e-mail on the details of the detected earthquake.

Caltech Pasadena and Berkeley UrEDAS started observations in July 2000 and February 2001, respectively. These observations are supported by Caltech and UC Berkeley, respectively. After earthquake detection, these UrEDAS sends an e-mail in real time. A formaer type of Pasadena UrEDAS was at work from September 1993 to August 1999, and observed the 1994 Northridge Earthquake.

The Berkeley UrEDAS is located just above the Hayward fault while the Pasadena UrEDAS is surrounded by faults, like Fig. 13.28. Despite the influence of faults, the Pasadena UrEDAS shows higher-accuracy estimation of focal parameters than Berkeley UrEDAS, and performance of earthquake detection is effective at both stations.

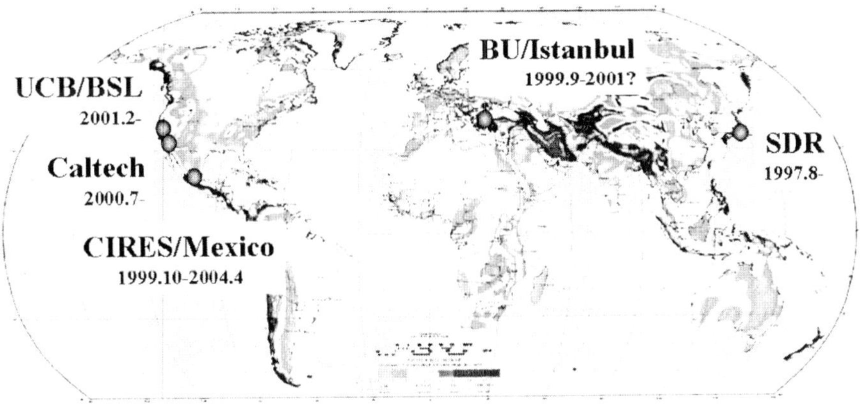

Fig. 13.27 Research UrEDAS in the world.

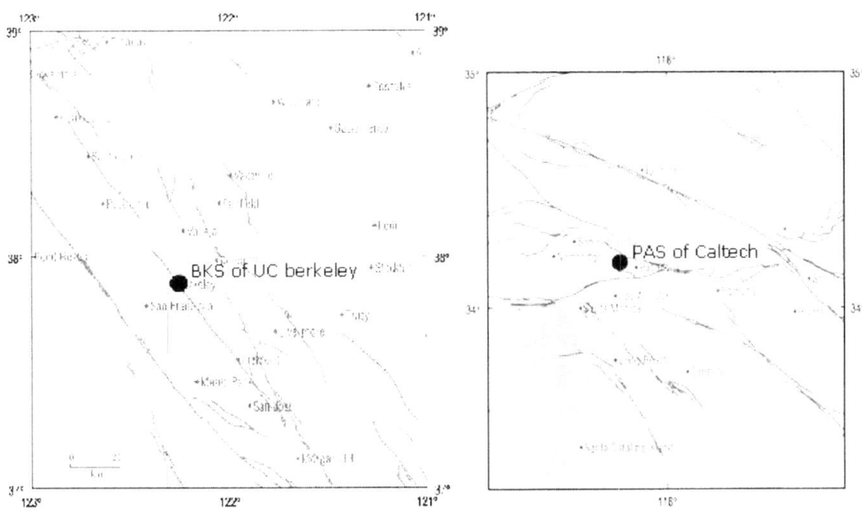

Fig. 13.28 Location of Pasadena and Berkeley UrEDAS.

13.6 Challenges for Earlier Estimation with High Accuracy

Figure 13.29 shows comparisons between UrEDAS real-time estimation results and USGS results. The three figures on the left are the results estimated by using three seconds of initial motion; the right-hand side shows the results of estimation by using one-second data of initial motion. From these figures, the accuracies of three-second estimations are not so good for the epicentral azimuth, within 0.5 for magnitude, half to twice as great for hypocentral distance, albeit excluding earthquakes over several hundreds km away. Contrasting with this, the accuracies of one-second estimations are within 30 degrees for epicentral azimuth, within 0.5 for magnitude, and half to twice as great for hypocentral distance, including distant earthquakes over 1000km. These results are interesting.

13.7 Conclusion

It is necessary that the measuring and processing functions of an earthquake disaster prevention information system be constructed in fully automated fashion and operated automatically.

The alarm in the expected damage area is of prime importance. The independent on-site warning system should be constructed with the awareness that network information from the source region is useful if it comes. Sometimes a robust, inexpensive and reliable system for disaster prevention should be built for early warning and indicating dangerousness even if rack of some accuracy. Efforts have been made to develop and disseminate small, low-cost on-site earthquake alarm instrumentation as a supporting tool for independent disaster prevention. It is expected to distribute many of this kind of equipments cooperating with existing information systems to reduce damage by disaster.

The necessary qualities for early warning systems may be summarized as follows:

- **Fully Automated:** As the time margin is limited, the facility should be directly controlled without human judgment.
- **Quick and Reliable:** As there is limited time to respond to earthquake motion, this kind of system is required to be quick and reliable.
- **Small and Cheap:** To install easily, the system must be small and cheap.
- **Independence:** To issue fail-safe alarms, the system must be independent of other systems.
- **Easy to Connect Network:** To deliver the earthquake information, the system must be easy to connect network.
- **Accuracy is Better:** For the alarm, accuracy of the information is not such a serious problem.

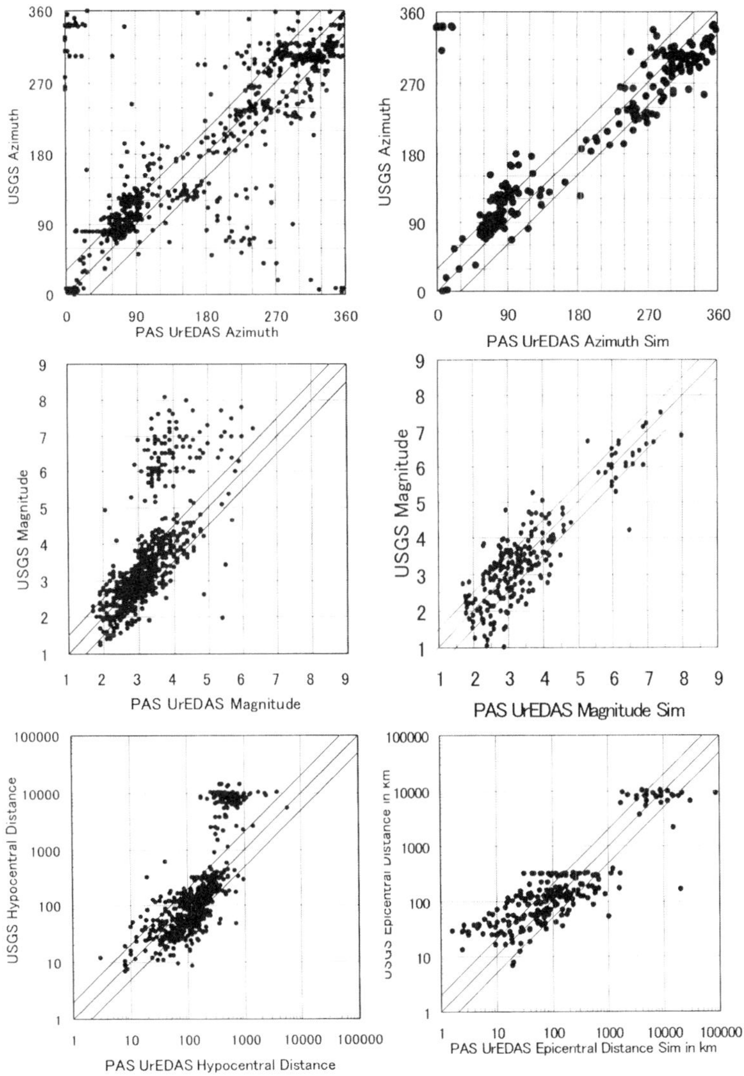

Fig. 13.29 UrEDAS simulation for next generation.

13.8 Acknowledgment

The authors would like to sincerely express their utmost appreciation and gratitude to the following people for their friendly cooperation and great contribution to setting up UrEDAS in each place, checking the condition of such systems, and fruitful discussion of the results observed.

- Prof. Hiroo Kanamori, California Technical Institute
- Prof. Fumiko Tajima, Hiroshima University
- Dr. Robert A. Uhrhammer, University of California, Berkeley Seismological Laboratory
- Prof. Mustafa Erdik, Kandilli Observatory and Earthquake Research Institute, Boğaziçi University
- Mr. Juan Manuel Espinosa Aranda, Centro de Instrumentacion y Registro Sismico
- Mr. Samuel Maldonado Caballero, Centro de Instrumentacion y Registro Sismico

UrEDAS in practical use was realized by the following companies and organizations:

- Central Japan Railway Company, Co., Ltd.
- West Japan Railway Company, Co. Ltd.
- East Japan Railway Company, Co. Ltd.
- Wakayama prefecture
- Tokyo Metro Co. Ltd.

The authors thank the hyper rescue teams of Tokyo Fire Department who opened up the new field of the early warning system.

References

Nakamura Y (1996) Research and Development of Intelligent Earthquake Disaster Prevention System UrEDAS and HERAS. Journal of Structural Mechanics and Earthquake Engineering, Japan Society of Civil Engineers 531(I-34):1-33

Nakamura Y (1988) On the Urgent Earthquake Detection and Alarm System (UrEDAS). 9th World Conference on Earthquake Engineering, Vol. VII, pp 673-678, August 2-9, 1988

The Chugoku Shimbun (2001) "Is Shinkansen Safe?" and "No clincher to countermeasure for derailment", reported on 1st and 2nd May, News Special 2001, May 2001

Proceedings of the Study Meeting for damage of recent earthquakes, the Miyagiken-Oki and the Algeria earthquakes (2003) Japan Society of Civil Engineers, 21st August 2003 (in Japanese)

Nakamura Y (2004) UrEDAS, Urgent Earthquake Detection and Alarm System, now and future. 13th World Conference on Earthquake Engineering, paper #908, 2004

Nakamura Y (2004) On a Rational Strong Motion Index Compared with Other Various Indices. 13th World Conference on Earthquake Engineering, paper #910, 2004

Appendix

San Francisco Daily Evening Bulletin of 3rd November 1868
Earthquake Indicator

EDITOR BULLETIN. Since the Japanese magnet indicator has proved a failure, we are now obliged to look for some other means of prognosticating these fearful convulsions, and I wish to suggest the following mode by which we may make electricity the means, perhaps, of saving thousands of lives in case of the occurrence of more severe shocks than we have yet experienced. It is well known that these shocks are produced by a wave-motion of the surface of the earth, the waves radiating from a center just as they do in water when a stone is thrown in. If this center happens to be far enough from this city, we may be easily notified of the coming wave in time for all to escape from dangerous buildings before it reaches us. The rate of velocity, as observed and recorded in Dr. J. B. Trask's work on Earthquakes in California from 1800 to 1864, is 61.5 (six and one fifth) miles per minute, or a little less per hour (40 miles) than the tidal wave is reported to have traveled across the ocean to this port from the Sandwich Islands or Japan.

A very simple mechanical contrivance can be arranged at various points from 10 to 100 miles from San Francisco, by which a wave of the earth high enough to do damage, will start an electric current over the wires now radiating from this city, and almost instantaneously ring an alarm bell, which should be hung in a high tower near the center of the city. This bell should be very large, of peculiar sound, and known to everybody as the earthquake bell. Of course nothing but the distant undulation of the surface of the earth should ring it. This machinery would be self-acting, and not dependent on the telegraph operators, who might not always retain presence of mind enough to telegraph at the moment, or might sound the alarm too often. As some shocks appear to come from the west, a cable might be laid to the Farallone Islands, 25 miles distant, and warnings thus given of any danger from that direction.

Of course there might be shocks, the central force of which was too near this city to be thus protected, but that is not likely to occur once in a hundred times.

J.D. COOPER, M.D.

14 State of the Art and Progress in the Earthquake Early Warning System in Taiwan

Yih-Min Wu[1], Nai-Chi Hsiao[2], William H.K. Lee[3], Ta-liang Teng[4], Tzay-Chyn Shin[2]

[1] Department of Geosciences, National Taiwan University, Taipei, Taiwan
[2] Central Weather Bureau, Taipei, Taiwan
[3] U. S. Geological Survey (retired), Menlo Park, California, USA
[4] Southern California Earthquake Center, University of Southern California, Los Angeles, California, USA

Abstract

We report here the recent progress and future development of real-time seismic monitoring in Taiwan, especially on earthquake early warning system (EWS) developed at the Central Weather Bureau (CWB), using the telemetered signals from strong-motion instruments. Utilizing a quick magnitude determination based on the first 10 sec of signals from a virtual and sub-network which is configured automatically, we were able to reduce the earthquake rapid reporting time to about 30 sec or less. This represents a significant step towards a more realistic earthquake early warning capability. This early warning system has been in operation at CWB since 2002. Comprehensive earthquake reports have been issued mostly in less than 30 sec, with an average of about 22 sec from the origin time. At 3 km/sec for a typical crustal shear-wave velocity, the present operation is not useful if an earthquake occurs less than about 70 km from a city, but the lead time will increase to more than 10 sec for cities at distances greater than 100 km from the earthquake source. In the latter case, a lead time of several seconds will allow pre-programmed emergency response to take place prior to the arrival of strong ground shaking. In order to offer earlier warning within an epicentral distance of less than 70 km, a τ_c and Pd method was experimented. We used the Taiwan Strong-Motion Instrumentation Program (TSMIP) accelerograms recorded within epicentral distances of less than 30 km from Mw > 5.0 crustal earthquakes. This method uses the first 3-sec length records from the onset of P-waves and could issue an on-site early warning within 10 sec after the event origin time, reducing the radius of the blind zone from 70 to 25 km from the epicenter.

14.1 Introduction

Located on the western circum-Pacific seismic belt with a plate convergence rate of 8 cm/year (Yu et al. 1999), Taiwan has experienced many destructive earthquakes with severe casualties and property losses. For examples, on March 17, 1906 a damaging earthquake (M=7.1) occurred in Chiayi (Hsu 2003), in 1935, a disastrous earthquake (M=7.1) in the Hsinchu-Taichung area (Hsu 2003), and the 1999 Chi-Chi earthquake (Mw = 7.6) occurred in Nantou County (Teng et al. 2001, Shin and Teng 2001). The potential earthquake hazard will continue to increase along with population growth. Therefore, it is essential for Taiwan to seek means through scientific research to reduce future earthquake hazards.

Due to the extreme complexity involved in the earthquake processes, reliable earthquake prediction is not currently possible (Kanamori 2003). Present technological advances in seismic instrumentation and in digital communication and processing permit implementation of a real-time earthquake monitoring system (Lee 1995). In terms of seismic hazards mitigation, an early warning system (EWS) is a practical and promising tool to reduce loss caused by a damaging earthquake (Nakamura 1988, Espinosa-Aranda et al. 1995, Lee et al. 1996, Kanamori et al. 1997, Teng et al. 1997, United States Geological Survey 1998, Wu et al. 1998, Wu and Teng 2002, Allen and Kanamori 2003, Lee and Espinosa-Aranda 2003).

The main purpose of this paper is to summarize the state of the art and progress in the earthquake early warning system in Taiwan. In a recent workshop on earthquake warning systems held on July 13-15, 2005 at the California Institute of Technology, Pasadena, Teng et al. (2005) summarized the development of earthquake rapid reporting and early warning systems in Taiwan; Hsiao et al. (2005) presented the current status of earthquake rapid reporting and early warning systems at the Central Weather Bureau (CWB) in Taiwan; and Lee et al. (2005) proposed a plan to the CWB to integrate tsunami warning with earthquake rapid reporting and early warning.

14.2 Physical Basis for Earthquake Early Warning and its Benefits

The physical basis for an earthquake early warning system is simple: (1) strong ground shaking from a damaging earthquake is caused by shear (S) waves and the following surface waves, (2) typical crustal P-wave velocity is about 6-8 km/sec, whereas S- and surface waves travel at about half the speed of the P-waves, and (3) the seismic wave velocities are much slower than electromagnetic signals transmitted by telegraph, telephone, or radio at about 300,000 km/sec.

Cooper (1868) was the first to propose an earthquake early warning system more than one hundred years ago. In the mid-nineteenth century, there were frequent earthquakes near Hollister, California, about 120 km southeast of San Francisco. Cooper proposed setting up seismic detectors near Hollister and when an earthquake triggered them, an electric signal would be sent by telegraph to San Francisco. This signal would then ring a big bell in the City Hall to warn citizens that an earthquake had occurred. Unfortunately, Cooper's scheme was never implemented. More than 100 years later, Heaton (1985) proposed a seismic computerized alert network for southern California. Nakamura (1988) implemented the single-station urgent earthquake detection and alarm system (UrEDAS) for the "bullet" train system in Japan. Espinosa-Aranda et al. (1995) established the first earthquake early warning system for large events in the Oaxaca coastal region several hundred km away; the system should broadcast to the public in the Mexico City. Subsequently, earthquake early warning systems have been implemented, experimented, and commented upon by several different groups as reported in Chapters 5.2 to 5.14 in Zschau and Kuppers (2003). More recently, an international workshop on earthquake early warning was held on July 13-15, 2005 at the California Institute of Technology, Pasadena (website address: http://www.seismolab.caltech.edu/early.html).

An EWS forewarns an urban area of forthcoming strong shaking, normally with a few seconds to a few tens of seconds of early warning time, i.e., before the arrival of the destructive S- and surface waves of a damaging earthquake. Even a few seconds of advanced warning time will be useful for pre-programmed emergency measures for various critical facilities, such as the deceleration of rapid-transit vehicles and high-speed trains to avoid potential derailment, the orderly shutoff of gas pipelines to minimize fire hazards, the controlled shutdown of high-technological manufacturing operations to reduce potential losses, and the safe-guarding of computer facilities to avoid the loss of vital data.

14.3 Progress in Earthquake Early Warning in Taiwan

Progress has been made towards earthquake early warning, for example, in Japan, Mexico, and Taiwan (see e.g., Nakamura 1988, Espinosa-Aranda et al. 1995, Lee et al. 1996, Wu and Teng 2002). Two approaches have been tried: (1) using an array of seismic stations for regional warning, and (2) using a single seismic station for individual onsite warning. In (1), there are two variants: (a) detect an earthquake and decide if it is "large" enough to issue a warning message to the public, and (b) detect and locate an earthquake, and estimate its magnitude and the expected ground motions in the monitored region, then issue a more informative warning message to the public. In (2), the beginning of the ground motion (mainly P-waves) observed at a site is used to predict the ensuing ground motion (mainly by S- and surface waves) at the same site; either an attempt is made for a single-station event location and magnitude estimation, or no attempt is made to locate the event and estimate the magnitude. The first approach is more comprehensive, but takes a longer time and cannot be used for early warning purposes at short epicentral distances. It has been used, for example, in Mexico and Taiwan. In contrast, the second approach is very fast, can provide useful early warning to sites even at very short epicentral distances where an early warning is most needed, and has been used, for example, in Japan and United States.

The approach of the EWS in Taiwan was motivated by the experience of the 15 November, 1986 Hualien, Taiwan, earthquake of M_L 6.8 (or M_W 7.3). Although the epicenter of that earthquake was located near Hualien, the most severe damage occurred in the Taipei metropolitan area, about 120 km away (Fig. 14.1). According to the travel times of past earthquakes in Taiwan, shear waves traveling over this distance should take more than 30 sec. Thus, if a seismic monitoring system can reliably estimate the location and magnitude within 30 sec of a large earthquake that could threaten a metropolitan area, then several or more seconds of advanced warning will be available for emergency response.

A continuous telemetered strong-motion network was installed in Taiwan, and a Rapid Earthquake Information Release System (RTD) was implemented by the CWB for monitoring earthquakes in real time since 1995 (Wu et al. 1997). In order to maximize the use of the data from this network, the CWB utilized its RTD system as a basis for the development of EWS capabilities.

This RTD system uses a real-time strong-motion accelerograph network that currently consists of 97 telemetered strong-motion stations distributed over Taiwan (Fig. 14.1), an area of 100 km x 300 km. Each station has 3-

component force-balanced accelerometers with signals being digitized continuously at 200 samples per sec per channel at 16-bit resolution, but telemetered at 50 samples per sec per channel using 4800-baud telemetry. The full recording dynamic range is ±2g, and has a sensitivity sufficient to record M > 4.0 events at distances of 100 km or more. Currently, RTD system can offer useful information about one minute after an earthquake occurs (Teng et al. 1997, Wu et al. 1997). Information includes earthquake location, its magnitude, and a shake map of Taiwan. Rapid damage assessment can also be determined and issued by the RTD system a minute or so later (Wu et al. 2002). By applying the sub-network approach (Wu et al. 1999, Wu and Teng 2002) and the M_{L10} method (Wu et al. 1998), the RTD system can achieve an earthquake reporting time of about 20 sec. Therefore, it can offer earthquake early warning for metropolitan areas located at 70 km from the epicenter and beyond.

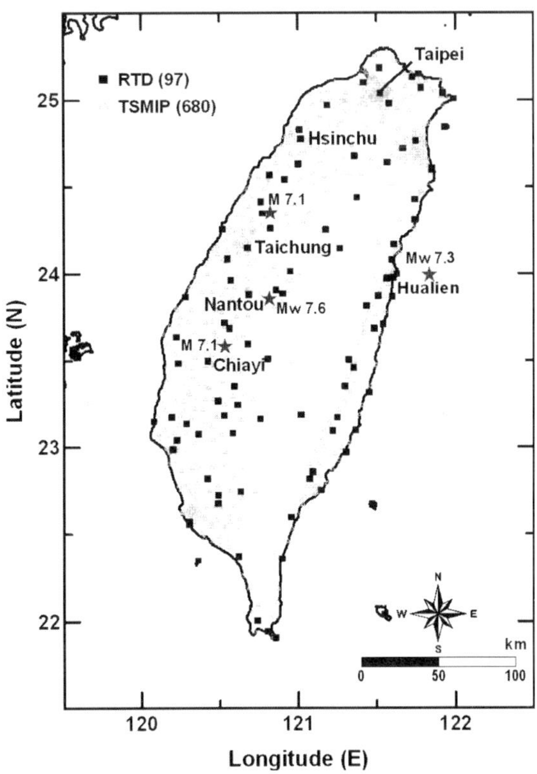

Fig. 14.1 Distribution of the 97 real-time and 680 free-field strong-motion stations in Taiwan.

The RTD system is a typical regional warning system for EWS. Recently, onsite EWS methods were also experimented in Taiwan (Wu and Kanamori 2005a, 2005b). It is possible to shorten the reporting time to about 10 sec. Therefore, this onsite method will play an important role for the future development of EWS in Taiwan.

14.4 Current Regional Warning System

The RTD system can routinely broadcast the location and magnitude of a strong earthquake as well as the intensity distribution in about 60 sec after the origin time of the earthquake. However, the 60-sec time lapse is too long to be practical for earthquake early warning purposes. Therefore, shortening of the reporting time of the RTD system is required in order to achieve early warning functions. Many studies include rapid event location, magnitude determination, sub-network, and virtual sub-network approaches were developed by the CWB in recent years (Lee et al. 1996, Wu et al. 1998, 1999, Wu and Teng 2002).

14.4.1 Rapid Local Magnitude Determination – M_{L10} Method

Two major requirements for an earthquake early warning system are the near real-time estimation of the earthquake location and of its magnitude. The first requirement on rapid location can be achieved readily in the 10-sec time window immediately following the first P-arrival. On the other hand, the second requirement of rapid determination of earthquake magnitude would be more difficult because the shear-wave trains may not be recorded completely within this time window, and, more importantly, since a moment magnitude (M_W) or its equivalence must be developed for large earthquakes. Thus, a method for quick M_W magnitude determination for large events needs to be developed. Although many researchers (e.g., Nakamura 1988, Grecksch and Kumpel 1997, Allen and Kanamori 2003, Kanamori 2005, Wu and Kanamori 2005a, Wu et al. 2006a, 2006b, Wu and Zhao 2006) have also tried to estimate magnitude from the initial portion of the seismograms, large uncertainties in magnitude are inevitable in these methods. Based on the current configuration of the RTD system and its monitoring area, we developed an empirical method to reliably determine earthquake magnitude within 20 sec after the first P-wave arrives at the nearest stations.

Twenty-three sets of strong-motion data from moderate earthquakes (M_L > 5.0) in the Taiwan area are used to achieve this goal. For earthquakes larger than M_L 5, epicenters can be reliably determined in about 15 sec after the arrival of the P-waves from the nearest stations. Figure 14.2 shows simulated Wood-Anderson seismograms from the 25 June, 1995 earthquake (M_L 6.5). In the first 10 sec, there are seven P phases and two S phases that can be used to locate the earthquake. The earthquake magnitude M_L cannot be determined in the same time frame due to the incomplete recording of shear waves at some stations. However, the magnitude based on the first 10 sec of the signals (M_{L10}) can be found to correlate with M_L (Fig. 14.3) as follows:

$$M_L = 1.28 * M_{L10} - 0.85 \pm 0.13. \qquad (14.1)$$

By applying this method for magnitude determination, the CWB system can determine hypocenter and magnitude with tolerable uncertainty in about 30 sec after the occurrence of an earthquake, and early warning is thus possible in Taiwan.

14.4.2 Sub-network Approach

In order to explore the feasibility of an earthquake early warning system for Taipei, a prototype early warning system was implemented in Hualien about 120 km away. Through previous studies (Wu et al. 1997), we concluded that using a dense sub-network under the RTD system is a good approach to shorten the reporting time, and thus gaining some earthquake early warning capabilities. Thus, a dense, real-time monitoring system with high-density station coverage in the Hualien area, and lesser density outside the Hualien area (Fig. 14.4) was developed for testing early warning capability from earthquake sources in the Hualien area. The high-density station coverage in the Hualien area was designed for recording shear waves to enhance magnitude determination, and the stations outside the Hualien area were designed to provide more P-arrivals to improve location quality.

For the 43 earthquakes with $M_L > 4.5$ that occurred from August 1998 to June 2005, this system successfully reported earthquake information in about 19 sec (Fig. 14.5) after the origin time with the location uncertainty under 10 km and the magnitude uncertainty below 0.3. Therefore, it provides about 15 sec of early warning time before shear-waves arrive in the Taipei urban area for earthquakes occurring in the Hualien area.

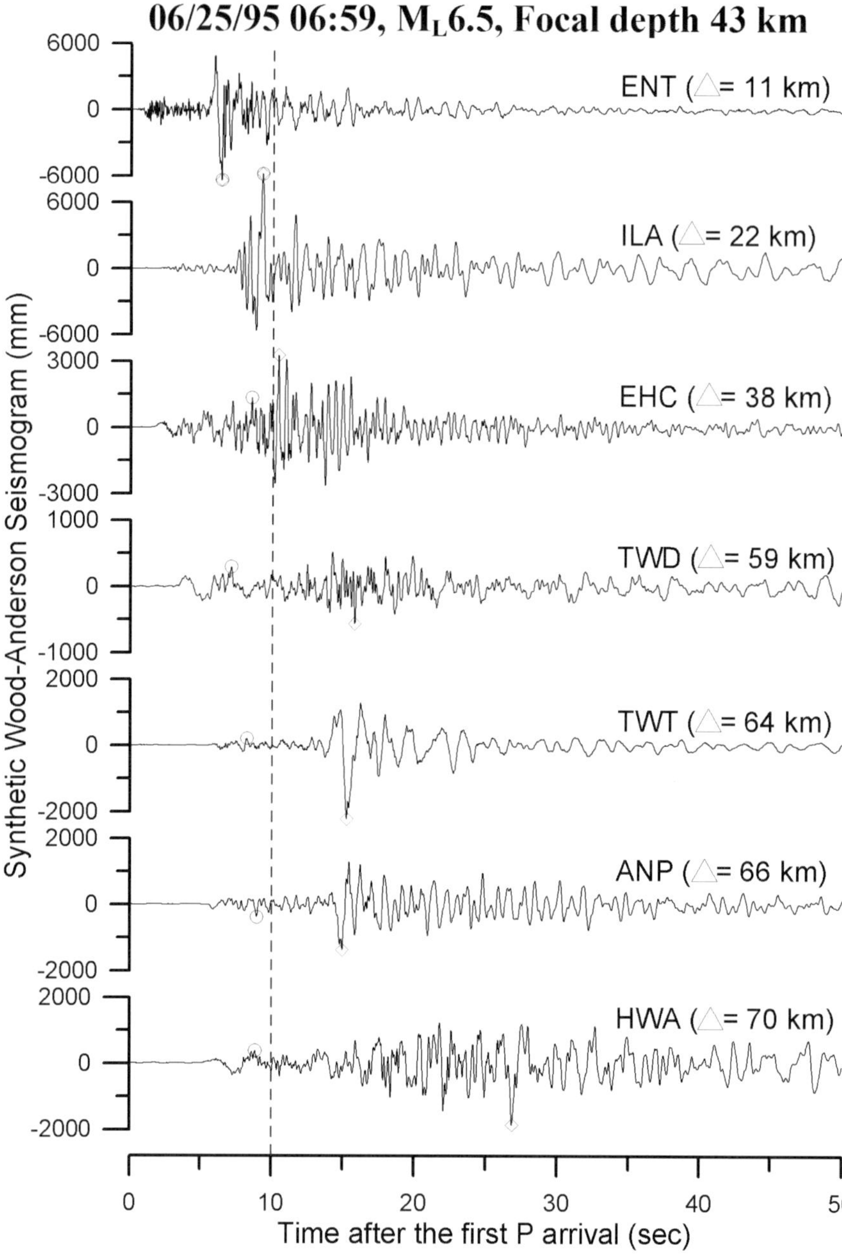

Fig. 14.2 Simulated Wood-Anderson seismograms showing peak readings in the first 10 sec (open circle) and 50 sec (open diamond) after the arrival of the P-wave at the nearest station.

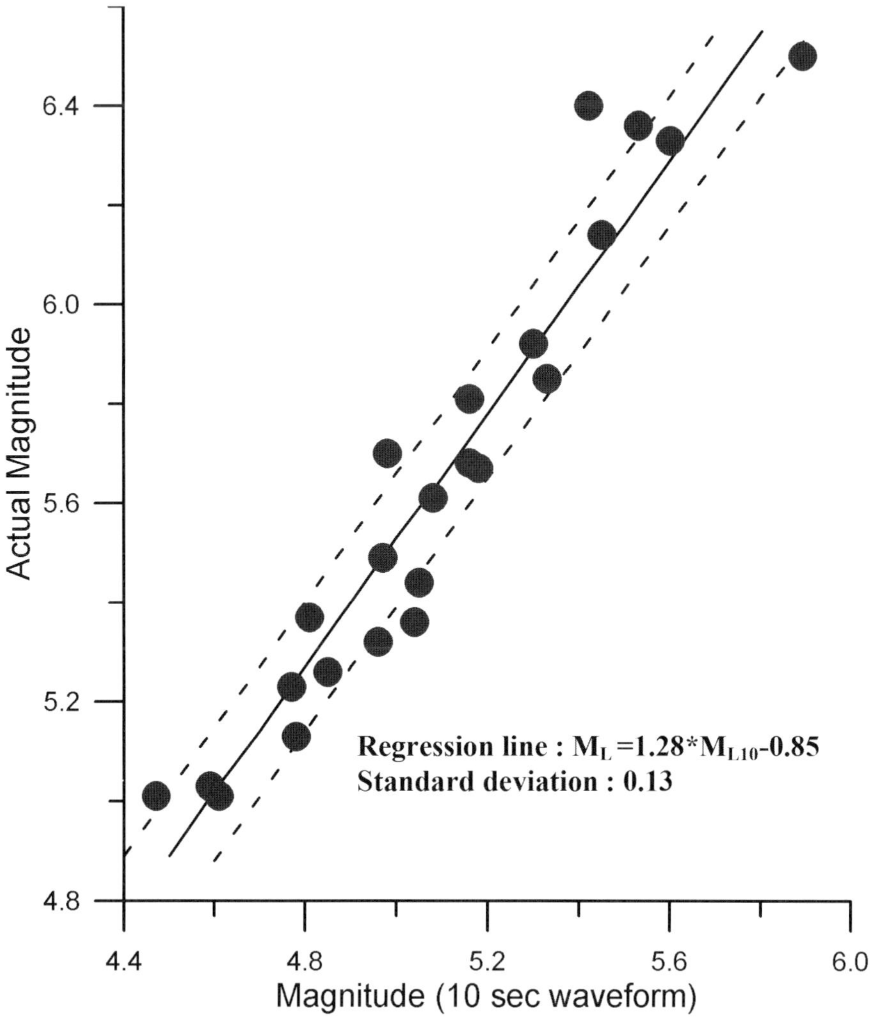

Fig. 14.3 Correlation between the actual M_L and tentative M_{L10} using the first 10-sec signals on an accelerogram after the arrival of the P-wave at the nearest station.

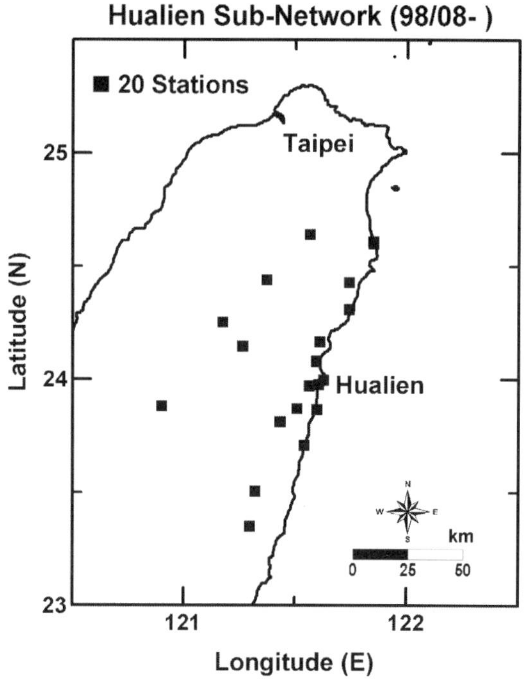

Fig. 14.4 Station distribution of the Hualien sub-network.

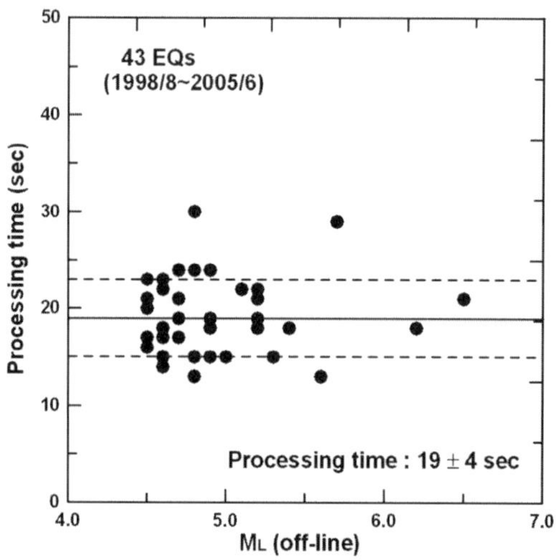

Fig. 14.5 Processing time of 43 events with magnitude larger than 4.5 from 1998 to 2005 being reported by the Hualien sub-network system.

14.4.3 Virtual Sub-Network (VSN) Approach

In the above experiment in Hualien, Taiwan, we demonstrated that earthquake reporting time can be significantly shortened by using a smaller network (Wu et al. 1997, 1999). This leads to the design and configuration of a VSN within the hardware system of the ongoing RTD network. The VSN, automatically configured by the monitoring system, is event-dependent and its configuration varies with time. By working with the VSN, we can substantially reduce the reporting timc such that an effective earthquake early warning capability is feasible to cover the entire Taiwan region.

In hypocenter and magnitude determinations, only stations close to the epicenter (less than 60 km away) contribute the crucial information. Within the framework of the RTD network, we chose to process only signals from a subset of the RTD stations that form a VSN network surrounding an event. As soon as the RTD is triggered by an event, the system automatically extracts a subset of the RTD input signal channels and configures a VSN with a 60-km radius centered on that event. Figure 14.6 also gives a number of possible VSN configurations; each normally consists of about a dozen stations. The extracted data stream for this event forms the basic VSN input data for the subsequent EWS work.

Signals of all stations within a 60-km radius are grouped and extracted through a Multi-IO-Board to form the VSN input, which will then be processed in parallel through the VSN software in a dedicated computer. We conducted a series of experiments to determine the optimum recording time for a 60-km radius network. Our results show that 10 sec is about the optimum. As soon as the 10-sec waveforms are presented at the VSN system, they will be immediately processed to give simulated Wood-Anderson seismograms for magnitude determinations that, in turn, will generate an equivalent moment magnitude. Further reduction of this recording time will cause a significant reduction in the reliability and stability of the magnitude determinations, since insufficient numbers of large S-wave amplitudes are available. An increase in the recording time, on the other hand, will severely cut into earthquake early warning time without significantly improving the magnitude determination. The VSN system is programmed to continue the recording of the waveforms up to 10 sec after the first P-arrival, after which hypocenter and magnitude determinations will be carried out. The results are disseminated automatically to the users.

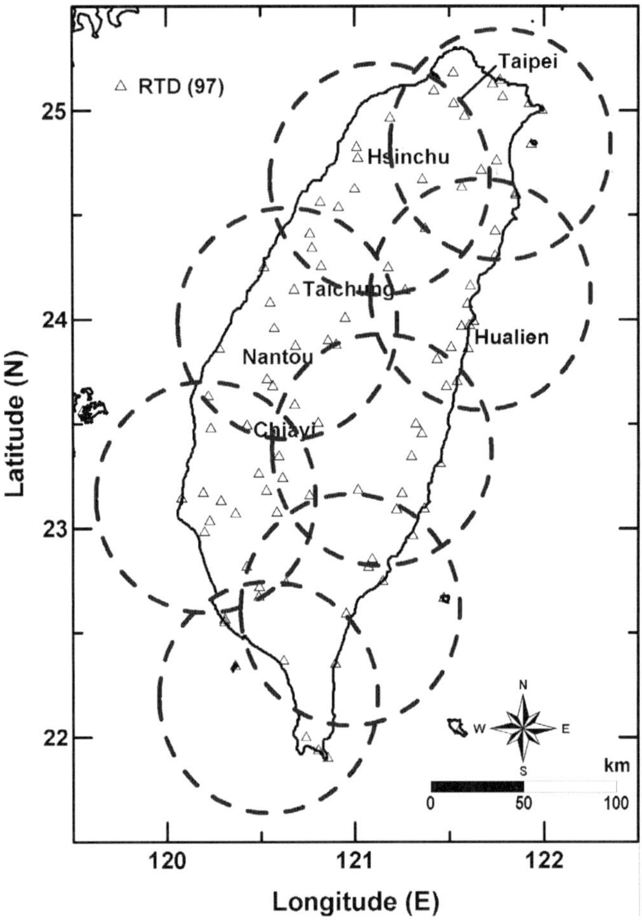

Fig. 14.6 Station distribution of the RTD system and sample VSN networks of 60-km radius configured by software from the RTD system. Each VSN network is centered at a hypothetical event in Taiwan.

We implemented the above VSN operation on the existing RTD network. During the period from December 2000 to June 2005, a total of 125 earthquakes of magnitude M_L ranging from 4.5 to 6.8 were detected, processed, and reported in real time. If we assume that the off-line manual measurements give the correct values, our VSN results give an average error of 6 ± 8 km in epicenter (Fig. 14.7) and an uncertainty of 0. ± 0.3 units in local magnitude. The reporting time is 30 sec or less, with an average of about 22 sec. Records of the CWB earthquake early warning system are given in Table 1.

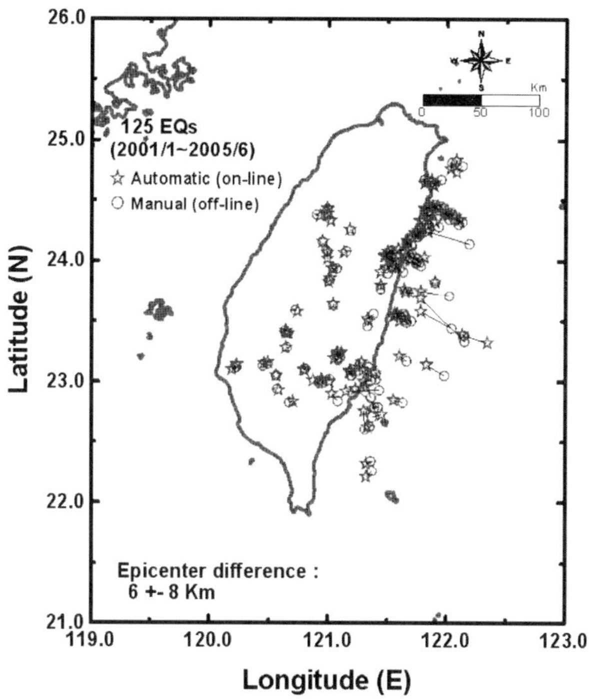

Fig. 14.7 Location differences between the results of the VSN network and those from manual locations.

The Chengkung earthquake (Mw=6.8) occurred on December 10, 2003 with a focal depth of 10.0 km reported by the Central Weather Bureau. The main shock occurred at 23.10°N and 121.34°E along the eastern coast of Taiwan near the town of Chengkung (Fig. 14.8). The EWS of the CWB reported this earthquake information for the same location and magnitude 6.6 at 22 sec after the occurrence. This is a representative case for the CWB early warning system.

For regional warning, the CWB achieved a short earthquake reporting time of about 20 sec (Wu et al. 1998, 1999, Wu and Teng 2002). This can offer earthquake early warning for metropolitan areas located 70 km from the epicenter and beyond. For an event with the same location as the September 20, 1999 Chi-Chi, Taiwan, earthquake, the Taipei metropolitan area is 145 km away, and would have more than 20 sec of early warning time. Figure 14.9 shows the expected early warning times for an event like the Chi-Chi earthquake for all parts of Taiwan. The small triangles in Fig. 14.9 give the locations of elementary schools, which essentially reflect the population density.

Table 14.1 Part of the typical record of the CWB Earthquake Early Warning System.

		Auto-picking				Manual-picking				
Date	Origin time	Lat. (N)	Lon. (E)	Depth (km)	M_L	Lat. (N)	Lon. (E)	Depth (km)	M_L	Reporting time
10/23/04	14:04:27	25.01	121.58	10.0	4.0	25.02	121.57	9.2	4.1	15
10/26/04	08:20:45	22.88	121.32	23.9	4.6	22.91	121.25	21.9	4.3	17
10/28/04	02:05:32	23.93	121.50	53.5	4.7	23.89	121.64	49.5	4.2	21
10/30/04	01:31:52	24.51	121.82	64.7	4.8	24.53	121.80	60.8	3.7	22
11/03/04	14:11:06	23.39	120.48	7.4	3.7	23.39	120.45	8.8	3.9	13
11/07/04	02:47:52	23.96	121.42	13.7	4.0	23.94	121.44	14.5	3.9	29
11/07/04	14:57:55	23.79	121.08	26.6	4.2	23.78	121.06	29.5	4.2	32
11/09/04	01:07:47	24.58	122.00	56.9	4.9	24.58	121.89	54.3	4.0	22
11/10/04	11:00:32	23.08	121.76	12.1	4.2	22.95	121.74	11.7	4.3	24
11/11/04	02:16:44	24.36	122.18	20.7	5.7	24.31	122.16	27.1	6.1	21
11/12/04	07:06:46	24.19	121.69	9.6	3.9	24.19	121.68	4.9	4.1	16
11/12/04	08:35:14	24.47	121.88	16.6	4.2	24.46	121.89	15.6	4.0	20
11/13/04	15:22:04	24.00	121.70	33.8	4.4	24.00	121.68	29.2	4.2	16
11/19/04	06:24:48	24.06	121.41	2.5	3.6	24.02	121.48	18.8	3.7	16
11/27/04	19:27:28	24.00	121.67	21.0	3.9	24.01	121.67	19.7	3.7	15
12/08/04	11:32:34	22.85	121.44	20.5	4.9	22.89	121.39	19.3	4.6	21
12/22/04	00:18:07	23.38	121.48	28.6	4.8	23.38	121.47	30.4	4.9	18
12/22/04	00:28:47	23.36	121.51	23.5	4.5	23.37	121.46	29.9	4.3	20
12/24/04	20:56:48	24.04	121.57	10.9	3.6	24.03	121.62	8.6	3.7	13

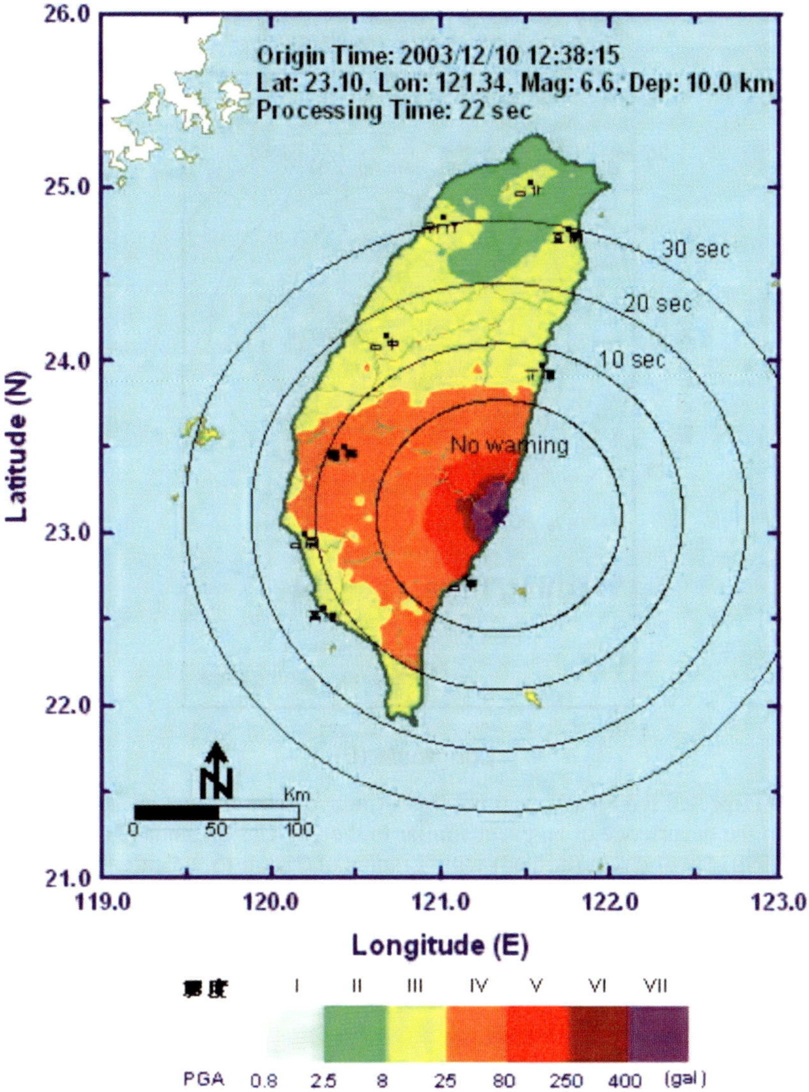

Fig. 14.8 Expected early warning times of the December 10, 2003 Chengkung, Taiwan, earthquake.

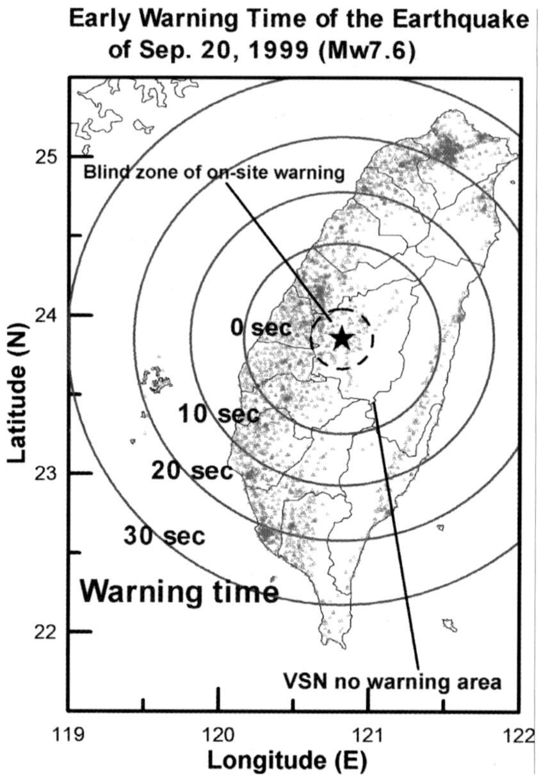

Fig. 14.9 Expected EWS early warning times (indicated by circles) in Taiwan with respect to the occurrence of an event similar to the Chi-Chi earthquake of 20 September 1999. Small circle (dashed) with a radius of 21 km to indicate the boundary of the blind zone of on-site warning method. Triangles give the location of elementary schools, which can be regarded as the population density of Taiwan.

14.5 Onsite Warning Methods

Any earthquake warning system will have a "blind zone" that receives no warning from an earthquake closer than a certain distance. The Taiwan EWS has its "blind zone" defined by a circle with a radius of approximately 70 km. To provide early warning to areas close to the source, a complementary use of an on-site EWS is desirable. Recently, Kanamori (2005) extended the method of Nakamura (1988) and Allen and Kanamori (2003) to determine a parameter, called τ_c, that reflects the size of an earthquake from the initial 3 sec of the P-waves. Results of Wu and Kanamori (2005b) showed that Pd (the peak amplitude of displacement during the first 3 sec of P-waves) correlates well with the peak ground-motion velocity (PGV) at the same location. By using the Pd information, we can predict the expected shaking intensity for earthquake early warning purposes. We explored the use of τ_c and Pd methods to complement the front-detection Taiwan EWS system.

A total of 208 strong-motion records from 26 $M_W > 5.0$ events in Taiwan are used for this study. The selection criteria are: $M_w > 5.0$ and focal depth < 35 km listed in the Harvard CMT catalog (http://www.seismology.harvard.edu/CMTsearch.html). All the events were well recorded by the Taiwan Strong-Motion Instrumentation Program (TSMIP) Network (Fig. 14.1). These events occurred during the period from 1993 to 2003 and were widely felt in Taiwan.

The vertical-component records of the closest eight stations within the epicentral distance of less than 30 km are used. The acceleration signals are integrated to velocity and displacement. We apply a 0.075 Hz high-pass recursive Butterworth filter to remove the low frequency drift after the last integration. Figure 14.10 shows the relation between PGV and Pd for the data set. The solid symbols indicate the average for each event. For the average values, we obtain a regression relation:

$$\log(\mathrm{PGV}) = 0.832 \log(\mathrm{Pd}) + 1.481 \tag{14.2}$$

where PGV is in cm/sec and Pd in cm.

Wu et al. (2003) obtained the following relation between the Taiwan Intensity I_t and PGV.

$$I_t = 2.138 \log(\mathrm{PGV}) + 1.890. \tag{14.3}$$

Combining Eq. 14.2 and Eq. 14.3, we can estimate I_t from Pd as

$$I_t = 1.779 \log(\mathrm{Pd}) + 5.056 \tag{14.4}$$

where Pd in cm.

The parameter τ_c is determined by

$$\tau_c = 2\pi / \sqrt{r} \tag{14.5}$$

where

$$r = \left[\int_0^{\tau_0} \dot{u}^2(t)dt \right] / \left[\int_0^{\tau_0} u^2(t)dt \right] \tag{14.6}$$

where $u(t)$ and $\dot{u}(t)$ are the ground-motion displacement and velocity, respectively; τ_0 is the duration of record used, usually 3 sec, and τ_c can be computed from the incoming data sequentially. τ_c reflects the size of an earthquake and in principle is independent of the distance at least to the first order. Figure 14.11 shows τ_c for all events (open symbols) and the average τ_c (solid symbols) as a function of Mw. The τ_c values generally increase with Mw and are useful for magnitude determinations (Kanamori 2005).

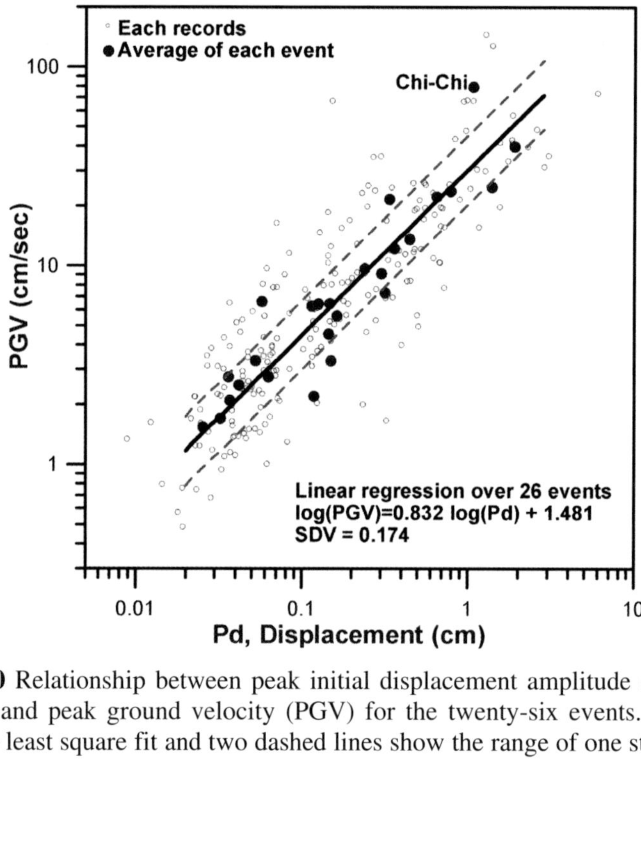

Fig. 14.10 Relationship between peak initial displacement amplitude (Pd) measurements and peak ground velocity (PGV) for the twenty-six events. Solid line shows the least square fit and two dashed lines show the range of one standard deviation.

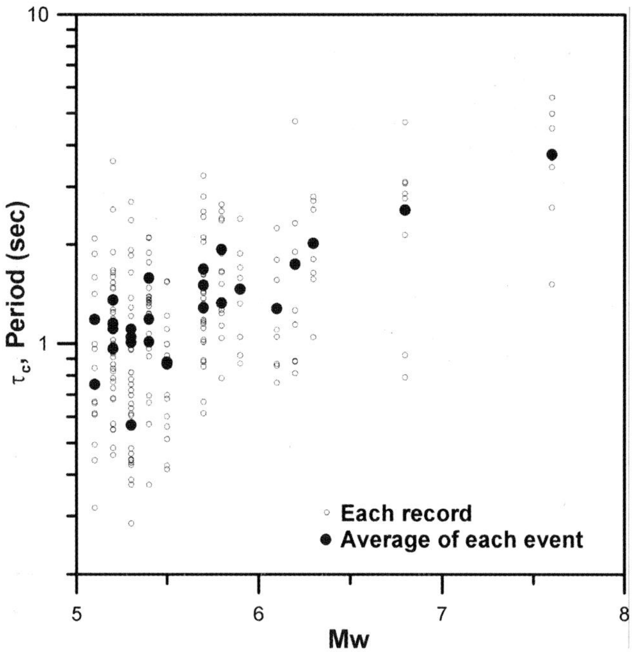

Fig. 14.11 Relationship between period parameter (τ_c) for the twenty-six events and moment magnitude (Mw).

Although $\log \tau_c$ increases approximately linearly with magnitude, the scatter is large for events with $M < 5.5$. The large scatter is primarily due to the low signal-to-noise (*S/N*) ratio when the amplitude during the first 3 sec is very small. Thus, we removed all the data with Pd < 0.1 cm and used only the first 8 stations within $\Delta < 30$ km to determine τ_c. Figure 14.12 shows the results for which the first 8 records with Pd > 0.1 cm are used. Determinations were made for the 12 events for which such data are available. The relationship between the average τ_c and M_w is given by,

$$\log \tau_c = 0.221 M_w - 1.113 \tag{14.7}$$

or, conversely,

$$M_w = 4.525 \log \tau_c + 5.036 \,. \tag{14.8}$$

By using the τ_c and Pd methods we can estimate the magnitude and shaking intensity from the initial three seconds of the P waves.

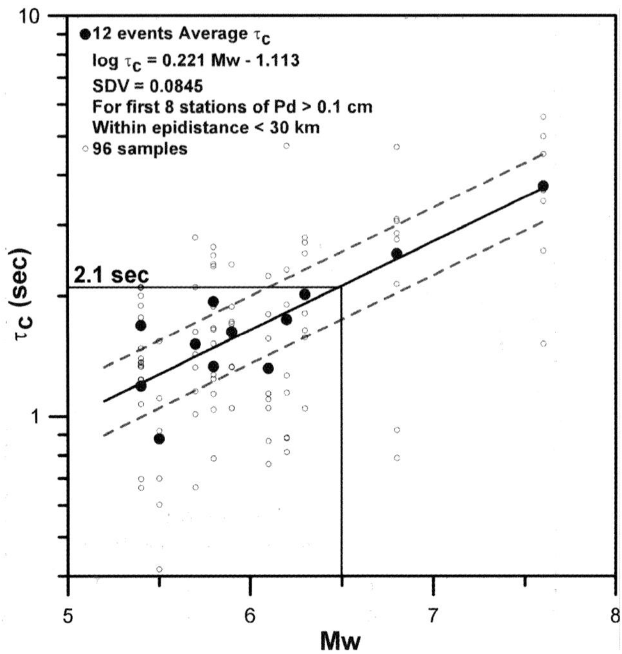

Fig. 14.12 Relationship between the averaged τ_c and Mw for the twelve events from the closest eight records with Pd > 0.1 cm.

Figure 14.13 shows the travel time of P- and S-waves for a reference focal depth at 10 km. Currently the Taiwan EWS can provide rapid reporting at 22 sec after the event origin time by using the VSN method. The current VSN method EWS provides a warning to areas at distances less than about 70 km from the epicenter (i.e., the radius of the blind zone is 70 km). In our experiment on the τ_c method, we used P-waves from the strong-motion records within 30 km from the epicenter. Figure 14.13 shows that P-waves need about 6 sec to reach 30 km. As this method needs only 3 sec of waveform data, the total procedure needs about 9 sec. At that time, S-waves have just propagated to about 25 km from the epicenter and we can ultimately reduce the radius of the blind zone to 25 km (Fig. 14.8).

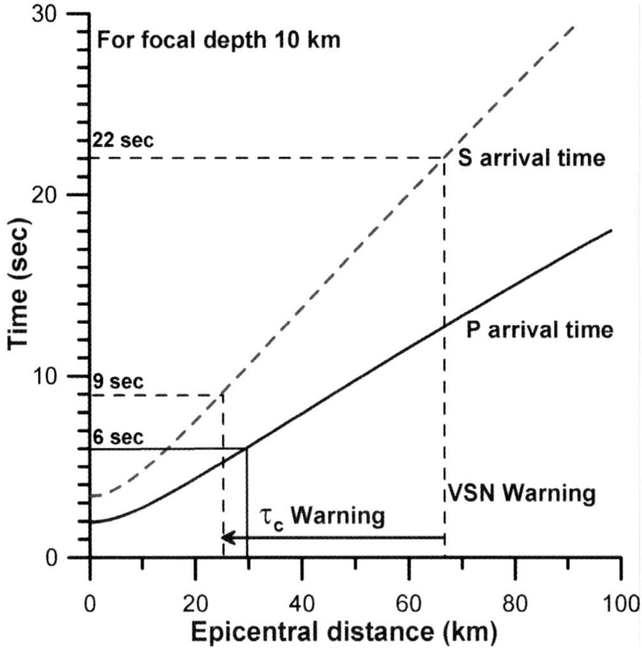

Fig. 14.13 Rapid reporting time and warning distance of τ_c and the VSN methods.

14.6. Prospects

Currently, EWS of the CWB is a reliable regional earthquake warning system. It can report information at about 20 sec after the occurrence of a large earthquake. In the future, by applying the τ_c and Pd method it is possible to shorten the reporting time to about 10 sec with necessary hardware improvement. The τ_c and Pd method is an onsite warning approach. Since it estimates earthquake parameters from the initial portion of P-waves, large uncertainties in the estimates are inevitable. Thus, a hybrid of both onsite and regional warning approaches should be built into a future system. Because of the disastrous Sumatra earthquake and tsunami, Lee et al. (2005) proposed a plan to CWB to integrate tsunami warning with earthquake warning. This integration plan is being studied for implementation in the near future.

14.7 Acknowledgements

We wish to thank Prof. Hiroo Kanamori for his stimulating ideas and encouragements. This research was supported by the Central Weather Bureau and the National Science Council of the Republic of China.

References

Allen R, Kanamori H (2003) The potential for earthquake early warning in South California. Science 300:786-789

Cooper JD (1868) Letter to Editor. San Francisco Daily Evening Bulletin, Nov. 3, 1868

Espinosa-Aranda J, Jiménez A, Ibarrola G, Alcantar F, Aguilar A, Inostroza M, Maldonado S (1995) Mexico City seismic alert system. Seism Res Lett 66: 42-53

Grecksch G, Kumpel HJ (1997) Statistical analysis of strong-motion accelerogram and it application to earthquake early-warning systems. Geophys J Int 129:113-123

Heaton TH (1985) A model for a seismic computerized alert network. Science 228:987-990

Hsiao NC, Lee WHK, Shin TC, Teng TL, Wu YM (2005) Earthquake rapid reporting and early warning systems at CWB in Taiwan. Poster presentation, Earthquake Early Warning System Workshop, July 13-15, 2005 at the California Institute of Technology, Pasadena, California (website address: http://www.seismolab.caltech.edu/early.html)

Hsu M T (2003) Seismological observation and service in Taiwan (up to 1970). In: Lee WHK, Kanamori H, Jennings PC, Kisslinger C (eds) International Handbook of Earthquake and Engineering Seismology, Part B, CD#2\79_15China(Taipei)\Tai70Hist.pdf, Academic Press, Amsterdam

Kanamori H (2003) Earthquake prediction: an overview. In: Lee WHK, Kanamori H, Jennings PC, Kisslinger C (eds) International Handbook of Earthquake and Engineering Seismology, Part B, pp 1205-1216, Academic Press, Amsterdam

Kanamori H (2005) Real-time seismology and earthquake damage mitigation. Annual Review of Earth and Planetary Sciences 33:5.1-5.20, doi: 10.1146

Kanamori H, Hauksson E, Heaton T (1997) Real-time seismology and earthquake hazard mitigation. Nature 390:461-464

Lee WHK (1995) A project implementation plan for an advanced earthquake monitoring system. Research Report of the Central Weather Bureau, Taipei, Taiwan, R.O.C., No. 448, 411 pp

Lee WHK, Espinosa-Aranda JM (2003). Earthquake early warning systems: Current status and perspectives. In: Zschau J, Kuppers AN (eds) Early Warning Systems for Natural Disaster Reduction, pp 409-423, Springer, Berlin

Lee WHK, Ma KF, Teng TL, Wu YM (2005) A proposed plan for integrating earthquake and tsunami warning at CWB in Taiwan. Poster presentation, Earthquake Early Warning System Workshop, July 13-15, 2005 at the California Institute of Technology, Pasadena, California (website address: http://www.seismolab.caltech.edu/early.html)

Lee WHK, Shin TC, Teng TL (1996) Design and implementation of earthquake early warning systems in Taiwan. Proc. 11th World Conf. Earthq. Eng., Paper No. 2133

Nakamura Y (1988) On the urgent earthquake detection and alarm system (UrEDAS). Proc. of the 9th world conference on earthquake engineering, Tokyo-Kyoto, Japan

Shin TC, Teng TL (2001) An overview of the 1999 Chi-Chi, Taiwan, earthquake. Bull Seism Soc Am 91:895-913

Teng TL, Tsai YB, Lee WHK (2001) Preface to the 1999 Chi-Chi, Taiwan, Earthquake Dedicated Issue. Bull Seism Soc Am 91:893-894

Teng TL, Wu YM, Shin TC, Lee WHK, Tsai YB, Liu CC, Hsiao NC (2005) Development of earthquake rapid reporting and early warning systems in Taiwan. Oral presentation, Earthquake Early Warning System Workshop, July 13-15, 2005 at the California Institute of Technology, Pasadena, California (website address: http://www.seismolab.caltech.edu/early.html)

Teng TL, Wu YM, Shin TC, Tsai YB, Lee WHK (1997) One minute after: strong-motion map, effective epicenter, and effective magnitude. Bull Seism Soc Am 87:1209-1219

United States Geological Survey (1998) A plan for implementing a real-time seismic hazard warning system – A report to congress required by public law 105-47. March 27, 1998, USA

Wu YM, Kanamori H (2005a) Rapid assessment of damaging potential of earthquakes in Taiwan from the beginning of P Waves. Bull Seism Soc Am 95:1181-1185

Wu YM, Kanamori H (2005b) Experiment on an onsite early warning method for the Taiwan early warning system. Bull Seism Soc Am 95:347-353

Wu YM, Teng TL (2002) A virtual sub-network approach to earthquake early warning. Bull Seism Soc Am 92:2008-2018

Wu YM, Zhao L (2006) Magnitude estimation using the first three seconds P-wave amplitude in earthquake early warning. Geophys Res Lett 33:L16312

Wu YM, Chen CC, Shin TC, Tsai YB, Lee WHK, Teng TL (1997) Taiwan Rapid Earthquake Information Release System. Seism Res Lett 68:931-943

Wu YM, Shin TC, Tsai YB (1998) Quick and reliable determination of magnitude for seismic early warning. Bull Seism Soc Am 88:1254-1259

Wu YM, Chung JK, Shin TC, Hsiao NC, Tsai YB, Lee WHK, Teng TL (1999) Development of an integrated seismic early warning system in Taiwan – case for the Hualien area earthquakes. TAO 10:719-736

Wu YM, Hsiao NC, Teng TL, Shin TC (2002) Near real-time seismic damage assessment of the rapid reporting system. TAO 13;313-324

Wu YM, Teng TL, Shin TC, Hsiao NC (2003) Relationship between peak ground acceleration, peak ground velocity, and intensity in Taiwan. Bull Seism Soc Am 93:386-396

Wu YM, Kanamori H, Allen RM, Hauksson E (2006a) Experiment using the τ_c and P_d method for earthquake early warning in Southern California. Submitted to Geophysical Journal International

Wu YM, Yen HY, Zhao L, Huang BS, Liang WT (2006b) Magnitude determination using initial P waves: A single-station approach. Geophys Res Lett 33:L05306

Yu SB, Kuo LC, Punongbayan RS, Ramos EG (1999) GPS observation of crustal deformation in the Taiwan-Luzon region. Geophys Res Lett 26:923-926

Zschau J, Kuppers AN (eds) (2003) Early Warning Systems for Natural Disaster Reduction. Springer, Berlin, 834 pp

15 FREQL and AcCo for a Quick Response to Earthquakes

Yutaka Nakamura, Jun Saita

System and Data Research Co. Ltd.

Abstract

There are two kinds of earthquake alarm. One is an "On-site Alarm" which is based on observations at the site of the objects under threat. The other is a "Front Alarm" which is the alarm based on observations near the epicentral area to spread the warning to potentially damaged areas. For each, there are two further kinds of alarm; one being the alarm exceeding a preset level (S-wave Alarm or Triggered Alarm), the other being the alarm during the preliminary motion (P-wave Alarm).

For P-wave Alarm, the new small instrument FREQL has been developed to shorten the processing time for issuing the alarm and combine the functions of UrEDAS and Compact UrEDAS. After P-wave detection, FREQL can issue the alarm within one second and estimate the earthquake parameters in one second. In 2005, FREQL was adopted for the hyper rescue team of the Tokyo Fire Department to protect the staff from the effects of large after shocks during their operation.

On the other hand, it is necessary for local facilities to capture immediately their "own" strong motion index for quick response. For this purpose, a simple seismometer was developed, known as "AcCo", Acceleration Collector. This unique palmtop seismometer has a bright indicator, memory, alarm buzzer and relay connecter.

Recently, seismometers have been installed in many locations and earthquake information is more easily obtained. Nevertheless, warning systems must issue alarms at their own risk, and the public information from authorities is only for the response after the earthquake. In this regard, FREQL and AcCo are surveyed for use in disaster prevention. FREQL is the fastest early warning system in the world to detect the P wave and issue the alarm: a single seismometer can issue four kinds of alarms; P wave alarms based on dangerousness and on earthquake parameters, and S wave alarms triggered by acceleration and real time intensity, RI. AcCo is a simple palmtop size seismometer to measure acceleration and intensity in real time. These instruments combined can achieve an effective warning system and indicate countermeasures for earthquake disaster.

15.1 Introduction

There are two kinds of earthquake alarm. One is an "On-site Alarm" which is based on observations at the site of the objects under threat. The other is a "Front Alarm" which is the alarm based on observations near the epicentral area to spread the warning to potentially damaged areas. For each, there are two further kinds of alarm; one being the alarm exceeding a preset level (S-wave Alarm or Triggered Alarm), the other being the alarm during the preliminary motion (P-wave Alarm).

As the first stage of an earthquake alarm program, a simple triggered alarm seismometer was only installed. This is the alarm seismometer, observing the strong motion just near the target of the alarm. When the ground motion exceeds a preset level, the alarm seismometer issues the alarm. In order to maintain the probability of false alarms as low as possible, the alarm level is kept quite high so that the alarm is issued almost at the same time as the severe strong motion. Nevertheless, the alarm seismometer is useful to shut down the gas supply or other systems automatically.

In order to extend the lead time before large motion arrival, some way of monitoring the earthquake near the focal area had to be considered. This idea was originally put forward in 1868 by Dr. J.D. Cooper in 1868 in California. Cooper proposed setting up seismic detectors near the earthquake source area. When an earthquake was triggered by the detectors, an electric signal would be sent by telegraph to San Francisco to ring a bell in the Town Hall to advise the citizens that an earthquake was to occur. More than 100 years after this original idea, the first system realizing Cooper's idea of "Front Alarm" was developed to protect the Tohoku Shinkansen railway line in 1982. Subsequently, SAS, Sistema de Alerta Sísmica, started operations for Mexico City in 1991.

A new system was then devised based on the detection of the initial part of the earthquake motion to issue an alarm based on the risk of earthquake damage. The first P-wave detection system for practical use, UrEDAS, was created as the front alarm system for the Tokaido Shinkansen line in 1992, and then almost the same system was installed for the Sanyo Shinkansen line in 1996.

The Great Hanshin disaster stimulated the development of an earlier P-wave alarm system as it was felt necessary to issue an on-site P-wave alarm. The Compact UrEDAS was thereby developed, being installed for the Tohoku, Joetsu and Nagano Shinkansen lines and the Tokyo metropolitan subway. Subsequently, the Wakayama Prefecture decided to install

UrEDAS for their own tsunami disaster prevention system and started test operations in 2000.

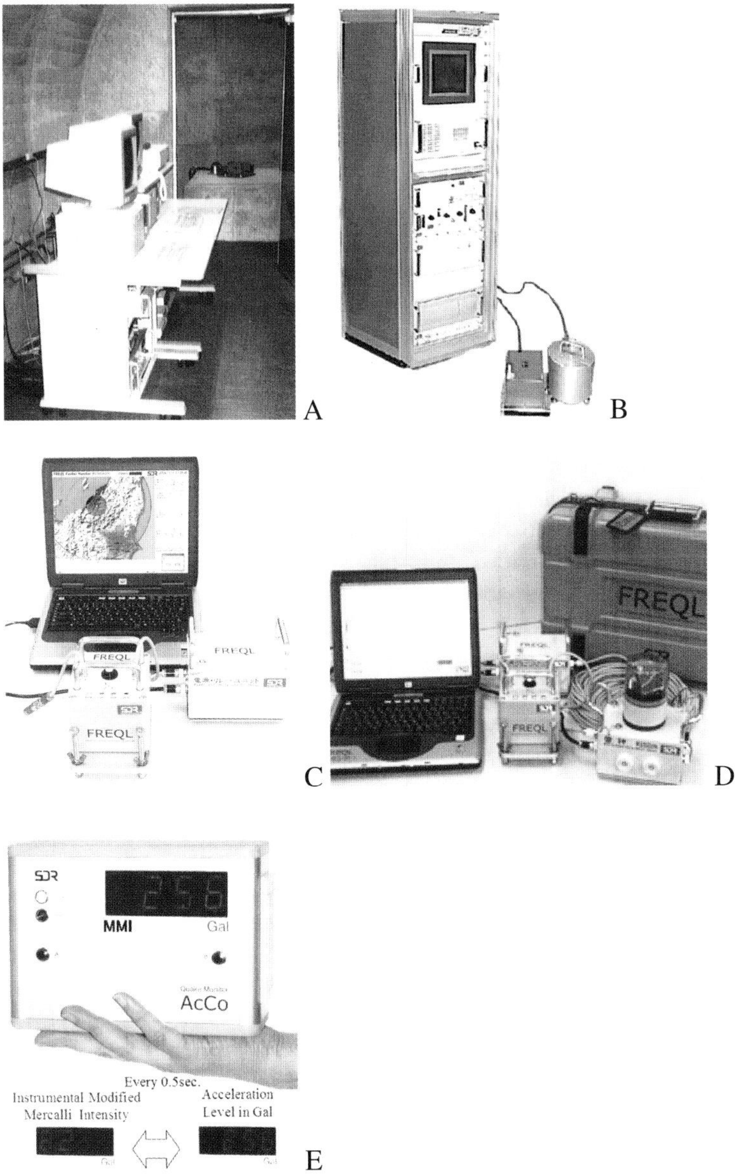

Fig. 15.1 UrEDAS, Compact UrEDAS, FREQL and AcCo. (A) UrEDAS. (B) Compact UrEDAS. (C) FREQL, Installation type. (D) FREQL, Portable type. (E) AcCo.

As the next generation of UrEDAS and Compact UrEDAS, the new small instrument FREQL has been developed to shorten the processing time for issuing the alarm and combine the functions of UrEDAS and Compact UrEDAS. After P-wave detection, FREQL can issue the alarm within one second and estimate the earthquake parameters in one second. In 2005, FREQL was adopted for the hyper rescue team of the Tokyo Fire Department to protect the staff from the effects of large after shocks during their operation. Famous for rescuing children from the landslide after the 2004 Niigataken Chuetsu Earthquake, the hyper rescue teams were afraid of the hazard caused by the aftershocks at that time.

On the other hand, it is necessary for local facilities to capture immediately their "own" strong motion index for quick response. For this purpose, a simple seismometer was developed, known as "AcCo", Acceleration Collector. This unique palmtop seismometer has a bright indicator, memory and alarm buzzer and relay connecter.

An on-site alarm is more important than a network alarm, because a network alarm is sometime missed during data communication. Hence it is not enough to receive information like the Earthquake Early Warning Information (EEWI) from JMA, the Japan Meteorological Agency. By contrast, FREQL has both functions of UrEDAS and Compact UrEDAS for on-site and network alarms. AcCo also has simple alarm functions for an on-site alarm. UrEDAS, Compact UrEDAS, FREQL and AcCo are shown in Fig. 15.1.

15.2 Real-time Seismology and Real-time Earthquake Engineering as a Disaster Prevention Tool

The difference between real-time seismology (RTS) and real-time earthquake engineering (RTEE) appears to consist in how they contribute indirectly or directly for practical uses, just like the difference between science and engineering. The former makes the countermeasure soon after the earthquake rational and prompt by sending useful information to the public at large. The latter sends information to certain customers as a trigger for the countermeasures against the earthquake disaster. In terms of the time domain, the former is required for rational action after the earthquake has terminated and the latter is necessary for immediate response just after earthquake occurrence or earthquake motion arrival.

RTS is required highly accurate but not immediate information, so it is possible to utilize effectively the knowledge and experience in seismology, as well as infrastructures such as observation networks. The challenge is to make the information on the earthquake observation as accurate as possible and to convey it rapidly to everyone concerned.

By contrast, as the most important aim of RTEE is to reduce the degree of the disaster or the possibility of disaster occurrence, it is necessary to issue the alarm rapidly and assuredly. For this purpose, the first concern must be to install one's own observation system for the alarm, without relying on information from other authorities. It is then possible to use other information if it can be received. How the alarm is issued and used necessarily depends on the situation for each company and field. Again, it is risky to rely on information from other authorities alone, using the data transmission network in the earthquake situation.

In Japan, the Japan Meteorological Agency (JMA) has started delivering the Earthquake Early Warning Information, EEWI, on trial. It is clear that EEWI belongs to RTS. Hence it is only an outcome of earthquake observation and must be delivered to the public across the board with no restriction on reception. Although for some cases it may be used the information as an alarm, generally EEWI is mainly for undertaking sound countermeasures after earthquake termination. It must be used for canceling the alarm quickly if the alarm is not needed. In this regard, the most important aspect is accuracy and the delay for a few seconds is not a problem, since the error in this kind of information may cause serious confusion. It is enough if accurate information is delivered within one or two minutes after the event. The alarm must be quitted rationally and EEWI may play an important role as one of the useful tools for this.

It is necessary to capture the distribution of earthquake motion at an early stage. The earthquake disaster prevention is expected to evolutes with the combination of rather late public information and locally dense, rapid information.

15.3 Proposal for a Reasonable Earthquake Index for the Alarm

15.3.1 DI Value and the Other Strong Motion Indices

Here the strong motion indices such as the DI value will be surveyed. The concept of DI was proposed in 1998 (Nakamura 1985). The P-wave alarm system based on the PI value, the maximum DI of the P-wave part, was operative for Tohoku Shinkansen in the same year.

The inner product of the seismic force and velocity is the power of the earthquake motion. Thus the inner product of a response acceleration vector **a** and a response velocity vector **v** can be considered to the generated power of the earthquake motion. Therefore, a way to calculate the power in direct was derived from using the acceleration record. Preventing the value from becoming high, DI was defined by taking the logarithm of power as below:

$$DI = \log(|\Sigma(a \cdot v)|)$$

The DI value is defined as the maximum value of DI. On analyzing DI changes in time and the DI value, DI increases when the seismic wave arrives and shows a good relation between the DI value and Ijma, seismic intensity of JMA, with a constant shift. The amount of the shift is 0.6 when DI is calculated with the acceleration of earthquake motion in Gal (cm/s/s) and velocity in 0.001cm/s within a frequency range of between 0.1Hz and 5Hz.

The influence of vertical motion to calculate DI was considered with the dataset of 314 sites including the 2003 Miyagiken-Oki Earthquake (M7.0), the 2003 Miyagiken-Hokubu Earthquakes (M6.2), and the 2003 Tokachi-Oki Earthquake (M8.0). Between the DI value calculated from three components, and omitting the vertical motion component, the average only differs by 0.013 and standard deviation by 0.035. Therefore the DI value is calculated by two horizontal components in this paper.

Nowadays, maximum acceleration observed at ground surface is widely publicized, but frequency range is not defined. Maximum acceleration tends to increase in the case of wide frequency range observations. Due to the recent improvement in observation techniques, the observed frequency range is spreading to higher frequencies. Hence the observed large maximum acceleration does not cause damage, as would have been expected from previous experience. By contrast, JR has already limited the frequency range to between 0.1 Hz and 5Hz for acceleration to alarm.

The datasets in this article are waveforms observed by K-NET, KiK-net and also by the Taiwan Meteorological Agency. The dataset range for magnitude and Ijma are M3.8~8.0 and 0.6~6.6, respectively. The main samples of earthquakes are the 1995 Hyogoken-Nanbu earthquake (M7.2), the 2000 Tottori earthquake (M7.3), the 2001 Geiyo earthquake (M6.7), the 2003 Miyagiken-Oki earthquake (M7.0), the 2003 Miyagiken-Hokubu earthquakes (M5.4-M6.2), the 2003 Tokachi-Oki earthquake (M8.0), the 2004 Niigataken-Chuetsu earthquake (M6.8) and the 1999 Chi-Chi earthquake of Taiwan(M7.6). The published peak ground acceleration (PGA) based on these data is calculated by two horizontal components or three waveform components without filtering and including up to 30Hz. To distinguish PGA from maximum acceleration with a JR alarm characteristic, "5HzPGA", 5 Hz low passed peak ground acceleration, will be used.

Ijma is calculated by each organization based on the definition of JMA. The SI value has several variations, but here the definition of the Tokyo Gas Company, acknowledged as the SI sensor, is adopted. This SI value is slightly smaller than the SI value defined by Incorporated Administrative Agency Public Works Research Institute (PWRI).

15.3.2 Relationship between Ijma, SI Value and 5HzPGA

In this section, three types of practical major indices, Ijma, SI value and 5HzPGA will be mutually compared. Figures 15.2 and 15.3 show relationships between Ijma and 5HzPGA and between Ijma and SI value, respectively. According to these figures, the range of dispersion of SI value corresponding to Ijma shows almost same as 5HzPGA. Figure 15.4 shows wide dispersion on the relationship Ijma and PGA. These figures show that acceleration generally does not have good relations with damage, but 5HzPGA filtering high frequency component shows conformity with SI value which shows good relation with damage. Here, to discuss on same frequency range are proposed because maximum acceleration depends on it.

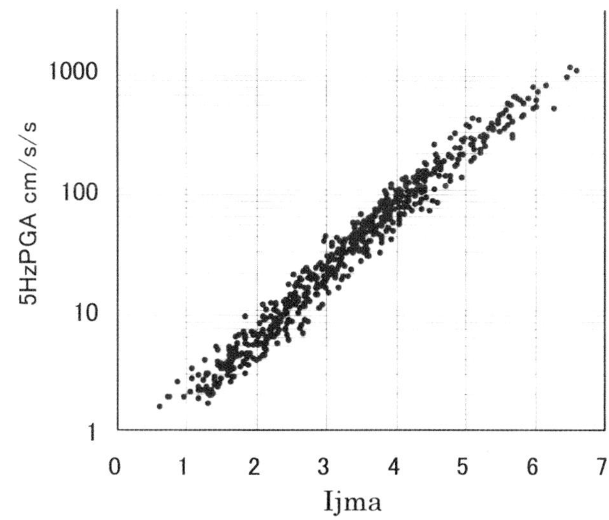

Fig. 15.2 Relationship between Ijma and 5HzPGA.

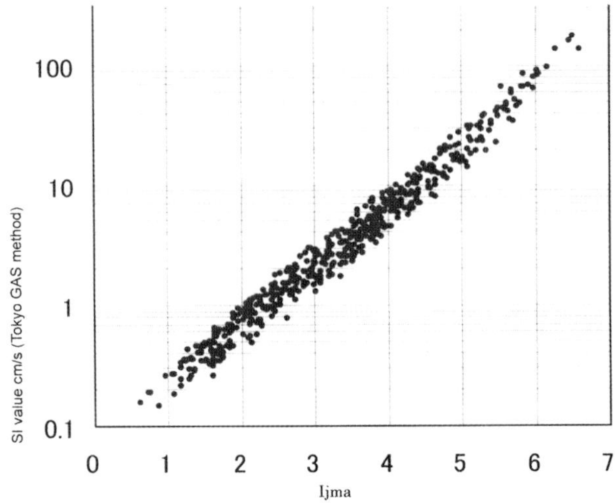

Fig. 15.3 Relationship between Ijma and SI.

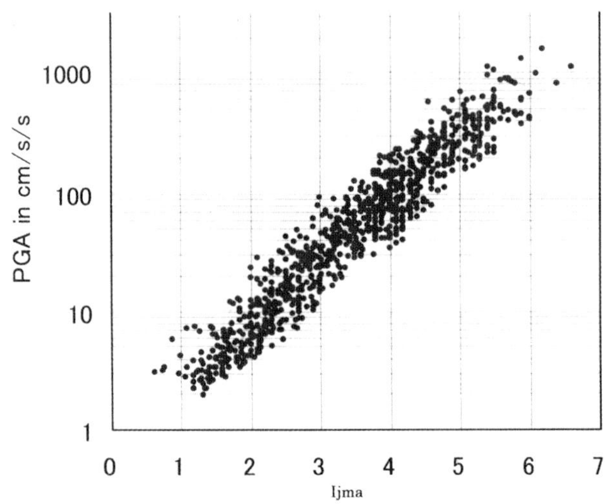

Fig. 15.4 Relationship between Ijma and PGA.

15.3.3 Proposal of RI Value and Relationship with Ijma

The RI value corresponding to Ijma will be defined as below. In the event of using the unit cm/s^2 and cm/s for acceleration and velocity, respectively:

$$RI = DI + 2.4$$

and using the unit cm/s^2 and 1/100 cm/s for acceleration and velocity, respectively:

$$RI = DI - 0.6$$

Figure 15.5 shows the relationship between the RI value, maximum RI and Ijma. The dataset used here has a magnitude range of M3.8 to M8.0 and Ijma range of 0.6 to 6.6. The number of strong motion data is 910, the average difference between Ijma and the RI value is 0.050, and the standard deviation 0.134. Thus it may be stated that both the RI value and Ijma are, in practice, the same. Figure 15.6 shows the relationship between the RI value and Ijma for magnitude. The RI value tends to be a little smaller than Ijma when $M<7$, a little larger than Ijma when $M>7$. Since the predominant frequency becomes smaller for large magnitudes, the RI value becomes smaller than Ijma when the predominant frequency is high, and when the predominant frequency becomes lower it becomes larger than Ijma. From a trend of recent relations between earthquake damage and earthquake motion, this RI characteristic seems to be better for the strong motion index.

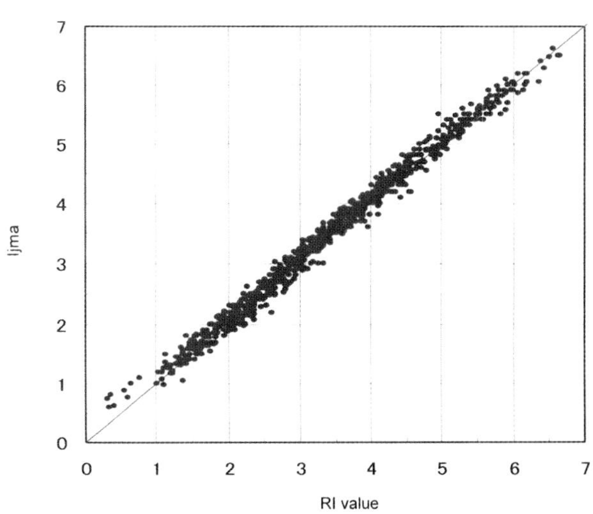

Fig. 15.5 Relationship between RI and Ijma.

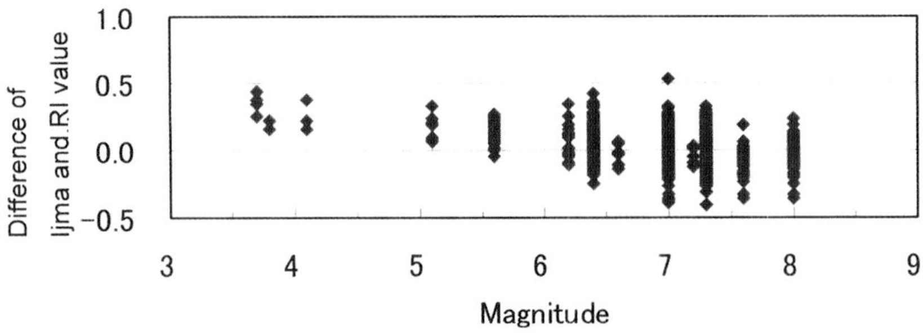

Fig. 15.6 Relationship between magnitude, RI and Ijma.

15.3.4 Proposal of Instrumental MMI

In the previous section, the RI value related to Ijma was defined on the basis of DI. In this section, instrumental MMI will be proposed. Scales of Ijma and MMI are 0 to 7 and 1 to 12, respectively. At first the RI value corresponding to Ijma is uniformly increased to MMI, and then the validity is verified by comparison with maximum acceleration and maximum velocity. The proposed formula is:

$$MMI = \frac{11}{7} RI + 0.50 = \frac{11}{7} DI + 4.27$$

Figure 15.7 shows the comparison between the MMI calculated by the formula above and PGA. In this figure, the relationship between MMI and PGA proposed by Richter (1958), Bolt (1993) and Wald (1999) are shown. After Richter's definition, PGA corresponding to MMI increases and the relationship between instrumental MMI and PGA is including the defference of definitions of Richter, Bolt and Wald.

Figure 15.8 shows the relationship between instrumental MMI and 5HzPGA, and this relationship shows good matches except for the newest relationship. These figures can be interpreted as follows; the maximum acceleration is able to measure for high frequency range with new instruments, since high frequency does not contribute to damage, PGA including a high frequency range deviated from seismic intensity. Generally, strong motion higher than 5Hz is not effective for structural damage. In the past, the observation range was lower than 5Hz, so the relationship between the instrumental MMI and 5HzPGA is expected to be similar to the relationship between MMI and PGA in original.

Figure 15.9 shows that the relationship between instrumental MMI and maximum velocity PGV almost agrees with the relationship found by Bolt. Therefore, it is possible to say that the proposed instrumental MMI in real time is verified.

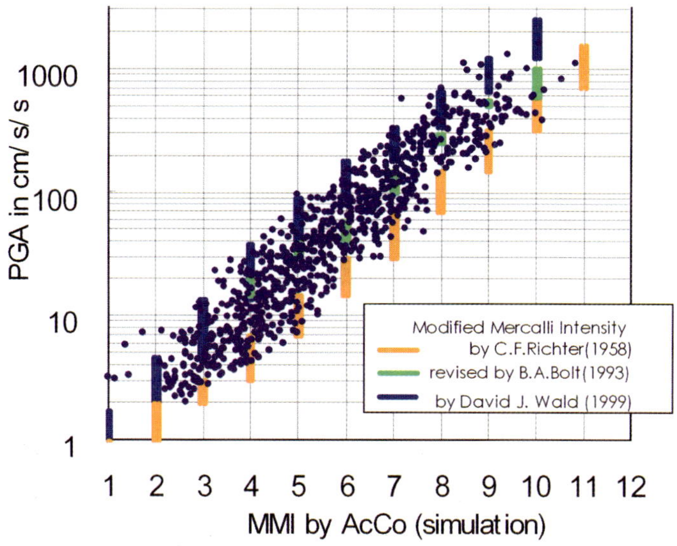

Fig. 15.7 Relationship between MMI and PGA.

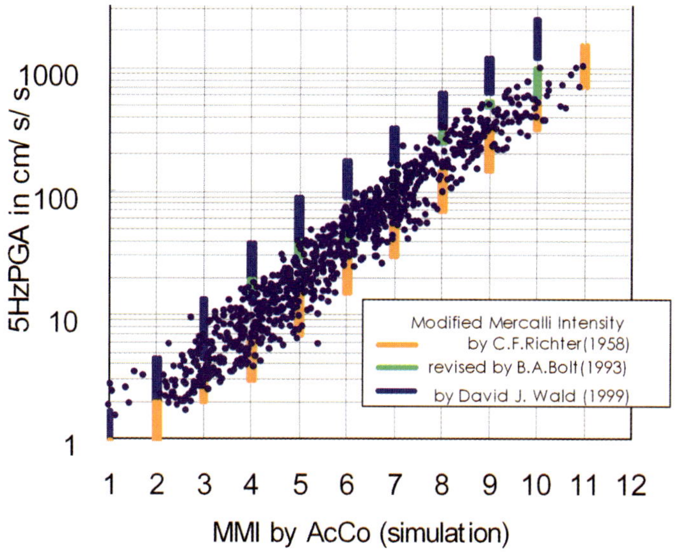

Fig. 15.8 Relationship between MMI and 5HzPGA.

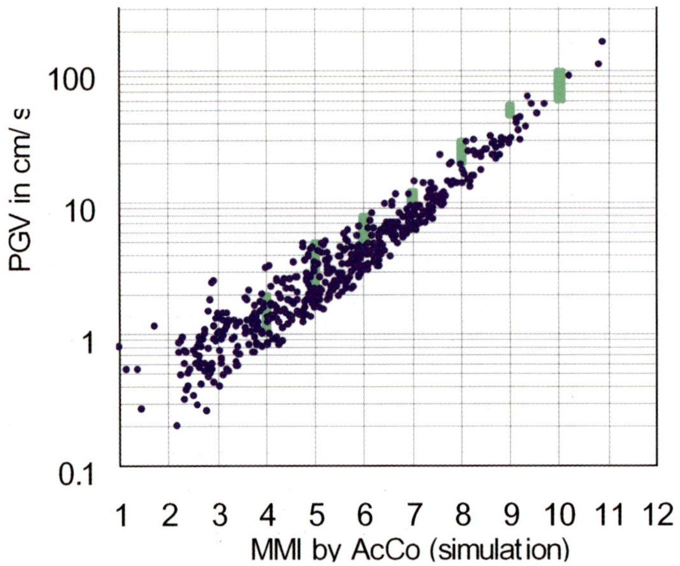

Fig. 15.9 Relationship between MMI and PGV.

15.4 AcCo

15.4.1 Overview of AcCo

Because the usual seismometers are expensive and required an installation and maintenance expert, they were installed for limited facilities. After the Kobe earthquake, the number of seismometers was increased but at most there are thousands of sets for the whole of Japan. This is not such a large figure, as it means one set per several tens of km^2 or per several ten thousand people. Nevertheless, there are many seismometers in Japan in comparison to other high-risk countries which have only a few seismometers. For such countries, it is difficult to take exact countermeasures against earthquake disasters because it is impossible to capture and analyze the damage based on strong motion records and to draw a plan of the city with a certain strategy.

AcCo, Acceleration Collector, is developed to realize a simple seismometer to issue alarms and record the strong motion at low cost. Since AcCo is just a palmtop size instrument, it can indicate not only acceleration but also the world's first real-time intensity. AcCo can thus issue an alarm with the trigger of both acceleration and intensity. AcCo indicates acceleration and intensity if the 5HzPGA (5 Hz low passed peak ground acceleration) exceeds 5 Gal. Intensity can be chosen from RI, MMI or

PEIS, Philippine Earthquake Intensity Scale. AcCo can output the digitized waveform via serial port and also record the waveform for the two largest events with delay memory. AcCo can work with AC power supply and backup battery for seven hours.

Because AcCo indicates acceleration as inertial force and RI as the power of the earthquake motion, it is useful to learn the meaning of acceleration and intensity from experience. This sense is required for the exact image against the earthquake motion. Table 15.1 shows the specification of AcCo.

AcCo is applied in many fields as a warning system, such as different school types, factories, railways and so on. Moreover, AcCo is used not only in Japan but also in the Philippines and Taiwan. Moreover, Tokyo Metropolitan Subway adopted AcCo as "Area Seismometer", for efficient re-operation after the earthquake alarm. More than 30 AcCo were installed in every 3 km mesh.

Table 15.1 Specification of AcCo.

Physical Specifications

Dimension	17cm(W) x 12cm(H) x 6.5cm(D) (not including projections)
Weight	0.8 kg (not including a battery)
Power Supply	AC100V-240V, 50/60Hz, and 4VA (AC adapter attached) Battery back up (*1)
Operating temperature	0-50°C
Operating humidity	Less than 80% (not dew)
Sensor type	Acceleration 2 direction
Measurement range	Each direction up to 2g (2 x 980Gal)
Noise level	Less than 1Gal-rms
Resolution	Approximately 1/6Gal
Sampling rate	1/100 seconds
Frequency range	DC-10Hz
Event Memory (option)	Record length 108 sec within pre-event memory 30 sec for max and next events
Display	

Frequency range	0.1-5Hz for PGA and intensity
Contents of a display	The maximum acceleration, RI or MMI(option)
Display type	4 characters of high luminosity 7 segment Light Emitting Diode
Alarm output	
Frequency range	0.1-5Hz
Alarm conditions	Exceeding preset level (*2)
Output method	Red LED luminescence, buzzer, relay and output via RS-232C
Output	
Communication	RS-232C (D-SUB 9 pin socket).
Contents of an output	Waveform and other information
Output method	The original protocol using RS-232C (D-SUB 9 pin socket). (Separately optional software and optional cable are required to communication with PC.)

The characteristics above mentioned changes not with a guaranteed value but with observation conditions.

(*1) 006P(9V) battery are required for back up.

(*2) selectable from 12 stages of 10, 15, 20, 25, 30, 40, 50, 80,100,120,150 and 200Gal.

15.4.2 Alarm Timing by a Simple Trigger Method

In the case of the system requiring earlier warning with no error or accidental warning, it is necessary to install a system with high reliability and sophisticated such as FREQL. However, in general, merely the simple warning system appears useful in cases where there are several alarms per year, even in the higher seismic activity areas of Japan. AcCo 10 Gal alarm or RI 2.0 alarm can play the role of this simple early warning. Figure 15.10 shows the relationship between the alarm timings.

Although the AcCo 10 Gal alarm or RI 2.0 alarm is triggered a little later than the P-wave alarm of FREQL, it is considerably earlier than the ordinary triggered S-wave alarm.

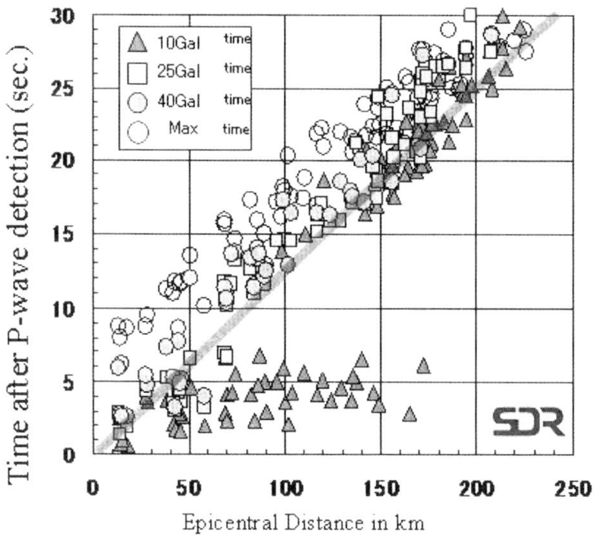

A

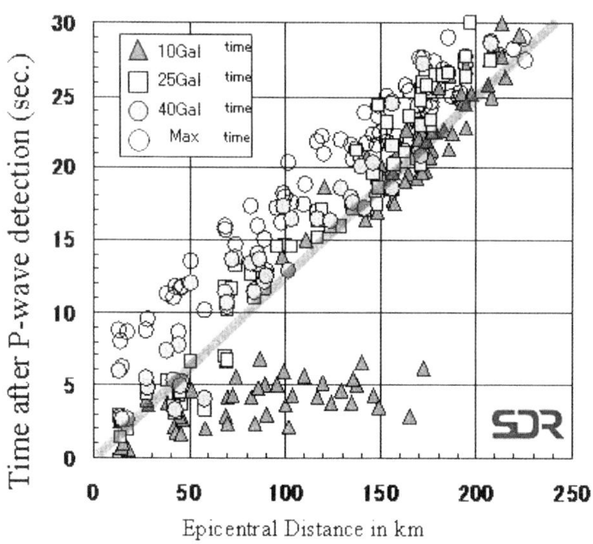

B

Fig. 15.10 Alarm timing by a simple trigger. (A) The 2000 Tottori-ken Seibu earthquake. (B) The 2001 Geiyo earthquake.

15.5 FREQL

15.5.1 Overview of FREQL

FREQL, Fast Response Equipment against Quake Load, integrates the functions of UrEDAS, Compact UrEDAS and AcCo. This means that FREQL can estimate the earthquake parameters one second after P-wave detection faster than UrEDAS, can judge the dangerousness of the earthquake motion one second after P-wave detection faster than Compact UrEDAS, and can output the information and alarm based on both acceleration and RI in real time in the same way as AcCo.

All the components of the seismometer (sensors, A/D converter, amplifier, CPU and so on) are put together in a small aluminum die cast 5 inches cube vessel, and the system is electrically isolated. Thus the FREQL is easy to install and the structure of FREQL is noise-proof. See Table 15.2 for the specifications of FREQL.

Table 15.2 Specification of FREQL.

Dimension	12cm(W) x 10cm(H) x 12cm(D) (not including projections)	
Weight	2.2kg	
Power Supply	AC100V-240V 50/60Hz (Battery unit with UPS for 3 hours)	
Water-proof	splash-proof	
Earthquake detection	P wave detection and S wave detection	
Observation	Acceleration and velocity	
Output	Acceleration, RI and alarm kind	
Observed component	Two horizontal components and one vertical component	
Sampling rate	1/100 seconds	
Alarm	P wave alarm	Based on the dangerousness Based on the estimated earthquake parameters
	S wave alarm	Triggered by acceleration level Triggered by intensity level
Communication	RS422	
Environment	0-40°C	

FREQL also has functions to omit the influence of electrical thunder noise and to detect the P-wave after a rather small pre-shock. Thus it can be stated that FREQL has solved the known problems of ordinary earthquake early warning systems. It is known that there was a pre-shock at the time the 1994 Northridge earthquake hit Los Angeles and the 1995 Kobe earthquake hit the greater Hanshin area. One drawback of the early warning system is that it is not possible to issue the alarm for large earthquake motion if there is a pre-shock just before the destructive earthquake because the pre-shock is recognized as a small event. It also seems difficult for the huge system to keep running perfectly under destructive earthquake motion. It is uncertain only with such remote systems due to the lack of information. This must be taken into consideration on installing the onsite warning system for an important facility.

15.5.2 Application of FREQL

FREQL is toward to the new field for the early warning system. Hyper rescue teams of Tokyo Fire Department decided to use FREQL for their activity under high risk of aftershocks (see Fig. 15.11).

Fig. 15.11 Rescue activities with FREQL.

Although the hyper rescue teams performed miraculously, their activity was always at risk of large aftershocks. After their activity at the damaged area of the Niigataken Chuetsu Earthquake, the Tokyo fire department approached us to adopt FREQL as a support system for rescue activity, taking note of the portability, rapidness and accuracy of the warning. FREQL for the Tokyo fire department consists of the FREQL main body, a power unit with backup battery for three hours, a central monitoring system and the portable alarm instrument with a loud alarm (>105dB) and rotary light.

The Tokyo Fire Department equipped the FREQL unit for two hyper rescue teams as from Spring 2005. At the time of their rescue activity after the 2005 Pakistan earthquake, they reported that FREQL worked soundly. This portable FREQL is spreading for many fire departments in Japan.

15.6 Conclusion

Recently, seismometers have been installed in many locations and earthquake information is more easily obtained. Nevertheless, warning systems must issue alarms at their own risk, and the public information from authorities is only for the response after the earthquake. In this regard, FREQL and AcCo are surveyed for use in disaster prevention. FREQL is the fastest early warning system in the world to detect the P wave and issue the alarm: a single seismometer can issue four kinds of alarms; P wave alarms based on dangerousness and on earthquake parameters, and S wave alarms triggered by acceleration and real time intensity, RI. AcCo is a simple palmtop size seismometer to measure acceleration and intensity in real time. These instruments combined can achieve an effective warning system and indicate countermeasures for earthquake disaster.

References

Nakamura Y (1985) Earthquake Warning System of Japanese National Railways. Railway Technology 42(10):371-376 (in Japanese)

Shabestari TK, Yamazaki F, Saita J, Matsuoka M (2004) Estimation of the Spatial Distribution of Ground Motion Parameters for Two Recent Earthquakes in Japan. Journal of Tectonophysics 390(1-4):193-204

Richter CF (1958) Elementary Seismology. W.H. Freeman and Co., pp. 136-140

Bolt BA (1993) Abridged Modified Mercalli Intensity, Earthquakes -New Revised and Expanded, Appendix C. W.H. Freeman and Co., p. 331

Wald DJ, Quitoriano V, Heaton TH, Kanamori H (1999) Relationships between Peak Ground Acceleration, Peak Ground Velocity, and Modified Mercalli Intensity in California. Earthquake Spectra 15(3):557-564

Nakamura Y (2004) UrEDAS, Urgent Earthquake Detection and Alarm System, now and future. 13th World Conference on Earthquake Engineering, Paper #908

Nakamura Y (2004) On a Rational Strong Motion Index Compared with Other Various Indices. 13th World Conference on Earthquake Engineering, Paper #910

16 Development and Testing of an Advanced Monitoring Infrastructure (ISNet) for Seismic Early-warning Applications in the Campania Region of Southern Italy

Emanuel Weber[1], Giovanni Iannaccone[1], Aldo Zollo[2], Antonella Bobbio[1], Luciana Cantore[2], Margherita Corciulo[2], Vincenzo Convertito[1], Martino Di Crosta[2], Luca Elia[1], Antonio Emolo[2], Claudio Martino[2], Annalisa Romeo[2], Claudio Satriano[2]

[1] Istituto Nazionale di Geofisica e Vulcanologia, Osservatorio Vesuviano, Napoli, Italy
[2] Dipartimento di Scienze Fisiche, Università di Napoli Federico II, Napoli, Italy

Abstract

In the framework of an ongoing project financed by the Campania Region, a prototype system for seismic early and post-event warning is being developed and tested, based on a dense, wide dynamic seismic network (ISNet) and under installation in the Apennine belt region.

This paper reports the characteristics of the seismic network, focussing on the required technological innovation of the different seismic network components (data-logger, sensors and data communication).

To ensure a highly dynamic recording range, each station is equipped with two types of sensors: a strong-motion accelerometer and a velocimeter.

Data acquisition at the seismic stations is performed using Osiris-6 model data-loggers made by Agecodagis. Each station is supplied with two (120 W) solar panels and two 130 Ah gel cell batteries, ensuring 72-h autonomy for the seismic and radio communication equipment. The site is also equipped with a GSM/GPRS programmable control/alarm system connected to several environmental sensors (door forcing, solar panel controller, battery, fire, etc) and through which the site status is known in real time.

The data are stored locally on the hard-disk and, at the same time, continuously transmitted by the SeedLink protocol to local acquisition/analysis nodes (Local Control Center) via Wireless LAN bridge. At each LCC site runs a linux *Earthworm* system which stores and manages the acquired data stream.

The real-time analysis system will perform event detection and localization based on triggers coming from data-loggers and parametric information coming from the other LCCs. Once an event is detected, the system will performs automatic magnitude and focal mechanism estimations. In the immediate post-event period, the RISSC performs shaking map calculations using parameters from the LCCs and/or data from the event database. The recorded earthquake data are stored into an event database, to be available for distribution and visualization for further off-line analyses.

The seismic network will be completed in two stages:

- Deployment of 30 seismic stations along the southern Apennine chain (to date almost completed)
- Setting up a carrier-class radio communication system for fast and reliable data transmission, and installation of 10 additional seismic stations.

16.1 Introduction

The Irpinia Seismic Network (ISNet) was designed in 2002 as an advanced research seismic network, and following its development, it is now in the phases of completion by the RISSC research group (a joint research group between the Physics Department of the "Federico II" University of Naples and the INGV-Osservatorio Vesuviano, Naples). It has been financially supported by the Campania Regional Center of Competence – Analysis and Monitoring of Environmental Risk. ISNet has a complete new communication and site infrastructure, and its fully digital and fast acquisition communication system renders it ideal for early-warning application studies.

ISNet is a highly dynamic, high density seismographic network under development in the southern Apennine chain. It is deployed across an area that has been struck by several destructive earthquakes over the last few centuries. The last of these occurred on November 23, 1980, with M=6.9, and it resulted in more than 3,000 casualties, and huge and widespread damage to the buildings and infrastructure of the whole region. In its final configuration, the network will consist of more than 40 highly dynamic seismic stations that are divided into physical sub-networks interconnected by a robust data transmission system.

In the early stages of the design of ISNet, it was conceived to monitor and analyze the background seismic activity produced by the regional fault system for purely research purposes. However, due to its high performance capabilities and within the framework of an ongoing project that is being financed by the Regional Department of Civil Protection, in 2006 the network will also become a prototype system for seismic early- and post-event warning which will be used to protect public buildings and infrastructure of strategic importance.

Implementation of a modern seismic network involves many different research and technology disciplines because of the sophisticated data management/processing and communication systems required to rapidly generate useful information. Due to this complexity, this paper provides a general technical overview of the ISNet architecture and implementation.

16.2 ISNet Architecture and Site Installation

ISNet covers an area of approximately 100 km x 70 km along the Campania-Lucania Apennine chain, and is deployed around and over the known active seismic system faults generating the 1980 Irpinia earthquake. ISNet does not follow a central site communication model for the transmission of seismic waveforms from a remote site, but an extended star topology that is designed to ensure fast and robust data analysis. The signals are acquired and processed at different locations in the network, and this leads to four fundamental network elements: the seismic stations, the local control centers (LCC), the central network Control Center (RISSC), and the data communication systems. Figure 16.1 illustrates the locations that comprise ISNet.

The stations are positioned along two imaginary concentric ellipses, with the major axis parallel to the Apennine chain. The average distance between pairs of stations on the outer ellipse is about 20 km, while the distance between the two ellipses is around 10 km. The inner ellipse is also filled homogeneously with seismic stations that have average inter-station distances of less than 10 km. Each seismic station is connected via radio bridge to a Local Control Center (LCC) (Fig. 16.1), which is itself linked to the RISSC Control Center by an E1 digital broadband (HDSL) wire line over frame relay. Through the use of permanent virtual circuits (PVCs), the frame relay allows the central site to use a single phone circuit to communicate with the multiple remote sites (the LCCs). The whole data transmission system is fully digital over TCP/IP, from the data-logger, through the LCC, to the control room in Naples.

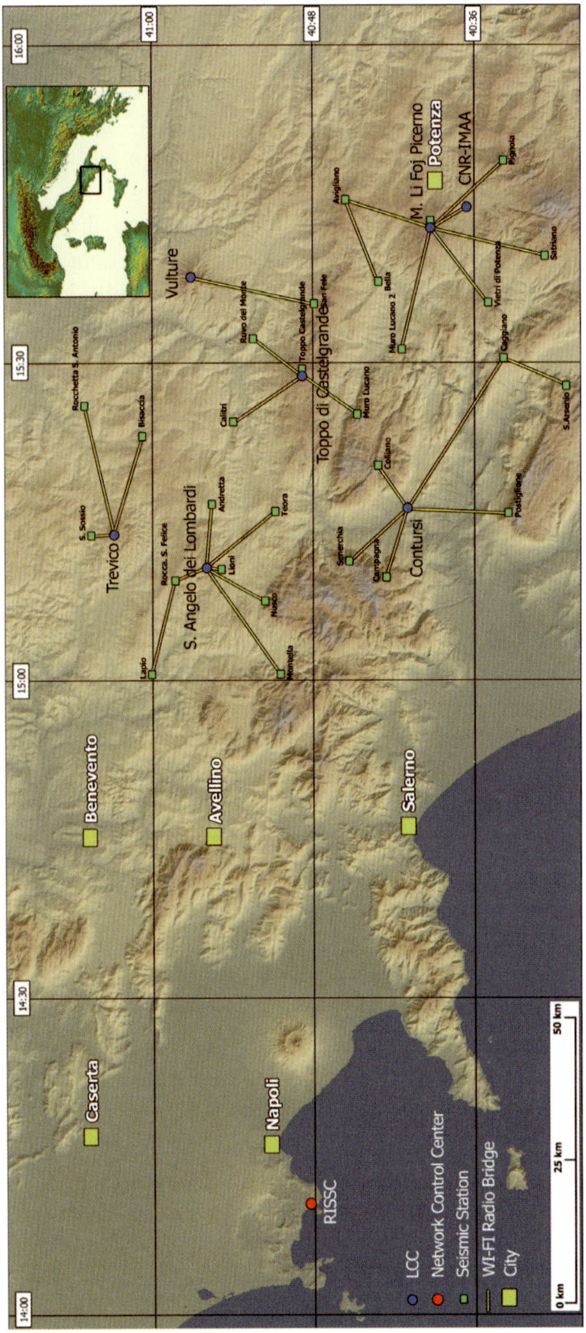

Fig. 16.1 Station map of Irpinia Seismic Network (ISNet). Symbols as given in the figure.

16.2.1 Seismic Stations and the Local Control Centers

ISNet consists of 30 seismic stations, each of which is connected with real-time communication to an LCC that is generally located in an urban area with a major communication backbone. The seismic stations are placed in 2 m x 2 m x 2 m shelters that are located inside 6 m x 4 m fenced areas. Each station is supplied with two (120 W) solar panels, two 130 Ah gel cell batteries (which avoids freezing damage) and a custom switching circuit board between the batteries. With this configuration, 72-h autonomy is ensured for the seismic and radio communication equipment. Each site is also equipped with a GSM/GPRS programmable control/alarm system connected to several environmental sensors (door forcing, solar panel controller, battery, fire, etc) and through which the site status is known in real time. This alarm system has standby power for at least three weeks. The GSM modem is connected to the data-logger and can be used as a backup data communication line. With an SMS (short message), and through the programmable GSM controller, the seismic equipment can be completely reset with a power shutdown. The GSM also controls the device start/stop release procedure when the battery goes over/below a predefined level.

The six LCCs collect and store the incoming data from the seismic stations to which they are connected via digital radio. The LCCs are positioned near small towns (in a shelter) or in existing buildings with an AC power supply and fast communication connections. In some sites, the LCC is also a seismic station. In this case, the sensors are outside in a shallow well, at a depth of 1 m to 1.5 m. The data-logger and other equipment are located inside an adjacent building. Each LCC has gel batteries for 320 Ah, a GSM remote-control system, a Cisco router, and an HP Proliant server with a 320 Gbyte hard-disk. All the instruments are connected to the batteries and 72-h standby power is guaranteed. Today, all the LCCs use the Earthworm (Johnson et al. 1995) linux version for data collection and processing, and event detection. The LCCs are connected to each other through the frame relay PVCs.

16.2.2 Sensors and Data-logger

Within each seismic site, all the sensors are installed on a 1 m^3 reinforced concrete base at least 0.8 m inside the soil. To ensure a highly dynamic recording range, each station is equipped with two types of sensors: a strong-motion accelerometer and a velocimeter. Twenty-five sites are equipped with the Guralp CMG5-T and the short period (T_0 = 1 s) Geotech S13-J.

The remaining five sites have Guralp CMG5-T and the broad-band Nanometrics Trillium (0.033-50 Hz band). After the first site installations, we noted a strong temperature dependency of the Guralp accelerometer. To minimize the thermal noise in the last sites to be built, a 30-cm hole was made in the concrete base. The hole contains the accelerometer and it is filled with sand to isolate the sensor. Before installation, the sensor/data-logger pairs are fully calibrated for single-channel responses by an automated process. This covers the entire frequency spectrum using an originally developed LabVIEW/MatLab software package that provides the transfer function in graphical mode and in terms of poles and zero.

Data acquisition at the seismic stations is performed using Osiris-6 model data-loggers made by Agecodagis (www.agecodagis.com). Some of the specifics of the Osiris include a Σ-Δ 24 bit A/D converter, a 100 MHz ARM processor with embedded Linux and open source software, on-site data storage (through one removable 5 Gbyte microdrive), serial and TCP/IP connectivity, GPS time tagging, an integrated SeedLink server, and simple/flexible configuration via a web interface (HTTP). The data-loggers have six physical and up to 24 logical channels, and each waveform can be analyzed with different sampling rates at the same time, for different purposes. An overview of the Osiris-6 data-logger has been provided by Romano and Martino (2005).

The external GPS receiver (NMEA/RS-232) guarantees a tagging accuracy that is better than 1 μs. A complete health status is available and helps in the diagnosis of station component failure or data-logger malfunction. The data-logger is remotely controlled from the IP configuration, sampling rate, gain, application of calibration signal to the resets of disks, GPS, etc. The data-loggers store the data locally on their microdrives or send it via SeedLink to Earthworm in the nearest LCC in 1 s packets. The real-time analysis system performs event detection and location based on triggers coming from the data-loggers, and parametric information provided by the other LCCs.

A *PostgreSQL* developed data-base tracks the network general configuration such as recorded channels, sampling rates for each channel, gain, sensor type, data-logger and other network devices, IP addresses, station position and serial number for each installed device.

At present, ISNet records real-time data from 33 stations with six channels, and in the near future we plan to add another 10 seismic stations. Twenty-eight of the stations installed have an infrastructure as described above, two are located in affiliated university buildings, and one is outside near a dam in the southern part of the Campania region. All the sites have the same sensors and data-loggers.

16.2.3 Current Data Communication Configuration

ISNet has a distributed star topology and at present it uses several different transmission systems. The seismic stations are connected via spread spectrum radio bridges to the LCCs (Fig. 16.1). To transmit waveforms in real time from the seismic sites to the LCCs, two outdoor 1310 G Cisco Wireless LAN bridges, operating in the 2.4-GHz ISM band, are used for each link. Each LCC is connected through different technology and media type to the Control Center in Naples, as shown in Table 16.1.

Table 16.1 Specifications of the data communication links between the LCCs and the Control Center in Naples.

Type	Frequency (GHz)	Bandwidth (Mbps)	# Number of Station	# Number of LCC	Comments
Spread spectrum	2.45	54	30	—	Throughput around 20-24 Mbps for links between 10-15 km (based on Ethernet packets with an average size of 512 bytes).
Satellite	Ku band	0.512	1	—	From outside to the nearest building through WI-FI bridge, then via satellite with a shared broadband internet solution to RISSC Control Center.
Internet	—	1.024	1	—	
SHDSL over Frame Relay	—	2.048	—	2	At the central site (RISSC) the CIR[1] has a maximum of 1.6 Mbps, depending upon the number of PVCs[2]. At the remote sites (the LCCs), the bandwidth is 640/256 kbps with a CIR of 64kbps for upload and download, over ADSL with ATM ABR service class.

Microwave SDH	7	54	—	2	Nera Networks carrier-class microwave link. Connecting LCC pairs. Truly full bandwidth available. Software upgradable to 155 Mbps (STM-1).
Microwave HyperLAN/2	5.7	54	—	1	The true usable maximum throughput of HiperLAN/2 is 42 Mbps.
Satellite	Ku band	0.512	—	1	Shared broadband internet solution to RISSC Control Center (512/256 kbps).

[1] CIR: committed information rate.
[2] PVCs: permanent virtual circuits.

The two primary backbone data communication systems of the central site use Symmetrical High-speed Digital Subscriber Line (SHDSL) technology over a frame-relay protocol. Frame relay offers a number of significant benefits over analog and digital point-to-point leased lines. With the latter, each LCC requires a dedicated circuit between the LCC and the RISSC Control Center. Instead, the SHDSL frame relay is a packet-switched network, which allows a site to use a single frame relay phone circuit to communicate with multiple remote sites through the use permanent virtual circuits. The frame-relay network uses digital phone circuits that can support up to 1.5 Mbit/second throughput for each single twisted-copper wire pair. In the present SHDSL configuration, two twisted-copper wire pairs are used to ensure 2 Mbps transfer. The monthly costs of long-distance digital packet-switched technology for fast data communication are much lower than for leased phone lines. With virtual circuits, each remote site is seen as part of a single private LAN, simplifying IP address scheme maintenance and station monitoring.

Several seismic sites, located well away from the earthquake source zone and outside the core of ISNet, use the Internet to transmit data directly to the Control Center site (RISSC). These sites are located inside affiliated university buildings in the main city of the region.

The satellite and SHDSL backbone are provided by commercial service providers who are regrettably unaccustomed to meeting the exacting demands of research teams involved in network infrastructures that require high reliability and complete data communication line control. There are thus short daily outages that require troubleshooting by our technical staff for failure detection. In this initial phase of the ISNet startup, and for further development towards new, fast, real-time applications, we test mainly the backbone links. These tests include line delay, system latency, availability and reliability. After this initial development, we are confident that we can reach 99.99% (53 min outage per year) of network availability. To date, we have never had radio link failure due to adverse weather conditions. Snow and heavy rain during the winter only have a slight effect on radio communication, without interrupting it.

Each seismic site has a real-time data flow of 18.0 kbps (at 125 Hz sampling rate for each physical channel), and the overall data communication bandwidth that is needed is around 540 kbps for 30 stations. ISNet supports this throughput under the worst conditions and it has been designed to guarantee further developments, such as the adding of further seismic or environmental sensors, without needing a larger economic and technological investment. The currently used data transmission protocol is TCP/IP, but for early-warning-application data acquisition, we intend to adopt the connectionless UDP/IP protocol to avoid unwanted overheads and handshaking between sending and receiving transport-layer entities before sending data segments. In early-warning waveform analysis, where single-packet error/missing does not critically affect decisions, this protocol is much faster and simpler to handle than TCP/IP.

16.2.4 RISSC Network Control Center

The seismic waveforms are stored locally on the data-logger and in real time at the nearest LCC. In its present condition, only selected signals are transmitted to the RISSC Control Center in Naples, and this is done manually and only for research purposes. We are developing a storage system at the Control Center that acts after the triggering of an LCC Earthworm or a station, which will have fully automated capabilities. We are also considering the data storage at RISSC, for redundancy of the whole datasets coming from each station, following a central site model. Hence we plan to install a large storage cluster at the Control Center on which all incoming waveforms can be loaded. The Network Control Center tracks the seismic events and monitors the entire network, including the data-loggers and radio links, through commercial network and bandwidth monitoring software.

16.3 Early-warning Prototype

Increased data processing capabilities and, above all, very fast reliable data communication systems today allow the warning of parts of the region that strong ground motion has been measured elsewhere in the region and that seismic waves may arrive shortly. The warnings include estimations of the earthquake origin and magnitude, and other source parameters, which must be continuously updated to improve data reliability while the earthquake is ongoing.

ISNet has been developed with modern communication, data acquisition and processing technology to be able to perform early warning and to have regional post-event capabilities, although it will need some infrastructure additions.

ISNet will have three new infrastructures integrated into it (Fig. 16.2):

1. an integrated seismic network which covers the Campania region outside the source area;
2. a proprietary data communication system connecting the LCCs to each other;
3. a second proprietary data transmission system that connects some LCCs to the Control Center in Naples.

16.3.1 New Seismic Stations

The function of additional seismic network stations will be to supply the data required for rapid building of regional shaking maps after large earthquakes. These stations will be located in urban and rural zones, but outside the earthquake source area (along the Apennine chain of southern Italy), and therefore will not be used for real-time localization and magnitude estimations. The 10 new seismic installations will have the Osiris data-logger with a strong-motion accelerometer and velocimeter. They will be deployed in buildings that will cover, in a wide mesh structure, the rest of Campania. The idea is also to record data close enough to urban conglomerations so that the waveforms observed can be correlated to eventual damage to structures. The data will be retrieved from these sites through a GPRS/UMTS communication system that will be provided by commercial service providers, and will be automatically downloaded and processed only after triggering from ISNet.

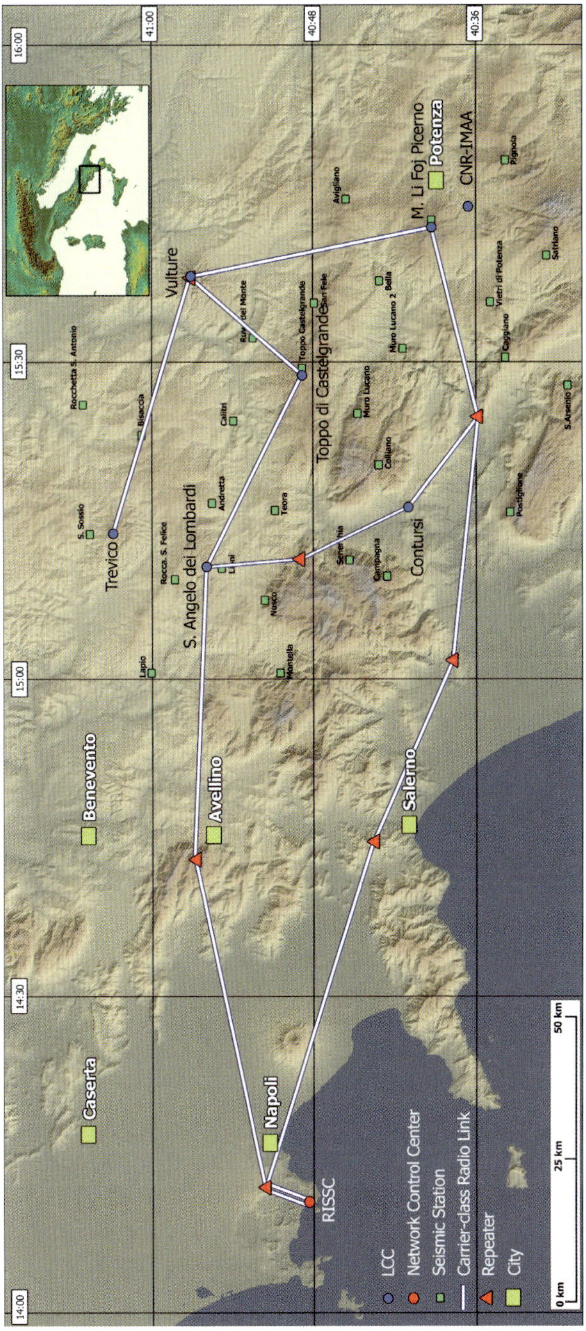

Fig. 16.2 Map showing the radio connections used in ISNet.

16.3.2 Data Communication Enhancements for Early-warning Purposes

The data communications system for early-warning applications is one of the fundamental points of this whole system. Based on the troubleshooting and experience of ISNet development, we are convinced that the complete control and management of the telecommunication systems, from the data-loggers to the Control Center, is one of the goals that must be achieved during experimentation of early-warning applications.

To understand latency, delay, failure and weak points in complex data communications structures gives quickly answers where can be earned fundamental time and give indications on future technological evolutions of the entire early warning system.

To avoid losing the entire network in the case of failure of one or more of the links, and to control the whole communication system without the need for commercial service-provider links, ISNet will be upgraded with a multiple radio-path data communication network. The redundancy in path determination and the percentage of unavailability acceptable for the system (outage) are fundamental parameters to be considered when the network enhancements are planned. Two rings have been created in the network: the first interconnects the six LCCs and the second connects the first to the Control Center in Naples. We are convinced that by selecting suitable technology and radio devices, we can achieve an overall system availability of 99.99% in the first development step, and 99.999% (five min outage/year) when the network is fully up and running. To reach this high availability rate, we have selected a carrier-class telecommunications device for the radio link. The first link is still being completed and will be used for the first main test purposes (Fig. 16.2). The radio device has the capability of an ethernet mapped over Synchronous Digital Hierarchy (SDH) device, with 155 Mbps (STM-1) throughput in a licensed 7 GHz frequency band.

Combining different technologies, such as satellite, radio and digital wire lines, will make it easier and faster to accomplish these high availability rates, and we will evaluate the use of this multiple communication technology in three LCCs. We have planned the new data communication system also taking the following constraints into account: system reliability and redundancy, low or no system damage during a strong earthquake, overall transmission delays of less than 100-200 ms, and data security.

16.3.3 General Overview of Network Management

As shown in Figs. 16.3 and 16.4, the ISNet data and information flow can be managed on three different levels:

– data-logger at the recording site;
– at the LCCs;
– at the network Control Center in Naples.

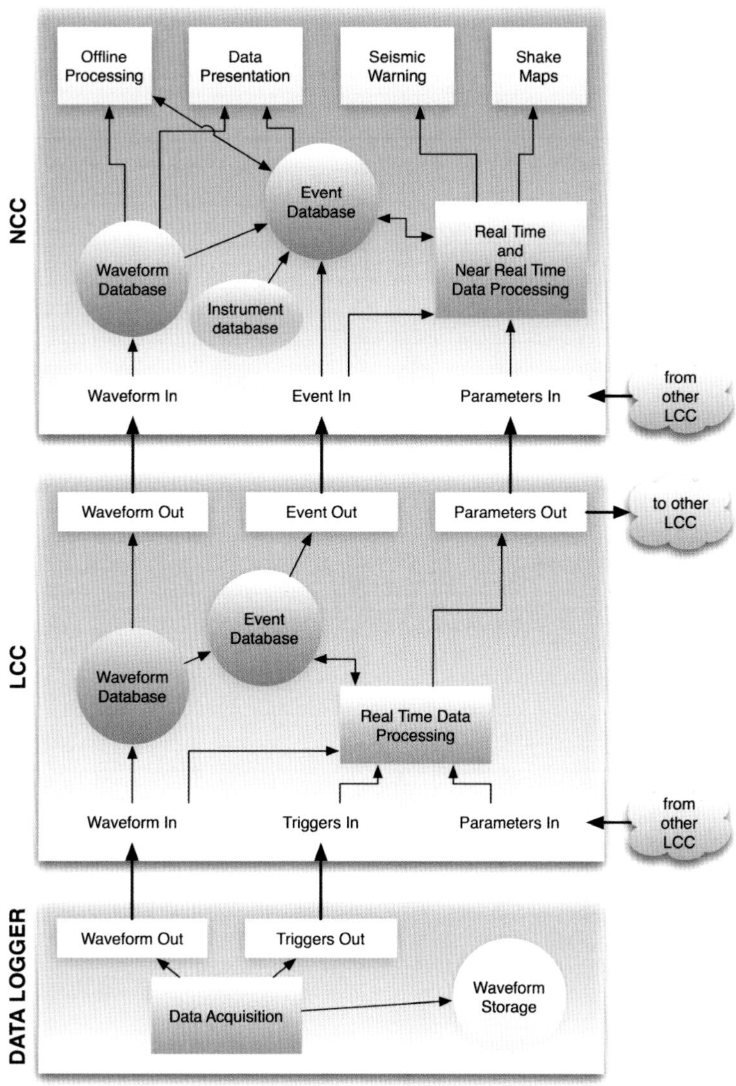

Fig. 16.3 Representation of the data flow in ISNet.

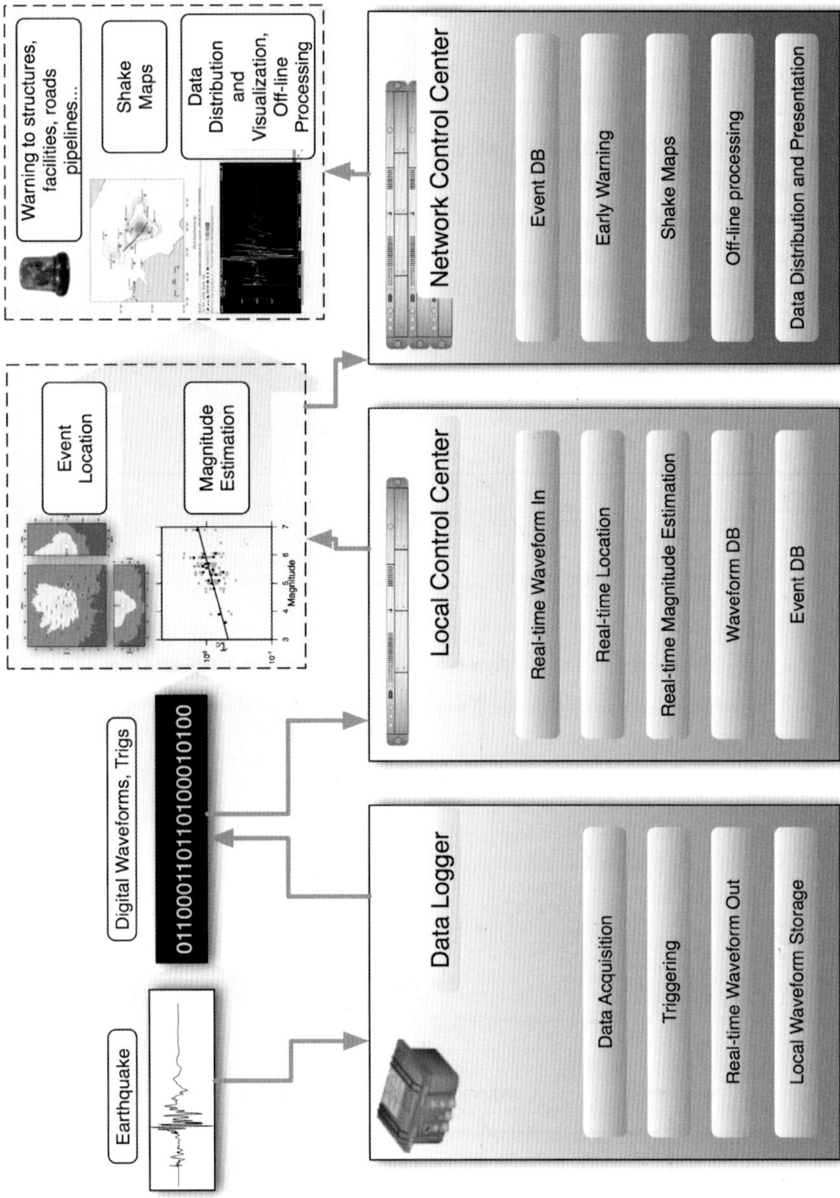

Fig. 16.4 Representation of the information flow in ISNet.

In particular, the data-loggers perform several functions: data acquisition, storage on hard disk, and transmission (data and trigs) over the network to the LCCs using the SeedLink protocol. The LCCs run the Earthworm real-time seismic processing system and they each keep a complete local database of waveforms from the seismic stations directly connected to them. The real-time analysis system performs event detection and localization based on triggers from data-loggers and parametric information from the other LCCs. Once an event is detected, the system performs automatic magnitude and focal mechanism estimations. The results of these analyses are used to build a local event database, and at the same time, they are sent to the other LCCs and to the RISSC Control Center. In the immediate post-event period, the RISSC performs shaking map calculations using parameters from the LCCs and/or data from the event database. The recorded earthquake data are stored in an event database, to be available for distribution and visualization for further off-line analyses.

As indicated in the previous section, ISNet has a complex infrastructure and needs to be accurately managed for proper real-time functionality. The most important information is the station and transmission network availability. To check the availability status of the network, a real-time cross-correlation between different information sources is needed. A large network is made up of many vendor devices for which the health status cannot necessarily be monitored by commercial software. To overcome these limitations, we are developing the ISNet Manager (ISNM). The ISNM will be a complete real-time monitoring suite with a notification system in the event of failure or critical status of any of the network elements (Fig. 16.5). The collected data from each device will be stored in a database. Considering the network complexity, various information sources will be used: SMS, E-mail, SMNP, ICMP (Ethernet), proprietary commands (Osiris), netflow (Cisco network flow-control protocol).

Today, in the first development step, the network manager is only static and serves as a devices data-base for the seismic network. It is made up of two fundamental design blocks:

- a database, managed by a server running PostgreSQL that is designed to store information related to sites, installed loggers, sensors, and generic and network hardware, along with their configuration and mutual connections.
- a user interface through a web application based on JavaServer Pages. The web server is powered by open-source standard Apache/Tomcat. A specific library of tags has been implemented to ease the interfacing of the web server with the database server.

The ISNM will have a server/client architecture. The data are collected at each of the LCCs and sent in parameterized form to the server, which analyzes the incoming information from all of the network elements.

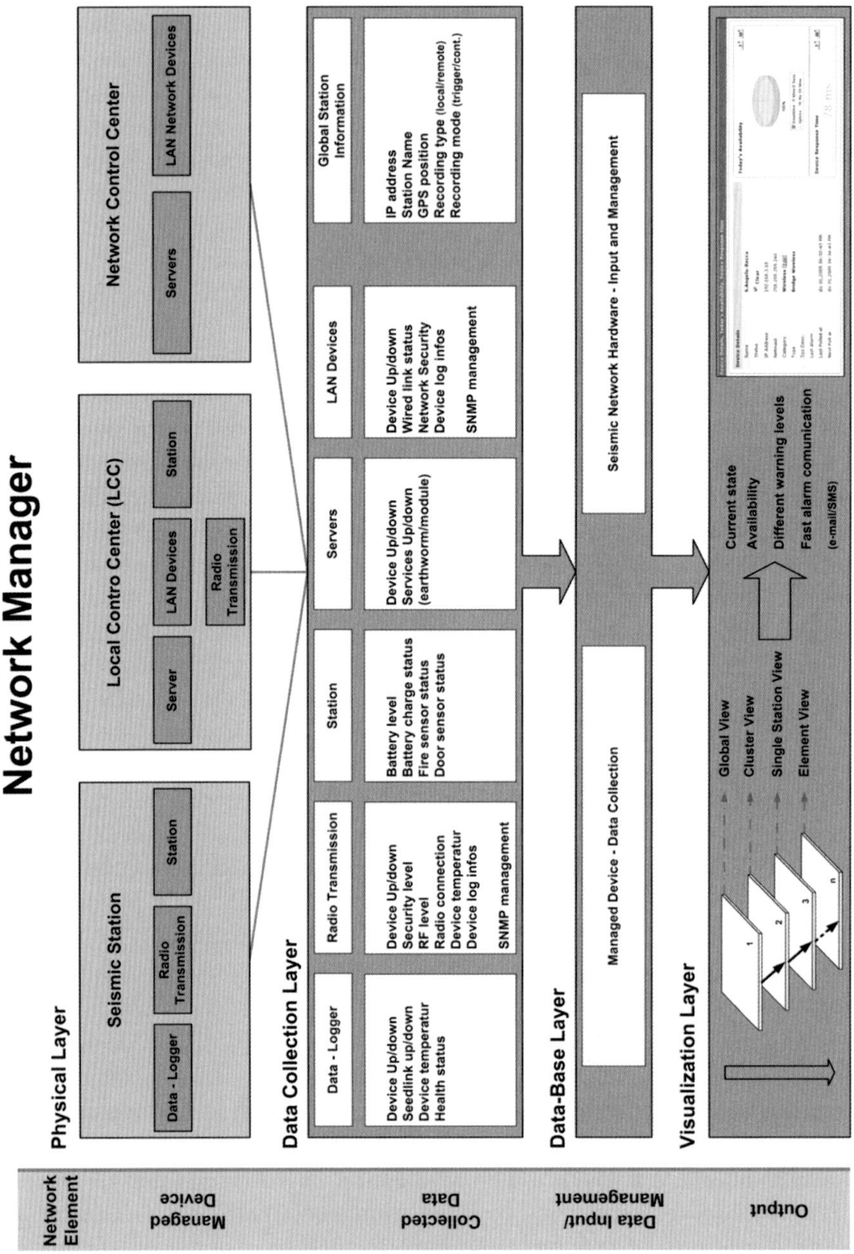

Fig. 16.5 Representation of the ISNet Manager.

Data- and network-security-related issues must be better investigated because of the sociological and psychological impacts of real-time alarms, especially if the general public is to have direct access to real-time shake maps. The network must be strongly protected against intrusion, which could intercept incomplete information that produce social panic, and/or generate false alarms.

References

Convertito V, De Matteis R, Romeo A, Zollo A, Iannaccone G (2007) A strong motion attenuation relation for early-warning application in the Campania region (Southern Apennines). In: Gasparini P, Manfredi G, Zschau J (eds) Earthquake Early Warning Systems. Springer

Hauksson E, Busby R, Goltz K, Hafner K, Heaton T, Hutton K, Kanamori H, Polet J, Small P, Jones LM, Given D, Wald D (2001) Caltech/USGS TriNet: Modern, Digital Multi-Functional Real-time Seismic Network for Southern California. Seism Res Lett 72:238

Johnson CE, Bittenbinder A, Bogaert B, Dietz L, Kohler W (1995) Earthworm: A flexible approach to seismic network processing. IRIS Newsletter 14(2):1-4

Romano L, Martino C (2005) L'acquisitore dati Osiris della rete sismica del CRdC-AMRA. INGV-Vesuvian Observatory OFR-5, 2005

Satriano C, Lomax A, Zollo A (2007) Optimal, real-time earthquake location for early warning. In: Gasparini P, Manfredi G, Zschau J (eds) Earthquake Early Warning Systems. Springer

17 An Early Warning System for Deep Vrancea (Romania) Earthquakes

Constantin Ionescu[1], Maren Böse[2], Friedemann Wenzel[2], Alexandru Marmureanu[1], Adrian Grigore[1], Gheorghe Marmureanu[1]

[1] National Institute for Earth Physics, Bucharest, Romania
[2] Geophysical Institute, Karlsruhe University, Germany

Abstract

Nowadays natural disasters phenomena as hurricanes, volcanic eruptions, tsunamis or earthquakes, are still difficult to prevent. Based on signaling of the phenomenon appearance in the destructive area, important human losses and material damages are avoided. For that reason, WARNING turns into a key objective, both in theoretical and practical research.

For the earthquakes, warning intervals are nevertheless very short – seconds to maximum one minute (Mexico City case). Even if the time window is reduced, automated decision measures are possible in case of a well organized system like important facilities.

In Romania, the major seismic risk zone is located in Vrancea region. The earthquakes occurring in this area are the main sources for the seismic hazard on Romanian territory. Seismotectonic characteristics of the Vrancea region offers the opportunity to create and develop a rapid seismic warning system. This system is simple, reasonably low-priced, robust and allows warning in an approximately 25 seconds time window for Bucharest. Warning signal obtained is sent to the responsible factors and specific users in order to control automated blocking of the installations and to carry out the required protection actions.

Romania is an earthquake prone area and it is of crucial importance to obtain quantitative information required for seismic risk mitigation and related public policies and seismic safety measures. The most damaging earthquakes in Romania concentrate in the Vrancea region, located at the sharp bend of the Eastern Carpathian Arc (Figs. 1a , 1b) in a well-confined focal volume at intermediate depths between 60 km and 200 km. Vrancea earthquakes have been documented for at least a millennium (since 985 AD) and present very peculiar characteristics. They are a permanent threat for large urban areas in Romanian territory and extended areas in Europe.

The Romanian capital Bucharest faces a significant earthquake hazard with a 50% chance of an event in excess of 7.6 moment magnitude every 50 years. Within the last 60 years Romania has experienced four strong earthquakes originating in Vrancea:

- Nov. 10, 1940 (Mw =7.7, 160 km deep);
- March 4, 1977 (Mw =7.5, 100 km deep);
- Aug. 30, 1986 (Mw =7.2, 140 km deep);
- May 30, 1990 (Mw =6.9, 80 km deep).

The 1977 event was catastrophic, producing the collapse of 35 high buildings and 1500 casualties, mostly in Bucharest.

The epicenters of the instrumentally well-located 60-200 km deep earthquakes are confined within a 30 x 70 km wide area (Oncescu and Bonjer 1997). The average epicentral distance to Bucharest is about 130 km. This concentrated location of earthquakes within a small source volume at a relatively fixed distance from Bucharest allows the design of an Early Warning System with a lead time of about 25 s for all the potentially dangerous intermediate depth earthquakes.

Therefore in Romania an Early Warning System could be developed similar to that implemented in Mexico, where strong earthquakes are located at the boundary of converging plates, off the Pacific coast, at a significant distance from Mexico City. A fairly constant warning time can be made available for both cities, although Bucharest has only one third of the time available to Mexico City.

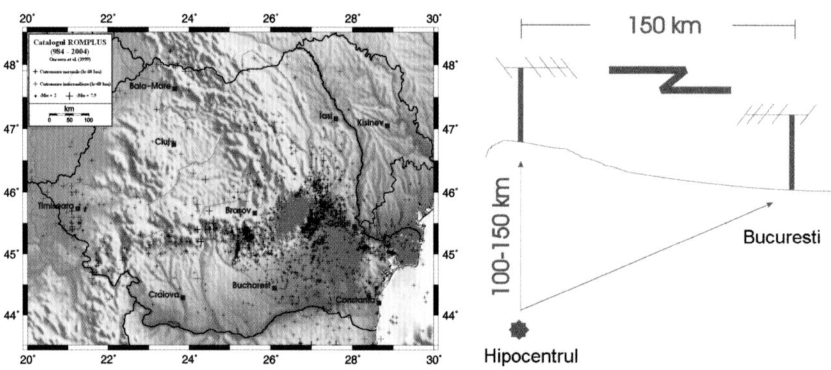

Fig. 17.1 The Vrancea epicenter seismic area and distance to Bucharest.

A group of civil engineers and seismologists from the National Institute for Earth Physics (NIEP)-Romania and Karlsruhe University-Germany has developed an earthquake Early Warning System (EWS) for the city of Bucharest and for industrial facilities. Simplicity and robustness of the system were sought in order to reduce the risk of false alarms, which is crucial for making the system cost-beneficial.

The design of a simple and robust Early Warning System to protect Bucharest from the Vrancea earthquakes is facilitated by the stationary epicenters, by the stability of the radiation patterns, and by the line-of-sight connection between the epicentral area and the capital city (Fig. 1).

In fact, besides the small focal volume, Vrancea earthquakes are characterized by extremely stable source mechanisms for larger and smaller events, such that a prediction of the level of ground motion to be expected in Bucharest can be based on the amplitude of the *P* wave in epicenter (Baresnev and Atkinson 1977, Wenzel et al. 1999) rather than on the cumbersome determination of magnitude and depth.

Real-time detection of the dangerous seismic waves from the Vrancea area is used as an Early Warning System. The EWS uses the time interval (that is about 25 seconds) between the moment when the earthquake is detected at the surface in the epicentre area (Vrancea) and the moment of arrival at the protected sites. This time interval depends on the earthquake depth and propagation conditions between focus and site (Fig. 2).

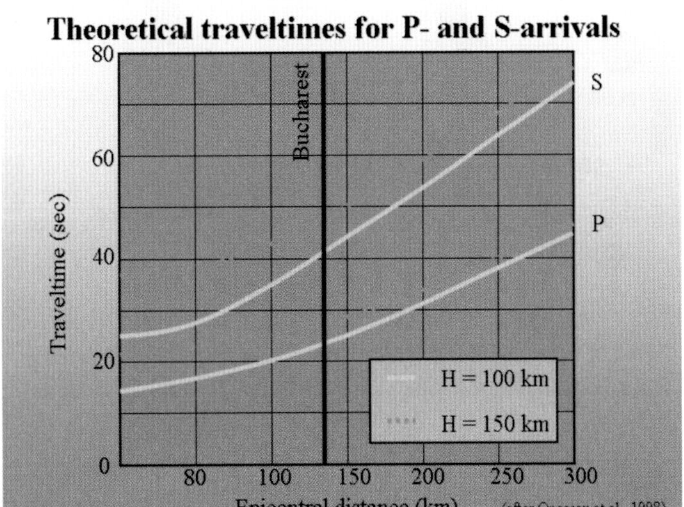

Fig. 17.2 Theoretical traveltime for P detected by borehole seismometer-and S-arrivals to user.

The EWS operates according to the following steps:

- detection of the P waves;
- processing of first arrivals (power spectra, fast Fourier analysis and coincidence on more than one accelerometer);
- alert generation;
- distribution of the alert to users (utilizing both SMS and e-mail).

The detection of the P waves in the epicentral area is the first step in the process. Triaxial strong motion accelerometers monitor ground motion at two different sites in the Vrancea epicentral area. The first site is situated at Vrancioaia Seismic Observatory (VRI) and uses an EpiSensor triaxial accelerometer installed in a vault. The other site is at Plostina (PLOR). It is equipped with two accelerometers: one is installed in a 50 m depth borehole (FBA23DH) and one on a vault at ground level (Fig. 3 left site). The distance between the two sites is about 8 km. These three strong-motion sensors form a small seismic network. The goal of this configuration is to reduce the false alarms generated by any accidental high noise (sonic bangs, explosions or dropping of heavy objects near accelerometers).

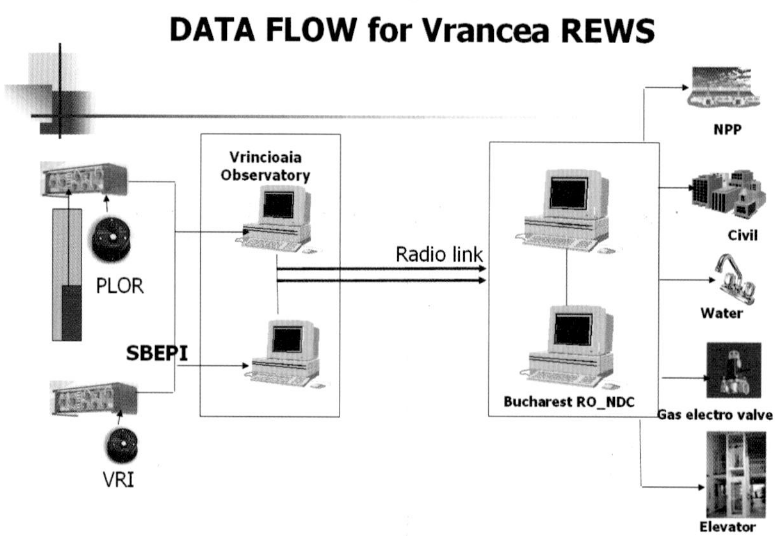

Fig. 17.3 Data flow chart for Vrancea real earthquake warning system.

The second step in the process is data processing.

The data provided by the accelerometers installed in the two sites are stored locally. The data stream is sent to the Vrancioaia Observatory where dedicated software performs real-time analysis.

One-second-interval data samples are processed by digital filters. Fourier spectra and power spectra are automatically computed in real time. Using the first seconds of the recording, preliminary estimation of the magnitude of the earthquake may be obtained. A special software module performs the analysis using a time window for all three accelerometers. The occurrence of an event is validated only when it is detected on more than one station. Eight levels of alarm are associated to acceleration values inferred from P waves. When an earthquake early warning alert is triggered, the following parameters are provided: time, peak acceleration value and level of triggering. An interactive processing module can modify the parameters of the alarm if in the next seconds the acceleration of the seismic waves is increased. Figure 4 shows the typical result of the early warning software analysis.

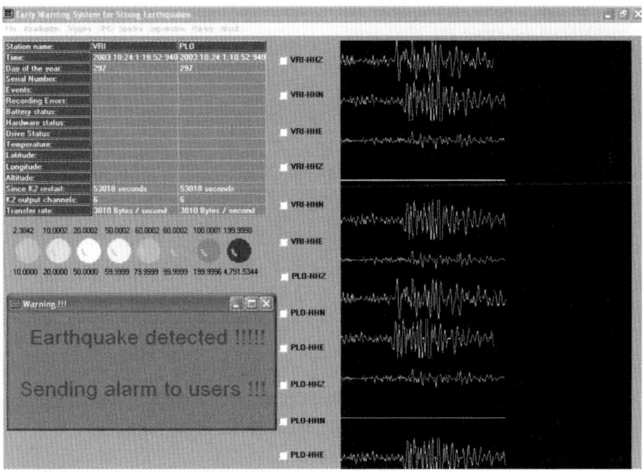

Fig. 17.4 Typical result of the early warning software analysis.

Distribution to users is the last step in the early warning process. At present this is ensured using dedicated connections. A special device designed by the NIEP staff connected to the Early Warning System's PC computer (by using TCP/IP communication) is the interface between a Central Station in Bucharest and the user's critical process. This device can be configured to trigger at eight different levels following the user's requirements (Fig. 5).

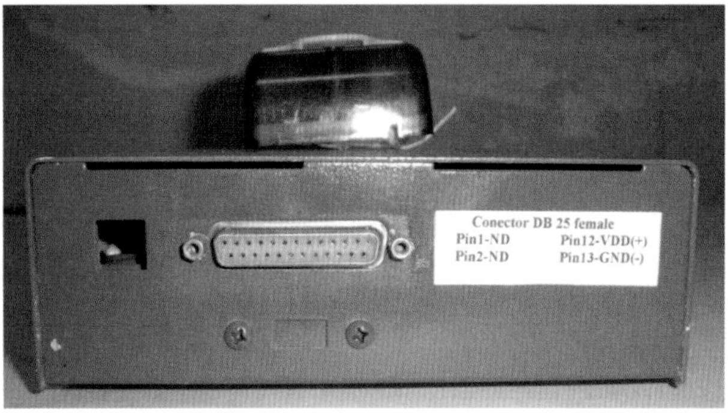

Fig. 17.5 Device configured to trigger at eight different levels following the user's requirements.

Communication is one of the most important parts of the Early Warning System. Redundant radio links ensure the communication between epicentral area (Vrancioaia and Plostina sites) and the Central Station in Bucharest.

At present the EWS is in the testing phase. It is however utilized at a nuclear installation by the Horia Hulubei National Institute of Physics and Nuclear Engineering (http://www.nipne.ro/) (Fig. 6) in order to place the nuclear source in a safe position prior to arrival of dangerous waves from the Vrancea area.

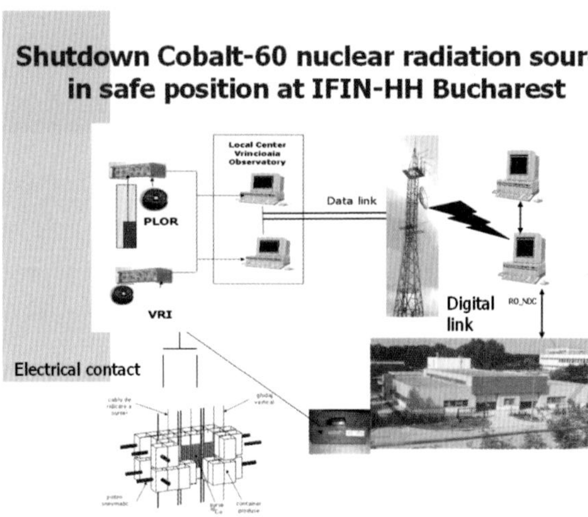

Fig. 17.6 EWS for an irradiator nuclear installation from Bucharest-Magurele.

An EWS can be interpreted narrowly as a technological instrument for detecting and forecasting impending hazard events and for issuing alerts. The Romanian EWS is one of the 20 Nominees for "THE EUROPEAN IST PRIZE" in 2005 (http://www.ist-prize.org).

The intention of NIEP is to start a project for installing, at the end of 2006, the early warning device at different users. The first EWSs will implemented at the surgery rooms of hospitals in Bucharest, chemical factories and railway companies.

References

Baresnev I, Atkinson G (1977) Modeling finite-fault radiation from the ω^n spectrum. Bull Seism Soc Am 87(1):67-84

Oncescu MC, Bonjer KP (1997) A note on the depth recurrence and strain release of large Vrancea earthquakes. Tectonophysics 272:291-302

Wenzel F, Oncescu MC, Baur M, Fiedrich F (1999) An Early Warning System for Bucharest. Seism Res Lett 70(2):161-169

Ionescu C, Marmureanu A (2005) Rapid Early Warning System (REWS) for Bucharest and Industrial Facilities. Presentation at Caltech University, July 2005

Printing: Krips bv, Meppel
Binding: Stürtz, Würzburg